Apollo 17

The NASA Mission Reports
Volume Two

Compiled from the archives & edited
by Robert Godwin

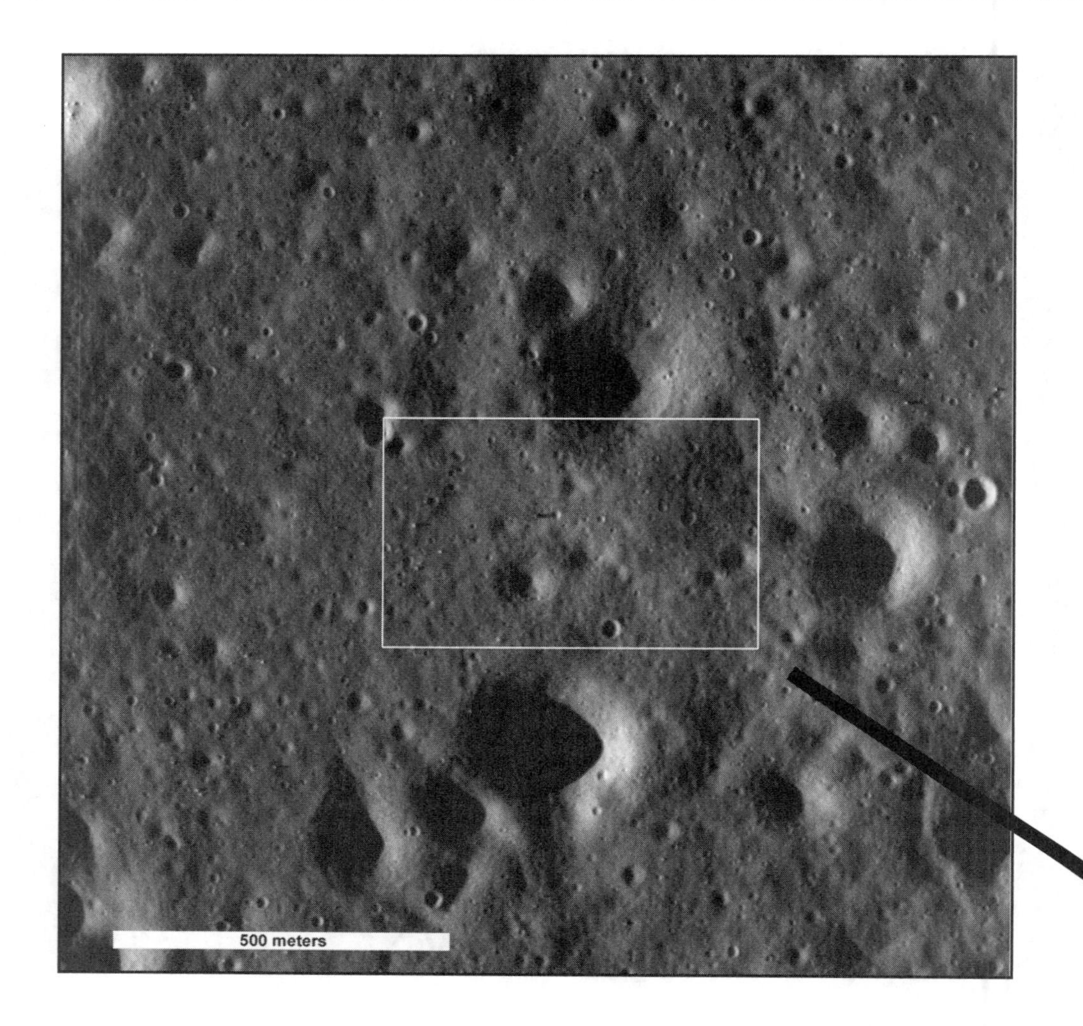

July 2009 Lunar Reconnaissance Orbiter Image of Apollo 17 Landing Site
Camelot crater is out of view at top left. Enlargement at right shows the
Lunar Module descent stage and EVA tracks.

Special thanks to:
Gene Cernan
Dr. Harrison Schmitt
Don Beattie
Rhett Turner

Dedicated to Ron Evans.

We acknowledge the financial support of the Government of Canada through the
Book Publishing Industry Development Program for our publishing activities.
Published by Apogee Books an imprint of Collector's Guide Publishing Inc., Box 62034, Burlington, Ontario, Canada, L7R 4K2
Printed and bound in Canada
Apollo 17 - The NASA Mission Reports - Volume Two
by Robert Godwin
ISBN 9781-926592-02-2

Apollo 17
The NASA Mission Reports
Volume Two
(from the archives of the National Aeronautics and Space Administration)

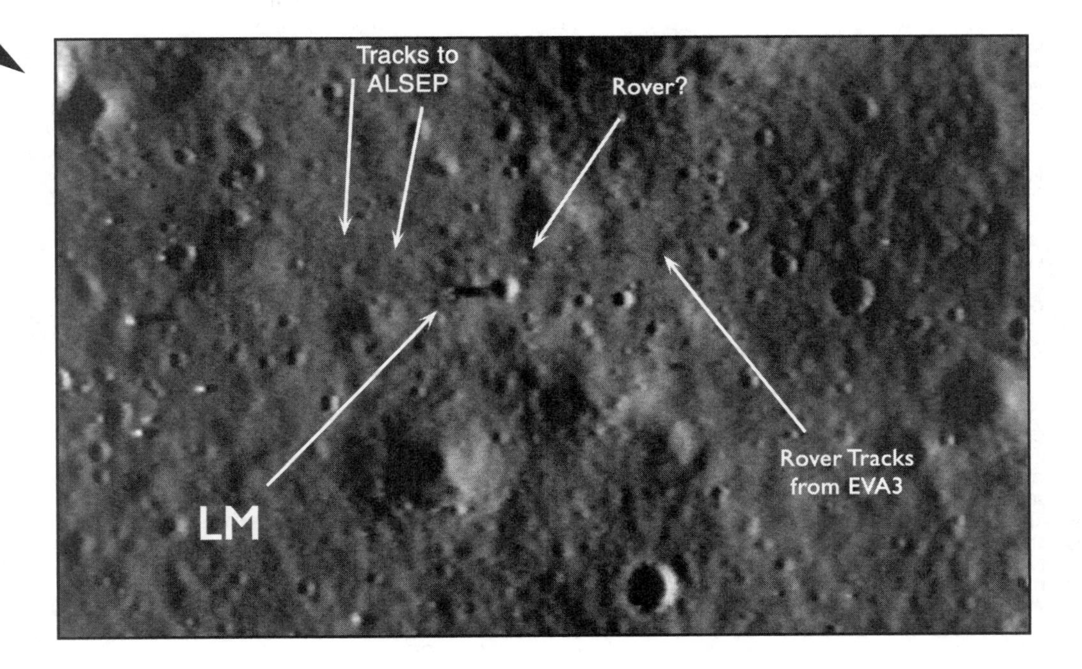

MISSION REPORT

TABLE OF CONTENTS

EDITOR'S INTRODUCTION

At the time of writing it is now over 36 years since humans have walked on the face of the moon. A frustrating span of time for those who were alive in 1972. Several generations, born and raised with the notion of living in space during the 20th century, have seen their dreams fade to nothing.

Our problems here on Earth have worsened since the heady days of Apollo. In the 1960s people like US Senator Edmund Musskie called for cutbacks in NASA's budget expressing the premise that the money should be spent to resolve problems like poverty, homelessness and a host of other social ills. NASA was an easy target, mainly because the mechanical marvels, executed routinely by the engineers and scientists, seemed to have no apparent benefit for the man in the street.

Ultimately those cut-backs would be made by the Nixon administration and NASA's budget, in real-terms, was slashed to a fraction of its 1966-high. The Space Shuttle program was then initiated to replace Apollo, on the promise of cheap, routine access to space. Despite many remarkable missions; including the construction of the International Space Station and the deployment and frequent repair of the Hubble Space Telescope; the Space Shuttle never came close to fulfilling that promise. The Shuttle was neither cheap, nor routine, and after almost three decades it is about to be scrapped, mainly because it has proven to be both expensive to maintain, and after two catastrophic failures, far from routine.

In 2009 NASA is now faced with the ignominious prospect of buying passage to the space station aboard Russian spacecraft, while the best and brightest at NASA are reverse-engineering Apollo-era spacecraft and propulsion systems. Should the Orion spacecraft fly, it will take NASA's astronauts to space in a system that is so alike to Apollo, it serves to only demonstrate the folly of the Nixon-era decision to cast aside the Apollo-Saturn system.

And what of all those other promises? To cut back NASA and thus fix all of our Earthbound problems?

NASA spent $274 Billion dollars (in 2009 money) between 1962 and 1972. That quarter of a trillion dollars employed as many as 400,000 Americans for much of that decade. Thousands of those jobs were high-technology jobs, but there were also truck drivers and janitors and accountants and every other imaginable type of employee. Compare those numbers to the current economic situation: the US unemployment rate has more than doubled between May 2007 and May 2009. It now sits at 14.5 million people (June 2009). The bailout number for the recession is a moving target, but some estimate it is currently above $8 trillion and still climbing. That's about $1 trillion a month, so far.

Using that *one month number*, in theory, NASA could employ 1.6 million Americans for ten years. There would be a statue of George Washington on Mars and the USA would again be the world's leading inventor and purveyor of high technology products to the rest of the world. Clearly people like Senator Musskie (and his philosophical counterparts) had no idea what a good deal they were getting from NASA in the 1960s.

I will not try to draw too many conclusions from the facts, they speak so clearly for themselves it seems almost pointless. I have therefore provided those facts in the form of a few graphs. All of the numbers have been adjusted for inflation. So they show everything in absolute terms, in 2009 dollars. With inflation factored in, it creates an obvious demonstration of what has happened in the last 47 years. I have not included *all* of the "sacred cows" in the US Federal budget, just some of those which were often used to hold NASA to ransom. Housing and Urban Development, the EPA, Health & Human Services, and The National Science Foundation. It was argued, and continues to be argued, that if NASA was shut down, the money could be used to reinvigorate these different agencies, but as you can see most of these have enjoyed year-over-year increases for decades (in REAL money i.e. adjusted for inflation) while NASA (and the EPA) have bumbled along with what amounts to static funding.

The truth is that NASA is one of the best bargains the American public has ever financed. The money spent on Apollo 17 and the rest of the space-race returned wealth in ways that are still being calculated. More than ever, it is NOW that the American people need to reinvest in a future of high technology.

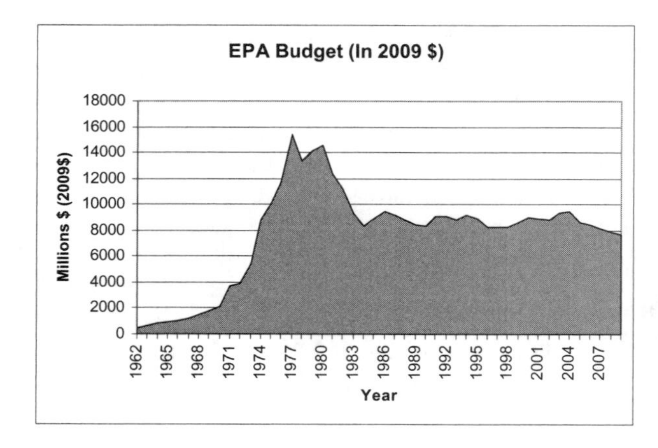

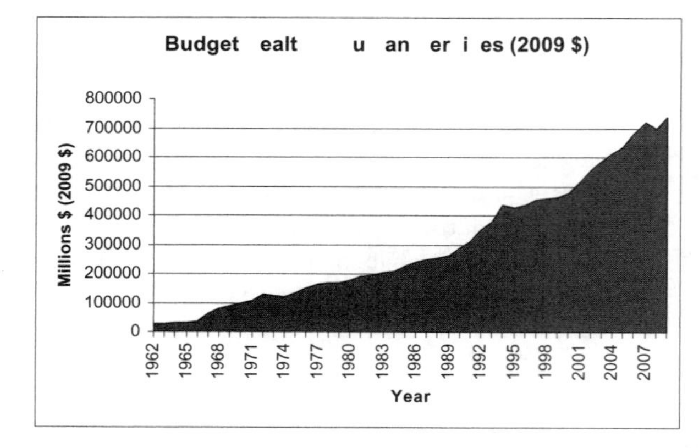

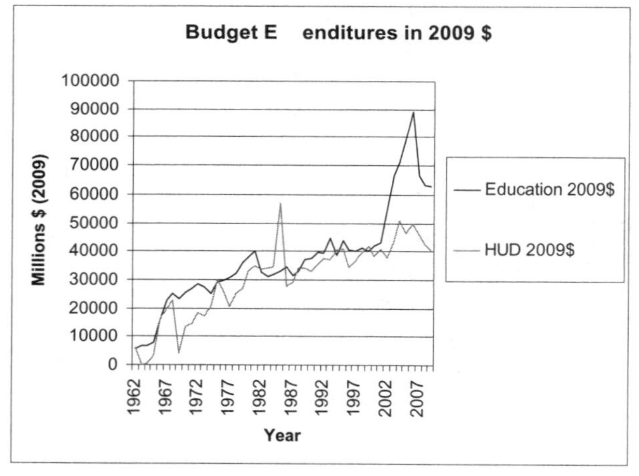

In the pages of this book you will see the results of just one geological expedition to the moon. But the real wealth was created back here on Earth, in technology, jobs and a robust economic future. That $274 billion *was spent in America*, not on the Moon. Just imagine how many jobs and what wealth could be created with just one month of the current bail-out expenditures.

Apollo 17 represented the end of an era of scientific expansion for the United States. No single peaceful undertaking in human history has produced so many benefits for a nation.

Sadly, NASA is being asked to recreate its finest hour with no substantial increase in funding, and this time NASA will not be building the next generation of launch vehicles on the back of massive military spending. Redstone, Atlas, Titan and Saturn were all spawned from military programs. Even the Space Shuttle system had considerable input from the DOD before being finalized. Aries is slated to use components of both Saturn and Shuttle systems but it is a peculiar hybrid and is already presenting unforeseen problems.

Based on past history, it might be more logical to once again lean on the current military heavy lifters and benefit from all of that DOD money already spent to get the Atlas and Delta families up to speed.

Meanwhile, this book begins with a reprint of the text from NASA publication EP-102, Apollo 17 At Taurus Littrow and concludes with three papers about Apollo's future plans written during NASA's heyday. It seemed like a logical way to start and end this book. The opening essay serves to revisit the wonder and excitement of the mission of Apollo 17. The closing papers, written by Farouq El-Baz, Noel Hinners and C.J. Byrne represent some of the many interesting discussions, that took place four decades ago, about potential future lunar landing sites. Places where people similar to Jack Schmitt and Gene Cernan may yet explore and learn more about our universe. One day, if smart minds prevail, we may see a young man or woman standing on the central peak of Copernicus Crater. They will be outfitted with the best new technology that the brightest minds in the world can concoct. New communicators, new gas flow systems, new cooling systems, new lighting systems, new materials and cloth, new radiation detection and protection, perhaps even new improved sun glasses. They will return to their base where they will use sophisticated new exercise devices, and then take their carefully engineered food out of improved refrigerators that are powered by superefficient solar panels.

None of these things are needed to send a robot to the moon. There are no biological developments when you send a robot to the moon. One has to wonder, what has the chess computer done for the game? What would a machine with a 200 mph fast-ball do for baseball; or a robot that can always drive a hole-in-one?

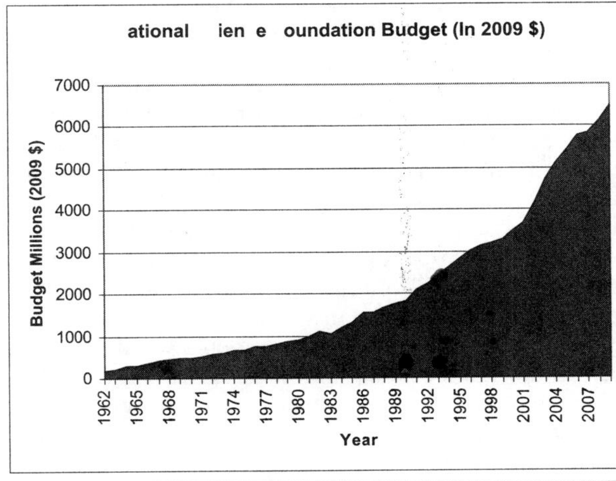

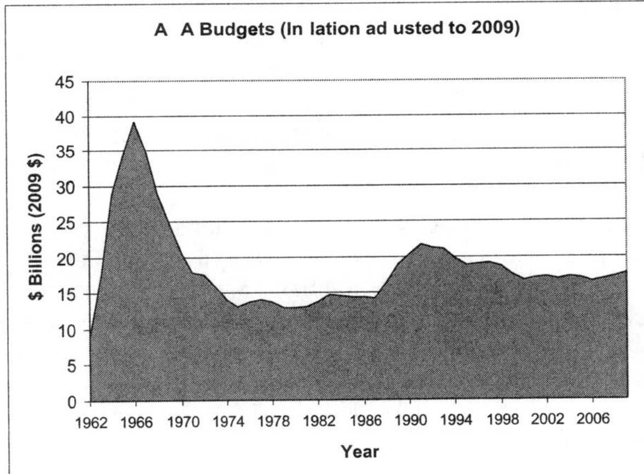

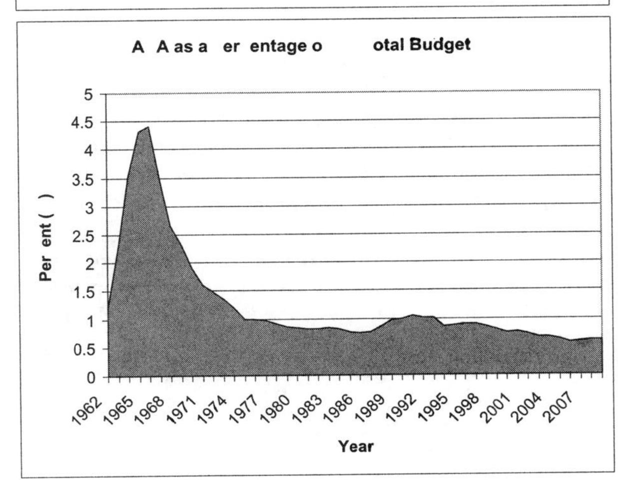

Would you pay to see a computer perform a concert? Are you really going to place more artistic value on a laser print than an oil painting? Do you prefer voicemail to a person on the other end of the phone? Is dropping a camera out of an aeroplane the same as skydiving? Would you be willing to give the swimming Gold to a speedboat, or to a Ferrari for the 100 meter dash? Why go on vacations to exotic places? Just turn on the National Geographic channel. Why not just send a robot to the top of Everest? Why go at all? Because extraordinary quests bring out the best in us. The smartest, toughest and brightest amongst us are the ones who improve the human condition for all of us.

When Lunar Orbiter I took a picture of Earth-rise barely anyone noticed, and then Bill Anders took the same picture from Apollo 8. It wasn't the information that affected us, it was the *experience*.

Maybe that future lunar adventurer standing at Copernicus will be a poet, instead of a scientist, a suggestion made by Jack Schmitt after his first view of the Earth from space. They might return to Earth with insights and ideas that might never occur to a test-pilot or a geologist, certainly never to a robot, but we'll never know just how far we can climb if we keep looking at our feet.

Robert Godwin
(Editor) (June 2009)

Apollo 17 At Taurus-Littrow

Preface

The Apollo 17 mission marked the end of the flight phase of lunar exploration. But in terms of analysis and understanding of the Moon, the Apollo 17 mission marked a point that was very near the beginning. For the past three years, experiments placed on earlier Apollo missions have transmitted their measurements back to stations on Earth, sending dots and traces that slowly have drawn a picture of the Moon. More than 23 billion data points had been received before the arrival of the Apollo 17 astronauts on the Moon; additional tens of billions will be added from the long lifetime experiments left by that mission. Years from now, those measurements will still be adding to the knowledge of the Moon, the phenomena that formed it, and the cataclysms that molded its early history. So after the centuries of philosophizing, of observing from the Earth, of assuming and hypothesizing and postulating and guessing, the first hard facts have become available with which to describe the evolution of another planet.

Introduction

Nothing can prepare you for the sights and sounds of an Apollo launching. Television, film, magnetic tapes and records are inadequate to convey the awe and the power of the experience. The towering launch vehicle looms hundreds of feet above its pad into the humid Florida sky. On top, almost 122 meters (400 feet) above the nearest stretch of sandy ground, is a tiny truncated cone, mounting a single launch emergency escape rocket attached by a tower structure. Inside the truncated form are three men in space suits, their provisions, and the instruments and equipment that will take them to the Moon, protect them on its surface, and bring them home again safely.

Thousands of items, tucked away in corners, built into complex installations, stowed in bags, worn, sat upon, and secured to the spacecraft structure cram the Apollo command module. In spite of all of this equipment, the crew's quarters provide six cubic meters (212 cubic feet) -two cubic meters (71 cubic feet) per man-of habitable volume. That's about one phone booth apiece, a mini-environment for work and eating and sleep during the trips out and back.

Below the command module is the cylindrical service module, the name for a catchall structure that holds fuel and a propulsion system in addition to scientific instrumentation, oxygen and hydrogen tanks, space radiators, helium tanks, batteries, and the other components that will service and support the crew and their mission until the last leg of the voyage homeward.

And below that sits the lunar module, closed in now by panels that conceal its angularity and its spindly legs. The two stages of the lunar module will make the descent onto the lunar surface and serve as housing for the two astronauts between their excursions on the Moon. It will carry with them to the Moon a lunar roving vehicle, a four-wheeled automobile that will be used by the two astronauts to move rapidly over long traverses on the Moon. The lunar module stands seven meters (23 feet) tall; halfway up is the front porch, where the first astronaut to emerge after lunar touchdown will pause before backing down the ladder and stepping off the footpad onto the Moon. The lunar module has two stages. Joined, they make the descent to the Moon. For ascent, after the lunar exploration has been completed, the upper stage separates and serves as a single spacecraft for ascent, rendezvous and docking with the command module in lunar orbit.

The weight of the lunar module alone is more than 16,000 kilograms (35,000 lbs), a phenomenal number to those familiar with the early days of spaceflight, when payloads were measured by a few kilograms, and as few as one. A large, loaded moving van weighs on the order of 16,000 kilograms (35,000 lbs), for comparison. Or think of it in terms of eight compact cars. It's a lot of weight to transport to the Moon.

So there they stand, stretching almost 25 meters (82 feet) from the base of the lunar module housing to the tip of the rocket motors on the escape system tower. And below them, for another 84 meters (275 feet), are stacked the three cylindrical stages of the launch vehicle.

Start nearest the spacecraft, with the third stage, the Saturn S-IVB, almost seven meters (23 feet) in diameter and just over 18 meters (59 feet) high. It is loaded with liquid hydrogen and liquid oxygen propellants for its single J-2 engine which will deliver 926,367 newtons of thrust in the final phase of the powered flight.

The third stage and the spacecraft will be

brought up to speed by the second stage, the Saturn S-II, a 10-by 25-meter (32.8- by 82-foot) cylinder with five J-2 liquid rocket engines delivering more than 5 million newtons of thrust. That performance seems almost insignificant when compared to the first stage, the mighty Saturn S-IC, again 10 meters (32.8 feet) in diameter, but peaking at more than 42 meters (138 feet) high. At its base are five F-1 rocket engines, four in a cruciform around a central unit, and each good for almost 7 million newtons of thrust at takeoff.

Ready to go, the whole assembly of spacecraft and launch vehicle towers 119 meters (363 feet) above the launching pad, and weighs almost 3 million kilograms (7 million lbs).

Night Launch

It is the warm, muggy evening of December 6 on a sandy cape area east of Orlando, Florida, where the Banana River winds through marshes, where herons, pelicans, and alligators search the waters. Stretching toward the sky at Pad A, Launch Complex 39, the huge rocket is lit by banks of searchlights, delineating its whiteness against the darkening skies. The countdown is routine, smooth. At 3 hours and 30 minutes before scheduled launch, there is a hold time built into the schedule. The clocks stop for one hour. Less than half an hour later, the doors of the Manned Spacecraft Operations Building open, and the Apollo 17 crew walk toward their transfer van, headed for their flight to the Moon. Eugene A. Cernan and Ronald E. Evans and Harrison H. Schmitt: white and bulky in their flight suits, carrying their portable breathing and cooling systems, their faces shielded by the clear visors of their near-spherical helmets. Flashguns fire brilliant lights against their bodies; Evans hesitates, raises his right arm, bumps his body toward the cameras.

From the observation site, this has been seen on television monitors that dot the area. In the grandstand, there are thousands of reporters, writers, researchers. Ahead of the grandstand, between it and the small body of water in whose black surface the Apollo rocket is reflected, stand dozens of cameramen. And off at remote sites there are other cameras and cameramen, television and movie, still and recording cameras, to make this last flight perhaps the most thoroughly documented of all.

The countdown goes on, flawlessly. Across the way perhaps a quarter of a mile, a parade of official vehicles, blue and red dome lights flashing, makes its slow way out toward the launch site. Cernan, Evans and Schmitt are on the way to the Moon by van and auto, rolling on rubber tires at a few miles per hour, gingerly, easily, toward the waiting spacecraft. Two hours and 40 minutes before the scheduled departure, they clamber aboard the command module. They "ingress" the module, in NASA's awkward but precise terminology. And then the checks of systems begin, right down to the final range safety checks. Two helicopters, one with a high-powered searchlight, patrol the beach area.

The searchlights at Launch Complex 39 bathe the scene in near-daylight, catching the wisps of venting oxygen as they drift slowly westward from the launch pad. The clock goes through T -5 minutes, and the count proceeds. Now at T -3 minutes, 6 seconds, the automatic sequencer starts to control the countdown. It performs each task, verifies that it is done, removes an interlock that prevents the next task from being done prematurely, or from being omitted, and does the next task. It does each, right down to 30 seconds, and the clock stops. Apollo control comes on the air to announce a hold, as it will do over the next two and one-half hours while the problem - a signal that the third-stage liquid oxygen tanks had not been pressurized - is solved by working out a safe alternate around that signal.

That alternate was developed on a "bread board" - a working duplicate of the complex control systems - by engineers at the Marshall Space Flight Center in Huntsville, Alabama. Their solution was a workaround which they checked and rechecked to verify its reliability. It worked, and passed the checks, and the communications links between the two NASA Centers carried the details of the alternate procedure to the control room at the Kennedy Space Center.

Finally, after a second prolonged hold at T-8 minutes, the final countdown begins, continues through the 30 second mark to cheers from the watchers, and approaches the T-10 second mark.

Apollo Saturn Launch Control: "Ten ... nine ... eight ... seven ... ignition sequence started ... all engines are started ... we have ignition. . . ."

There is a short, intense burst of white flame at the bottom of the Saturn first stage; it blossoms to a huge burning whiteness, bright as the Sun, spreading outward through billowing clouds of steam and is surrounded, but not swallowed, in bright orange sheets of fire from the five rocket engines.

"Two ... one ... zero ... we have a liftoff. We have a liftoff and it's lighting up the area, it's just like daylight here at Kennedy Space Center as the Saturn V is moving off the pad. It has now cleared the tower."

The enormous bulk of launch vehicle and spacecraft moved slowly, ponderously upward through the steam and flame. The only noise was the commentary from launch control and the cheering of the crowd in the grandstand The rocket began to climb into the night, above the searchlights now but needing no other light than the fire in its engines and the incandescent foxtail of flame streaming behind it. And then, perhaps fifteen seconds later, the noise hit, a blasting, snapping, crackling roar unlike any noise before. The physical impact stopped the breathing momentarily, and then the metal roof of the grandstand began to rattle and shake under the hammering of the shock waves from the engine. They sounded like a rapid-fire series of cannon shots from a few yards away, sharp explosive bangs that were startling, unsteady, high-pitched and roaring at the same time.

And through all the blasting the launch vehicle slowly climbed through the blackness, its five-forked flame streaming orange light, illuminating the upturned faces of the shouting crowd. It began to turn into its slow curving path downrange, and the flames began to open up behind the rocket like a pale orange umbrella of fire as the atmosphere thinned and reduced the external pressure on the engines. The individual rocket engines burned like bright dots of orange behind the veil of paler orange as the Saturn booster continued to turn downrange and the fiery umbrella kept opening outward, seeming to fill the sky and cover the Cape. Then the center engine was shut down automatically and a few seconds later, the outboard engines blinked out together. There was a brief pause, and then the white fireball of the second-stack engines igniting lit up the sky like a lightning stroke, and stabilized into a burning sphere of white light streaking toward the Moon like a rising star. Foxtail flame from five rocket engines in the Saturn lights the sky and the ground around the launch complex.

From The Earth To The Moon

The approach path to Taurus-Littrow valley is visible against the mountain bases at the left in this view from the lunar module. The Sculptured Hills are the bumpy terrain at the bottom. The command/service module is seen in the distance. About 12 minutes after liftoff, the on-board computers read out the maximum and minimum altitudes of the Earth orbit that had just been established. But this was simply a parking orbit for a short time, a time to catch the breath, figuratively speaking, and to get ready for the rocket engine firing that would launch the command and service modules across the long blackness toward the tiny Moon.

About 3 hours later, the Saturn S-IVB stage began its second burn, thrusting the spacecraft out of its parking orbit and heading it into space. For almost six minutes the rocket engine fired, adding about 3,000 meters per second (10,000 feet per second) to the velocity of the speeding spacecraft. Now at 10,440 meters per second (34,250 feet per second), the spacecraft engine stopped and the astronauts were on course and on speed for the Moon.

Separation and Descent

That, too, was a routine trip. But the sights were new to two of the astronauts -Evans and Schmitt, who had not been in space before-and they could not tear themselves away from the windows during the parking orbit, or the flight outward. After one pass around the world, one radioed back: ". . . we had almost a completely weather-free pass over Africa and Madagascar. And the scenery both esthetically and geologically was something like I've never seen before ... there were patterns like I haven't seen in textbooks. . ."

Schmitt, the geologist, became interested in the weather patterns that he could see developing and changing on the Earth.

"It looks like Mexico in general is pretty nice, although there is a band of east-west trending clouds that start from the Gulf of California across Sonora, and probably up through New Mexico and over into Texas as far around as I can see. Southern California looks like it's in pretty good shape today, but northern California looks like it's probably overcast. And a major system probably associated with that stretches into the north western United States. But a band of clear weather . . . stretches from Arizona right on up through - I would guess - through Colorado, Kansas and probably into the midwest pretty well."

And Capcom answered...... you're a regular human weather satellite." Later, Schmitt called in with this observation: "Bob, you always wish that you had a poet aboard one of these missions, so he could describe things that we're seeing and looking at and feeling in terms that might transmit at least a part of that feeling to everybody in the world. Unfortunately that's not the case, but . . . I certainly hope that some day, in the not too-distant future, the guy can fly who can express these things."

The long hours passed, and then the Moon, which had been looming larger and larger, became the primary target of their thoughts and views. It was the smoothest landing, closest to the planned touchdown point, of any of the Apollo missions. "Challenger," the lunar module, had undocked and separated from "America," the command/ service module, earlier, and its astronauts had completed the second descent orbit insertion maneuver without incident.

And now it was time to start the final descent toward the lunar surface. Astronauts Cernan and Schmitt, in constant contact with Mission Control at Houston, completed check after check preliminary to the actual firing of the descent engine that would lower them to the Moon. The checks completed and the go-ahead given, the final details of the pre-ignition sequence came over the voice link from Challenger:

"Ten seconds ... fuel ullage; we've got ullage ... two ... one ... ignition, ignition Houston ... attitude looks good. . . ."

The descent engine had fired, applying its thrust to brake the coasting speed of Challenger and move it out of the lunar transfer orbit toward the vertical descent and touchdown.

"Challenger, you're go for enter," said Mission Control, and the final procedures began on board the lunar module. They were closing in on the landing site, and recognizing its features: "Okay, I've got the South Massif ... I've got Nansen, I've got Lara and I've got the scarp ... oh, are we coming in ... oh, baby!"

Now just a hundred feet above the surface, the commentary came fast as Challenger eased toward touchdown.

"Eighty feet ... going down at three ... getting a little dust . . . very little dust ... stand by for touchdown ... down two . . . feels good ... twenty feet ... ten feet . . . contact!" Then the engine shutdowns, the switch closings, and the two astronauts had arrived at the surface.

"Okay, Houston, the Challenger has landed . . . tell America that Challenger is at Taurus-Littrow."

There was a brief time while the crew completed more checks and waited for permission from Mission Control to go on with the planned phases of the mission that would take them to the opening of the hatch and the exit onto the Moon.

"I'll check everything again. Let's just double-check. Okay. That hasn't changed ... it looks good ... the manifold hasn't changed ... the RCS hasn't changed ... ascent water hasn't changed ... the batteries haven't changed.

"Oh, my golly, only we have changed!" They had, just as other astronaut crews before them had changed. Somehow, the emotions of the moment, the enormity of the task just completed, the awe of being on the Moon, the stark beauty of its features, reach through the outer shell of objectivity and machinery that shields the astronauts. "Only we have changed!" He said it for all of them, and for all of mankind. Man is on the Moon, and we've changed.

EVA, The Sixth Lunar Surface Expedition

"The next thing it says is that Gene gets out." Astronaut Jack Schmitt, reading the checklist before their first moves outside the protective environment of the lunar module, had come to that line. This was the beginning of the first extra-vehicular activity (EVA), the NASA jargon for moving and working on the lunar surface. Cernan, on hands and knees now and backing out of the tiny tunnel opening in the lunar module, edges along carefully, wary of getting stuck or hung up.

Cernan: How are my legs? Am I getting out?

Schmitt: Well, I don't know. I can't see your legs . . . I think you're getting out though, because there isn't as much of you in here as there used to be.

Cernan: Okay . . . my legs are out ... -Houston, Commander is on the porch of Challenger."

And now Cernan is backing down the ladder.

"I'm on the footpad. And, Houston, as I step off at the surface at Taurus-Littrow, I'd like to dedicate the first step of Apollo 17 to all those who made it possible."

Then it was Schmitt's turn to descend the ladder and to step onto the lunar surface. After a quick walkaround inspection, and a first look at the surface around the landing site, Cernan and Schmitt freed the lunar rover from its berth, tucked into the lunar module. It came out, unfolding as it came, and stood ready for work. Sixteen minutes later, after a bit of difficulty getting the vehicle locked together, Cernan climbed aboard for a test drive.

"Can't see the rear ones, but I know the front ones turn, and it does move. Houston, Challenger's baby is on the road."

The test drive over, Cernan and Schmitt began to explore the area, Schmitt picking up rocks and commenting on their apparent structure. They fumbled, stumbled and fell, getting used to the strange feelings of moving in one-sixth the gravity force of the Earth. Cernan warned Schmitt: "You've just got to take it easy until you learn to work in one-sixth g," and Schmitt answered, "Well, I haven't learned to pick up rocks which is a very embarrassing thing for a geologist." And then, just before they attached the television camera, Cernan called over to Schmitt, who was working hard at loading the rover: "Hey, Jack, just stop. You owe yourself 30 seconds to look up over the South Massif and look at the Earth."

They assembled and flew the flag, and after the now-ritual salutes and pictures, Cernan talked to Houston: ". . . this flag has flown in the MOCR (Mission Operations Control Room) since Apollo 11, and we very proudly deploy it ... in honor of all those people who have worked so hard to put us here. . . ."

Then they unloaded the Apollo Lunar Surface Experiments Package (ALSEP), and assembled it into a barbell package for easier carrying. Schmitt made a gravimeter reading for calibration of that instrument, and they were off to deploy the ALSEP, Schmitt loping into the distance carrying the unit. But Cernan had knocked a fender off the rover by catching it with his hammer, and stayed behind to fix it. He made a temporary binding from an adhesive tape, and joined Schmitt at the ALSEP site.

It was Cernan's job to drill the two holes for the heat probe experiment. "Man, is that thing biting," he said about the drill. And later, he added, "I'm in something tough down there now ... must be in the mother lode.... Boy, the old fingers really suffer on these...... But he completed the job, with both probes driven to their planned depth. "Hey, Bob, just out of curiosity, what kind of heart rate has this drill been producing on me?" Cernan queried. And Capcom answered, ". . . you've been running at 120 pulse speed with peaks of 140 to 150 from time to time."

Cernan had to drill again, this time a ten-foot probe for the neutron flux experiment. He was beginning to feel the effects of the unaccustomed strain of working against the constraint of the pressure suit, and complained of arm pains when this job was done. The next task was a rover ride to a point near the crater Steno, named after Nicholas Stenonis, a 17th century Danish scientist. On the way, the rover stopped briefly at the site selected for the Surface Electrical Properties (SEP) experiment, and Schmitt placed the transmitter on the surface. On toward Steno they went, and collecting rock samples, taking photos, and making a gravimeter reading for another point on the gravity profile traverse.

Then it was back to the Surface Electrical Properties experiment (SEP) site for deployment of its transmitter antennas, placing them carefully in a set of right-angled tire tracks driven on the location earlier by Cernan. On the way back the broken tape gave way on the fender splice, and the fender was lost somewhere on the surface. "Oh, it pretty near makes me sick at losing that fender," Cernan said.

Back at the lunar module after these jobs, Cernan and Schmitt dusted off the rover, and were getting ready to re-enter the module when Cernan was startled by something. "Hey, something just hit here! What blew? Hey, what is that?" Schmitt had the answer. "It's the styrofoam off the high-gain antenna backing."

Capcom called with a request for a complete description of what happened to the rear fender, and the extent of the damage. They planned to think and work on the problem during the rest period for the astronauts, and to get back to them with some temporary fix if possible. For the rest of the stay on the surface, Cernan and Schmitt spent the time dusting themselves off. "How'd you get so dirty?" asked Cernan, and Schmitt retorted, "Huh, just wait until I show you a picture I took of you."

All of the experiments had been emplaced properly, and all were checked out as operative. They had made 6 readings with the gravimeter, placed 2 explosive charges for the seismic experiments, taken 229 color pictures and 197 black-and-white pictures, collected one deep drill core and 17 samples of rocks and rake fragments, a total of about 13 kilograms (about 29 pounds).

Then up the ladder and into the lunar module, ending the first EVA. They closed the hatch and repressurized the module. The time of the EVA was 7 hours, 12 minutes and 13 seconds, and there were two tired astronauts at the end of that day's work. "Oh, what a nice day ... there's not a cloud in the sky," said Schmitt as he started his trip down the ladder for the second EVA. He joined Cernan on the surface and, after the usual first-thing checks, the two astronauts attached their new fender. This history-making event-the first automotive repair on the Moon-had been figured out at MSC while the crew slept. Essentially it consisted of taping together four plasticized map sheets that would not be used, folding them once and fastening them to the fender rails with clamps cannibalized from the optical alignment telescope lamp.

The reason for the seriousness of this apparently minor problem was the dust plume thrown up by the lunar rover wheel. Without that rear fender, the surface dust was thrown onto the astronauts and the equipment on the rover, aggravating an already annoying situation. The fender fix worked, and the day's tasks began.

The first stop was scheduled to be Station 2, near the crater Nansen, named after Fridtjof Nansen, a Norwegian Arctic explorer. On the way, they drove south of the crater Camelot, and Schmitt described as they rode: "The surface is not changing in terms of the detail. The surface texture of the fine grained regolith still is a raindrop pattern.... Occasional craters show lighter colored ejectas all the way down to, say, half a meter in size.... Most of the brighter craters have a little central pit in the bottom which is glass lined. The pit is maybe a fifth of the diameter of the crater itself."

After more than an hour, during which they had stopped for samples and to set two more explosive charges, Cernan and Schmitt arrived at Station 2, lying at the base of the South Massif, at the contact area between the massif and the lunar mantle. Here the samples and observations were directed toward describing the massif bedrock and the light mantle. Cernan and Schmitt spent almost an hour here, gathering samples and photographically documenting them and the region . Then they drove the rover on toward Station 3, the crater Lara, named after the heroine of "Dr. Zhivago."

The same sampling routine followed, and they packed the specimens and headed on to Station 4, the crater Shorty, named after a character in Richard Brautigan's novel, "Trout Fishing in America." And there, almost five hours into the EVA, Schmitt was standing on the rim of Shorty and calmly describing its physical appearance. Finished, he said that he was going to take a panoramic shot with the camera. He moved to do so, and then his excited voice cut through the transmissions: "There is orange soil!" Cernan: "Well, don't move it till I see it."

> Schmitt: "It's all over, orange!"
> Cernan: "Don't move it till I see it."
> Schmitt: "I stirred it up with my feet."
> Cernan: "Hey, it is, I can see it from here!"
> Schmitt: "It's orange!"
> Cernan: "Wait a minute, let me put my visor up ... it's still orange!"
> Schmitt: "Sure is. Crazy. Orange ... It's almost the same color as the LMP decal on my camera."
> Cernan: "That is orange, Jack."

Quoting the words doesn't convey the sense of excitement that this discovery caused. And why? Because the presence of the reddish-orange material around the crater points to its volcanic origin, and suggests volcanic activity both before and after the formation of the light mantle.

Said Schmitt: "If I ever saw a classic alteration halo around a volcanic crater, this is it. It's ellipsoidal. It appears to be zoned. There's one sample we didn't get. We didn't get the more yellowy stuff, we got the center portion. . . ."

After this, the rest of the EVA was almost anti-climactic. They drove on to Station 5 at Camelot crater, collected samples, and then pushed on to the ALSEP site for another check of the lunar surface gravimeter. Indications were that the instrument was not working, and the scientists back in Houston wanted to make certain that it had been correctly installed and was level. Schmitt confirmed that. They arrived back at the lunar module, cleaned up the rover and turned off the television camera. It had been a second full day. They had traveled a distance of about 20 kilometers (12 miles) and had worked on the Moon for 7 hours, 37 minutes and 22 seconds. They had taken 218 color photographs and 627 black-and-white photographs, and collected 56 samples weighing about 36 kilograms (80 pounds). The scientific experiments were working properly, except for the lunar surface gravimeter.

In the final minutes before mounting the ladder and entering the lunar module, Cernan looked around. "I stand out here and I look at that flag, and I look at the rover, and I look at those feet. It's still hard to believe." And Schmitt answered by wondering what they had done to deserve being out there.

"It's about 4:30 Wednesday afternoon, as I step out on to the plains of Taurus-Littrow's beautiful valley," said Astronaut Cernan at the start of the third EVA. He and Schmitt did their usual routine checks and make-ready tasks before setting out in the lunar rover about a half-hour later. They drove toward Station 6 at the base of the North Massif, stopping on the way at about half the distance for sampling of the surface.

They paused again at Turning Point Rock, named because it was a prominent landmark on their course and the place where they were to turn to head toward Station 6. That rock, perhaps 6 meters high, was photographed and observed at close range before the trip continued.

They parked the lunar rover on the slope near the base of the massif and began their sampling and documentation. "Man, here's a big white clast, and there's one on top about a foot and a half across, and here's one must be two feet across ... three feet ... and that's in the blue-gray," reported Cernan, describing some of the local rocks of a type that geologists have labelled "breccias," or rocks which are made of fragments of specific materials. Bluegray, of course, refers to the color of the rock.

"Feel like a kid playing in a sandbox," said the exuberant Cernan, in the middle of his reporting. And then Capcom came on the air with instructions: "We're ready for you guys to leave this rock . . . and either get the rake soil and cores near that crater down below the rock ... or else go on to some other different variety of rock. . ."

Schmitt answered, ". . . going down to that crater is not a problem; getting back up is."

Capcom: ". . . it's not that vital to get to that crater ... if it's that much of a job to get down to it and back up."

Schmitt: "Okay. Bob, we don't move around from here too much. I tell you, these slopes are something else."

Before they left the site, Cernan called in an accident report: "Houston, we've got a couple of dented tires."

"What's a dented tire?" queried Capcom, and Cernan responded, "A dented tire is a little ... golfball size or smaller indentation in the mesh. How does that sound to you?"

"Sounds like a dented tire," said Capcom. But the damage was negligible and would not prevent the lunar rover from continuing to do its job. From Station 6 to Station 7, just off the slopes of the massif was a matter of a five-minute drive. First things first at the site; Capcom asked Cernan to dust off the mirror and the lens of the television. And Cernan came back with, "You know what? I'm getting tired of dusting. My primary tool's the dust brush. . . ."

By working at both stations 6 and 7, the astronauts had documented the structure of the base of the massif, first a distance up the slope and then a distance into the plain at its foot. At Station 8, another position at the base of the massif, they continued their geological field trip, with sampling and documentation. And Schmitt made some observations on lunar locomotion: "This is the best way for me to travel, up hill or down hill ... like this two-legged hop . . . Man, I can cover ground like a kangaroo!"

Cernan continued to comment on the dust and the way it got into everything. . . . we sure are giving this suspension system (of the rover) a workout ... I can't even see it . . . everything's getting awful dusty." Schmitt, confirming that, added, "Boy, everything is stiff. Everything is just full of dust." But Cernan had the last word: "There's got to be a point where the dust just overtakes you and everything mechanical quits moving."

They left Station 8 after 47 minutes of sampling and headed on to Station 9, on the rim of Van Serg crater and near the rims of three others: Shakespeare, Cochise and Gatsby.

After about a half hour of work in the area, Capcom advised the astronauts that it was time to go, and suggested that they leave ". . . immediately if not sooner to head for Station 10." There was a brief flurry of activity, and then Schmitt's voice broke through in excitement. "Come here, Gene, quickly. We can't ... we can't leave this. This may be the youngest mantle over whatever was thrown out of the craters." "Take pictures of it," answered Cernan, and then speaking to Capcom, he added, "Bob, we've got to take five more minutes ... what Jack's doing is he's dug a trench in the southwest-northeast direction and he discovered about three inches below the surface a very light gray material ... a possibility here is that ... this upper six inches of gray material . . . is the latest mantling in the area and the light colored debris may be what's left over from the impact."

Capcom, now concerned about the duration of their stay on the surface at this distance from the protection of the lunar module, came on again: "Okay, I copy. I understand. But we'd like to get you going, in case you didn't get the clue."

But Schmitt, ever the professional geologist, was reluctant to leave this exciting and unexpected find, and Cernan seemed to share his view. Schmitt took photos; they started to make ready to go back when Capcom reprieved them: "We've had a change of heart ... and we're going to drop station 10 . . . and we're going to get a double core here ... then we're going to leave and go back. . . ."

Cernan and Schmitt worked harder now, with Cernan starting to drive the core while Schmitt was still arguing with Capcom about whether or not it was even going to be possible. They got the core and then Capcom asked for a special sample from the shallow trench that Schmitt had dug earlier. Capcom added, "We'd also like to have you moving in 4 minutes. That's with wheels rolling in 4 minutes."

There was more argument, Cernan saying that he didn't know if that could be done, Capcom insisting, and Cernan saying that they could not get that special sample in 4 minutes. Capcom agreed to eliminate it, and Cernan said, "Let's roll."

They left Station 9 and headed back to the lunar module, calling in observations as they passed craters on the way back. At about 5 hours and 44 minutes after they had set out, they arrived back at the home base for the last time this trip. They still had work at that site: photography, a final check of the ALSEP, and the tasks of clean-up after the continuing problem with the lunar dust. But there were some other, non-work tasks that remained, too, and Cernan, as commander of the Apollo 17 mission, was its spokesman. He spoke to the International Youth Science Tour that was in Houston that day, watching some of the activity by television transmission from the lunar surface. Cernan told them that, ". . . Jack has picked up a very significant rock ... composed of many fragments, of many sizes and many shapes, probably from all parts of the Moon, probably billions of years old. But a rock of all sizes and shapes.... and even colors, that have grown together to become a cohesive rock outlasting the nature of space, sort of living together in a very coherent, very peaceful manner ... we'd like to share a piece of this rock with so many of the countries throughout the world. We hope that this will be a symbol of what our feelings are. . . ."

And Schmitt added that portions of the rock will be sent to museums or other agencies in each of the countries represented by the young people in Houston on that day.

Then Commander Cernan again, speaking for Apollo 17 and all the Apollo missions that came before: "We'd like to uncover a plaque that has been on the leg of our spacecraft ... I'll read what that plaque says . . . "Here man completed his first exploration of the Moon December 1972 A.D. May the spirit of peace in which we came be reflected in the lives of all mankind. "

It was a serious moment, but it wasn't the finale of the spectacle. That finale would have been the last routine checks before boarding the lunar module. But Capcom had some final words about the lunar surface gravimeter experiment, which still wasn't working; the scientists wanted to make one more try.

So Capcom told Schmitt "Tap sharply on the gimbal with the UHT (Universal Handling Tool) and then reverify the level."

Schmitt: "You mean tap on the thing that swings?"

Capcom: "That's what they say."

Schmitt: "You always wanted to do that, didn't you?"

Capcom: "Yeah, that's right."

Schmitt: "How much is sharply?"

Capcom: "Sharply is sharply. It's probably not heavily, but sharply. Fairly light, but sharply."

Schmitt: "Okay, here it goes. I did it ... that was sort of a moderate tap."

Capcom: "Go ahead and hit it harder."

Schmitt: "Okay. Okay? I can hit it harder yet."

That didn't do it, and Schmitt was released to work on other tasks. But a few minutes later, back came Capcom, with more instructions. "We'd like to return to the surface gravimeter, Jack . . we'd like you to rap even more sharply, more strongly on the gimbal another three times."

Schmitt: "Okay, Bob. Here come the raps. About three times. Okay."

Capcom: "Okay, Jack, that's really fighting it pretty hard."

And still there were no measurable results. And, after more talk and tasks, Capcom finally closed out the experiment attempt: "Okay, Jack, we're ready to leave the ALSEP."

"I hate to do that, Bob. I'm sorry about this gravimeter," said Schmitt.

And Capcom responded, "Well, you're not the only one ... there's a whole room full of people that are sorry."

For the last time, they closed out the check list, moved the sample bags into the lunar module, and prepared for the last steps up the ladder.

"Okay, you guys, say farewell to the Moon," said Capcom, and Cernan answered, "Bob, this is Gene, and I'm on the surface . . . as we leave the Moon at Taurus-Littrow, we leave as we came, and, God willing, we shall return, with peace and hope for all mankind."

Rendezvous With America

And now in lunar orbit, Challenger and America eased toward each other for the critical rendezvous in lunar orbit, the achievement of the concept that made the entire Apollo mission possible. The Challenger ascent stage moved into an elliptical orbit around the Moon, closing to within 16.2 kilometers (9.3 nautical miles) of the lunar surface and reaching a maximum distance of 88.7 kilometers (47.9 nautical miles). Astronaut Evans, in the command/ service module, was in a near-circular orbit 114.8 kilometers (62 nautical miles) above the surface. Challenger saw America first, at about 207 kilometers (112 nautical miles). Houston transmitted orbital corrections to Challenger, so that the two orbits would be brought together for the final phase of the lunar rendezvous. Then Evans' voice came through, confirming that he could see the Challenger spacecraft.

Challenger, behind America, closed rapidly on the spacecraft. Evans, in America, said: "I just started picking you up on the telescope ... I don't care what you look like, come on back. I was going to shave and look nice for you, but I didn't have time. . . ."

Soon after, both spacecraft moved behind the Moon and their signals were lost. Hidden from the monitors on Earth, Challenger's ascent engine again thrust-sharply and briefly, nudging the spacecraft out of its elliptical orbit into an intersecting path with the course of the command module. Then out of the shadow they came, checking in with Houston and getting down to the fine details of rendezvous.

"I can see your thrusters firing now, Ron."

More exchanges of closing data, and then Evans said, "Looks like Challenger's in good shape. I don't see anything hanging down. . . ."

"She's in excellent shape," Cernan answered, and added, . . . you look pretty." Then slowly, slowly the two vehicles closed the distance between them, Evans maneuvering to line up the spacecraft for the docking maneuver. "Okay, coming in nice and slow ... no problems." "You're looking good . . . looking good ... looking good ... must be a couple of feet away . . . about two or three feet is all."

The first touching placed the two spacecraft in soft lock, without final latching and locking accomplished. There were some minutes of maneuvering, backing off, recapture and then Challenger spoke: "Okay ... here he comes. Bang, two good old barber poles! ... Okay, Houston, we're hard docked." The barber poles, indicators on the command panels of Challenger, showed the latching. Challenger and America once again were a single spacecraft in lunar orbit. Evans removed the probe and drogue assemblies and passed them through the airlock to Cernan and Schmitt, along with a vacuum cleaner and a list of the items that needed to be transferred from Challenger to the command module. Emptied, Challenger's ascent stage was sealed off from the command module and the three astronauts were back together in the command module for another day of work in lunar orbit before the start of the long voyage homeward.

"Don't let us bother you, Ron," said one of the returning astronauts. "You just go about and do whatever you want to do. We'll just get clean for the next three days."

Then the Challenger ascent stage was separated and, behind the Moon, nudged into a trajectory that would take it down to do its last job: crash on the Moon to get one more set of data points on the seismic instrumentation.

Turning Toward The Planet Earth

"Bob, that's the most beautiful crescent Earth I've ever seen," radioed one of the astronauts just before the end of their 67th revolution around the Moon. They were nearing the end of the long exploratory work that had begun with their first visual sightings of the lunar surface, expanded through the work on the surface by Cernan and

Schmitt, and continued with visual observations of the Moon during the orbits after ascent and rendezvous. That phase was to continue for only eight more turns around the Moon, and then the trans-Earth injection burn would be done, in the shadow of the Moon, after the 75th revolution. As they emerged from behind the Moon, they would be headed back toward Earth and splashdown in the Pacific.

Now the last data were sent to update computers for the burn. "Okay, Jack, here's the numbers you've been waiting for," said Capcom. "TEI rev 75 SPS/G&N 36 372 plus 063 plus 086. Noun 33 is 236 42 08 35 plus 30398 minus 01850. . . ." The signals went on in lengthy procession, and soon after, the spacecraft moved behind the Moon and signal was lost. The times of signal loss always make for tension. The success or failure of operations done behind the Moon never is known until the spacecraft returns to the near side and radio contact can be re-established.

There was an expected dividend on this flight; it was to be only the second time that television from an Apollo spacecraft could show pictures of the far side of the Moon. The reason was that the burn on the far side would propel the spacecraft into a higher altitude, enabling contact with the Earth to be established earlier than usual. Then the on-board television camera could transmit pictures of portions of the far side.

The first signals from the returning America came via the 64-meter (210-foot) diameter antenna at Goldstone, California, and they were followed almost immediately by pictures of the far side, showing the crater Tsiolkovski. "Houston, do you read America?" queried the spacecraft, and, after Houston affirmed it, Commander Cernan said, "America has found some fair winds and following seas, and we're on our way home."

But even on that leg of the trip the work continued, with visual observations of the Moon. Schmitt reviewed some of the results: "... we have pretty good evidence as a result of the Apollo program that ... basalt flows ... some three to four billion years ago erupted on the Moon and filled many of the lower areas that existed at the time. Not an awful lot has happened to the Moon-except for the impact craters, some of the younger ones-since three billion years ago, which is one of the reasons it becomes so interesting to man.... The Moon's frozen in a period of history three billion years and older, which is a period of history we cannot recognize very readily on Earth because of the dynamic processes of mountain building and oceans and weathering that are taking place even at the present time. Understanding that early history of the Moon may mean an understanding of the early history of the Earth. And I think we're well on our way to a first-order understanding of that history as a result of the program."

Return, Re-entry And Recovery

The trip home was as routine as any trip back from the Moon could be. There was the excitement of Astronaut Ron Evans' excursion outside the command module to retrieve the film cassettes from the instrument bay in the service module, and his euphoric comments during that spectacular space walk. He got the film and returned it, and he commented on the view, the external condition of the spacecraft, spoke to his two children, sang, and generally behaved as others have who personally encounter space, freed of the shackles of a confining spacecraft.

The final test of the spacecraft and its systems came during the last minutes of the flight and the blazing re-entry through the Earth's atmosphere. The command and service modules separated, and Cernan maneuvered the command module to position it for the re-entry, its heat shield facing into the direction of flight. And about 304 hours after the spacecraft had lifted off the pad at Cape Kennedy, it slammed into the Earth's atmosphere, in a streaming, flaming dive toward the surface. Behind it trailed the tongues of flame from the instant combustion and ionization of the heatshield material.

Then there was loss of radio contact, as the spacecraft plunged into the blackout zone, where the ionization from the burning heat shield hides all radio transmission. After a long and agonizing three minutes, voice contact came through again and then the atmosphere began to slow the craft, the drag force pressing the astronauts into their contoured seats with more than six times their normal weight.

Then there was the welcome sudden deceleration as the three drogue parachutes streamed behind the spacecraft, slowing it further. And, less than a minute later, the three main braking parachutes streamed, and blossomed behind and then above the descending command module. Down it came, dropping slowly toward the calm blue Pacific, and splashed down into the pattern of threefoot waves within sight of the recovery ship.

Cernan, Evans and Schmitt were back home, back home on Earth, back home on a small blue world in a dark corner of the galaxy.

Science Results

The flight phase of the Apollo program now is over; but the scientific analysis of its data has hardly begun. For months and years to come, the measurements from Apollo 17 and its predecessor missions will continue to provide bases for understanding and argument about the Moon, the Earth, and their very ancient histories. For science moves deliberately-some might say slowly-to analyze data taken during experiments. The sheer volume of data-already measured in the tens of billions of data points-precludes any rapid results. Further, some of the experiments that were emplaced on the Moon are designed for long-term exposure and are intended to return raw data for several years. But there are preliminary results from some of the tests. The lunar seismic profiling experiment, for example, which uses explosive charges to explore the interior of the Moon, already has noted the impulse of the lunar module ascent engine firing, and the impact of the spent module after it impacted back on the Moon. The small explosive charges left on the surface were detonated after the astronauts had returned to lunar orbit, and showed that the sub-surface structure of the Apollo 17 site was definitely different from that of either Apollo 15 or 16.

That lunar module impact, plus the much earlier impact of the spent Saturn S-IVB stage, registered on the passive seismology experiment previously deployed on the Moon. And these two shocks, which doubled the amount of previous data, have shown the possibility that the Moon's crust is considerably thinner than the 60 kilometers (37.2 miles) earlier postulated. It's possible, said one scientist, that the thickness is only 25 kilometers. But there was prompt disagreement from another scientist, and the final results are yet to come.

The heat flow from the lunar crust appears to confirm the results of Apollo 15, which surprised the scientists with its high value. What this means is that the interior of the Moon may be warmer than expected, and this in turn hints at large quantities of radioactive materials in the lunar structure. There appears to be a much higher concentration of those materials on the Moon than there is on the Earth, leading to further arguments against the theory that the Moon was torn from the mass of the Earth. First results from the lunar traverse gravimeter show that there is a high-density material filling the valley between the North and South Massifs, with greater thickness than expected. The implication, then, is that the valley originally was a very deep one, prior to its flooding from the Mare Serenitatis.

Geologically, there was a feeling that everything planned had been accomplished. The surface samples, the detailed documentation and observation of large boulders and surface features, and the visual descriptions of the surface by the astronauts during their EVA and orbital times, added immeasurably to man's knowledge of the Moon. The orbital science experiments, oriented toward geophysics in contrast to earlier missions that concentrated on geochemistry, worked to near-perfection. Five continuous revolutions' worth of data were obtained by the laser altimeter, which added to altitude profile data, and the continuity of the data will help recover gravity information as well. The panoramic and mapping cameras, which recorded the lunar surface on several miles of film, should produce the best series of maps yet developed for the Moon.

The lunar sounder, designed to probe the depths of the Moon with electromagnetic signals, acquired that data along a 49,000-kilometer (38,000-mile) track and returned more than 500 million soundings. All of that data was recorded on film, and will have to be processed and analyzed after the return to Earth. The infra-red scanning radiometer produced coverage of about one-third of the surface area of the Moon, and made about 100 million independent temperature measurements, during both day and night on the Moon. These points, which will produce thermal maps of the Moon, will lead to a new technique for exploration of planetary surfaces that lack an atmosphere. The ultraviolet spectrometer experiment, whose prime function was to measure the lunar atmosphere, was unable to detect evidence of any major or even minor constituents of that atmosphere. In fact, all it recorded was one trace component of the lunar atmosphere. Said the principal investigator of that experiment, "What's behind us is the rather startling discovery that the Moon simply is not de-gassing; it has nothing left in terms of anything that would create an atmosphere."

Astronauts

Cernan, Evans and Schmitt, the crew of Apollo 17, were as typical or untypical as any other astronaut crew. In their late thirties, all three had long experience in their specialties. Cernan and Evans were from the Navy, and had been pilots; both had engineering degrees. Schmitt was the first professional scientist to go on a space mission, being a hard-rock geologist.

Eugene A. Cernan, born in Chicago in 1934, was an engineering graduate in Electrical Engineering from Purdue University, and holds a master of science degree in Aeronautical Engineering from the U.S. Naval Postgraduate School. He was commissioned in the Navy through the ROTC program at Purdue, entered flight training just after graduation, and has since logged more than 3,800 hours of flight time. He is a Navy Captain, and was one of the third group of astronauts, selected in October 1963. Space flight is not new to Cernan. He shared the Gemini 9 mission as co-pilot to command pilot Thomas P. Stafford, and completed a two-hour space walk outside that spacecraft. He was lunar module pilot for the Apollo 10 mission, the dress rehearsal for all that came after, and flew that spacecraft to within 14.8 kilometers (8 nautical miles) of the Moon's surface. On Apollo 17, he was crew commander.

Ronald E. Evans was born in St. Francis, Kansas, in 1933. His training included a bachelor of science degree in Electrical Engineering from the University of Kansas, and a master of science degree in Aeronautical Engineering from the U.S. Naval Postgraduate School. Commissioned after ROTC at the University of Kansas, he flew naval aircraft in tours of duty on carriers. When NASA notified him of his selection as an astronaut in April 1966, he was serving in the Pacific aboard the USS Ticonderoga, the same carrier that was to become the recovery ship for the astronauts of Apollo 17 more than six years later. This flight is his first into space; but he has been a backup and support crew member on earlier Apollo missions. He was Apollo 17 command module pilot.

Harrison H. Schmitt was born in Santa Rita, New Mexico, in 1935, the son of a mining geologist. His undergraduate work was done at California Institute of Technology, and after post-graduate studies at the University of Oslo in Norway, he received his doctorate in geology from Harvard University. He was selected as an astronaut by NASA in June 1965 and went through more than a year of flight training to learn the piloting skills necessary to the mission. With more than 1665 hours flying time logged, Schmitt was selected as the lunar module pilot for the Apollo 17 trip.

The Next Steps In Space

Apollo 17 has been called an end and a beginning. More accurately, it is an important way-point in the exploration of space by manned and unmanned spacecraft. Apollo 17 was the last of the manned Apollo missions; but it is far from being the last manned space flight effort. At this moment, the first of the Skylab orbital workshops and its launch vehicle are standing assembled in the huge Vehicle Assembly Building (VAB) at NASA's Kennedy Space Center. Due to be launched in mid 1973, the Skylabs will be used for a wide range of scientific and medical experiments, aimed primarily at understanding the requirements for long-term stays in space. In Skylab, the first emphasis will be properly placed on a series of medical experiments to determine how well man can adapt to the conditions of longterm exposure to the peculiar characteristics of space. Additionally, a number of solar astronomy experiments will be performed, studying the Sun through the uncluttered distance between an onboard telescope and our primary star.

Potentially the most valuable of the Skylab experiments, in their possible applications to the problems of everyday life, are a series of Earth surveys that will be made. Observations of the Earth, from Skylab heights, will continue to yield more of the kind of information that can be useful at ground level. Skylab draws on both the expertise and the hardware of the Apollo program. The Saturn V booster will place in orbit the workshop cluster, a structure consisting of the workshop itself, modified from a Saturn S-IVB stage, the Apollo telescope mount, an airlock and a docking adapter. This clusterweighing 82,237 kilograms (181,300 pounds) and spanning 35.97 meters (118 feet)-will be orbited at a height of 435 kilometers (235 nautical miles) in a nearly-circular path inclined at 50 degrees to the Earth's polar axis.

About a day after the launch of the workshop cluster, the first manned flight-Skylab 2-will be launched. Lifted into orbit by a Saturn IB booster, the three-man crew will make the trip to their orbital home in a modified Apollo command/service module. They will rendezvous and dock, and enter the workshop for a time period of 28 days to conduct the planned programs of experimentation. Their laboratory will give them considerably more room to move around than the Apollo astronauts enjoyed. The total work volume of the Skylab workshop cluster is 316.45 cubic meters (12,673 cubic feet) which is about the size of a four-bedroom house. At the end of their 28-day stay, the crew will ready the workshop for a twomonth dormant period. They will then enter the command/ service module, separate from the cluster, and return to Earth. After two months, the second crew will make its journey to the work shop, this time for a stay of 56 days. But, having studied most of the medical problems in the earlier stay, this time period will concentrate on the solar astronomy and Earth resources experiments. Skylab 4, the last of the planned programmed launches, will go aloft about a month after the completion of the second mission, and will also stay in orbit for 56 days. Earth resources will receive the major share of the attention on this trip.

You can see the evolution of the space program by studying the progress made from the early days of the one-man Mercury flights. A program as daring as the introduction of man into space must, of necessity, start at the first step. And so the first Mercury flights lofted astronauts into ballistic trajectories above the Earth to check initial problems and reactions to manned space flight. Stay time was increased as the orbital trips were added. Two-man Gemini teams explored the concepts of working together to achieve longer stay times and to perform more useful tasks.

Then came Apollo, with its long-term Earth orbits with a three-man crew. Apollo 11, as the first lunar landing, proved that the trip could be made safely. Apollo 12 proved that lunar landings could be made precisely. Later Apollo missions proved the usefulness of man as an explorer and scientist in space as he was on Earth. Then came Skylab, the logical next step, with its extended stay times in space and the opportunity to study Earth over extended periods. It's too early to predict, but the Skylab work may prove to be the most valuable space experiments ever undertaken, in terms of direct applications to everyday problems on Earth.

Beyond Skylab there are two program of extreme interest. The first is the joint Apollo-Soyuz test program, to be the first international space effort. American astronauts, in an Apollo space craft, will rendezvous, dock, and visit an orbiting Soyuz. In turn, Soyuz crewmen will pass through the docking module and return the visit to the Apollo. The target date for the launch is July 15, 1975. There is a deep and serious purpose behind this first planned experiment with Apollo and Soyuz. The first flight will test the designs, the ability to work together, and the ability to communicate and control a joint program. Later spacecraft of both nations will be designed so that they can rendezvous and dock with each other and with their space stations in an extension of that first cooperative joining in space.

Finally, the real benefits of space appear to be achievable only if some economical means of getting to and from orbit can be established. It is, at the moment, costly to use a huge launch vehicle to loft one spacecraft into space only to soon be discarded in a flaming plunge back into the Earth's atmosphere One answer is the space shuttle, an airplane-like orbiter roughly the size of a contemporary twin-jet airliner. Strap-on solid propellant rockets will launch the shuttle initially, and its own propulsion system will take over as the second stage. The spacious design of the shuttle will provide an enormous amount of space-equivalent to the passenger cabin volume of today's four-engined jet air transports-in which payloads can be carried to and from space. Imagine the convenience of carrying a complete scientific satellite, plus the crew to maintain it, to an orbit where it can be parked and watched. Or the ability to retrieve an important satellite communication link, and replace some critical component whose life is limited by time.

The shuttle, after its trip into orbit, can literally fly back to Earth, making a conventional landing at an airstrip. Two weeks later it can be ready for another trip. The payoff here lies in two areas: first, the convenience of the shuttle in getting to and from space easily and economically; second, in the economics of the shuttle. To take a pound of weight to orbit now costs about $1,000 or more, depending on the vehicle used. The shuttle costs are expected to reduce that figure to something less than $200, perhaps as little as $160 per pound in orbit. To put that into perspective, you can orbit the Earth once, right now admittedly at a much lower altitude and via commercial airlines-for about $10 per pound of your weight. The space shuttle economics may or may not ever permit fares like that. But one thing is certain: the space shuttle will provide the only way to make the benefits of space readily and cheaply available. As President Richard Nixon said early in 1972, the space shuttle "... will go a long way toward delivering the rich benefits of practical space utilization and the valuable spinoffs from space efforts into the daily lives of Americans and all people." Skylab will conduct space science, medical, Earth resources and other experiments in an Earth-orbiting laboratory.

Epilogue

... if ever there was a fragile appearing piece of blue in space, it's the Earth right now."

When much else has been forgotten about the Apollo program, people will remember the astronauts' descriptions of their views of the Earth from space, a small, blue, lonesome sphere sailing through a dark galaxy. It was the manned spaceflight program that first gave people an awareness of just how small and finite and limited are the resources of spaceship Earth. The problems were defined more clearly and more tellingly than ever before. Manned space flights awakened a worldwide consciousness that Earth, our home, was all we had to share among billions of people of all colors ar all philosophies. It underlined the fragility of this tiny blue sphere, the pitifully small amounts of water and rainfall, snow in the mountains and blue seas, of green pampas, broad steppes, hillside farms and rice paddies.

But Apollo did more than define the problem. It began to suggest solutions. With men as trained observers in space, people could watch over their own spaceship, could see the fouling of its water and the loss of its green lands, and could begin to plan solutions. Sensitive instruments, cameras, spectrometers, human eyes could see what needed to be done, and send back the coordinates of trouble. It has been said that you can't begin to understand a problem until you can look at it from a distance. The Apollo program, and the manned space flights that will surely follow, place mankind at that needed distance so that he can ponder solutions to his problems. So it is interesting that scientists will better understand the origins of the Moon, and therefore, of the Earth. But it is vital that man finally has seen his home from the depths of space, and has understood, after all, how frail an abode it really is.

Apollo 17 Time of major flight events 1972

Launch at Kennedy Space Center 12:33 a.m. December 7

Spacecraft enters lunar orbit 2:48 p.m. December 10

Lunar module lands at Taurus-Littrow 1:55 p.m. December 11

Liftoff from Moon 5:55 p.m. December 14

Trans-Earth injection 6:35 p.m. December 16

Splashdown 2:24 p.m. December 17

Apollo 17 EVA on Moon 1972

Start	Duration	Distance driven with LRV	Rocks collected
First EVA 6:55 p.m. Dec. 11	7 hr. 12 min.	4.4 km. (3 nmi.)	13 kg. (29 lb.)
Second EVA 6:17 p.m. Dec. 12	7 hr. 37 min.	20 km. (12 nmi.)	36 kg. (80 lb.)
Third EVA 5:23 p.m. Dec. 13	7 hr. 17 min.	11.6 km. (6 nmi.)	66kg (145lb.)

Note: All times Eastern Standard.

APOLLO 17 VEHICLE CHARACTERISTICS

VEHICLE DATA

STAGE/ MODULE	DIMENSIONS DIAMETER (FT)	LENGTH (FT)	WEIGHT DRY (LB)	AT LAUNCH (LB)
Launch Vehicle		363		6,530,820
S-IC Base	63.0			
S-IC	33.0	138.0	288,015	5,038,468
S-IC/S-II *IS (small)	33.0	1.9		1,359
S-IC/S-II *IS (large)	33.0	16.3		8,631
S-II	33.0	81.5	80,377	1,087,580
S-II/S-IVB *IS	33.0 Base 21.7 Top	19.0	7,000	8,019
S-IVB	21.7	59.3	25,084	265,938
IU	21.7	3.0		4,511
SLA	21.7 Base 12.8 Top	28.0		4,059
LM		16.9		36,237
SM	12.8	24.5	13,538	54,007
CM	12.8	11.1		12,844
LES	2.2	33.4		9,167

*Interstage

TOTAL PAYLOAD – 107,147 LB
PAYLOAD CAPABILITY – 108,565 LB

ENGINE DATA

STAGE/ MODULE	QTY	MODEL	NOMINAL THRUST (LB) EACH	TOTAL	BURNTIME (MIN)
S-IC	5	F-1	1,530,000	7,650,000	2.7
S-II	5	J-2	230,000	1,150,000	6.6
S-IVB	1	J-2	208,238	208,238	1st 2.48
			200,550	200,550	2nd 5.88
LM					
Descent	1	MIRA-10K	10,500	10,500	15.0 (TOTAL)
Ascent	1	8258	3,500	3,500	7.6 (TOTAL)
SM	1	AJ10-137	22,000	22,000	12.5 (TOTAL)
LES	1	LPC-A2	150,000	150,000	3.2 (SEC)

FLIGHT DATA

STAGE/ MODULE	EVENT	VELOCITY (MPH)	WEIGHT AT EVENT (LB)
S-IC	Engine Cutoff	6,112	1,857,873
S-II	Engine Cutoff	15,530	477,908
S-IVB	Earth Orbital Insertion	17,457	306,768
S-IVB	Trans Lunar Injection	24,239	143,313
CSM/LM	Lunar Orbit Insertion	3,585	69,971
S-IVB	Lunar Impact	5,683	31,200
LM	Lunar Touchdown	0-2	19,216
LM	Lunar Liftoff		10,688
LM Ascent	Lunar Impact	3,756	
CSM	Trans Earth Insertion	5,640	
CM	Earth Insertion	24,640	

NOTE: The above values are all approximations.

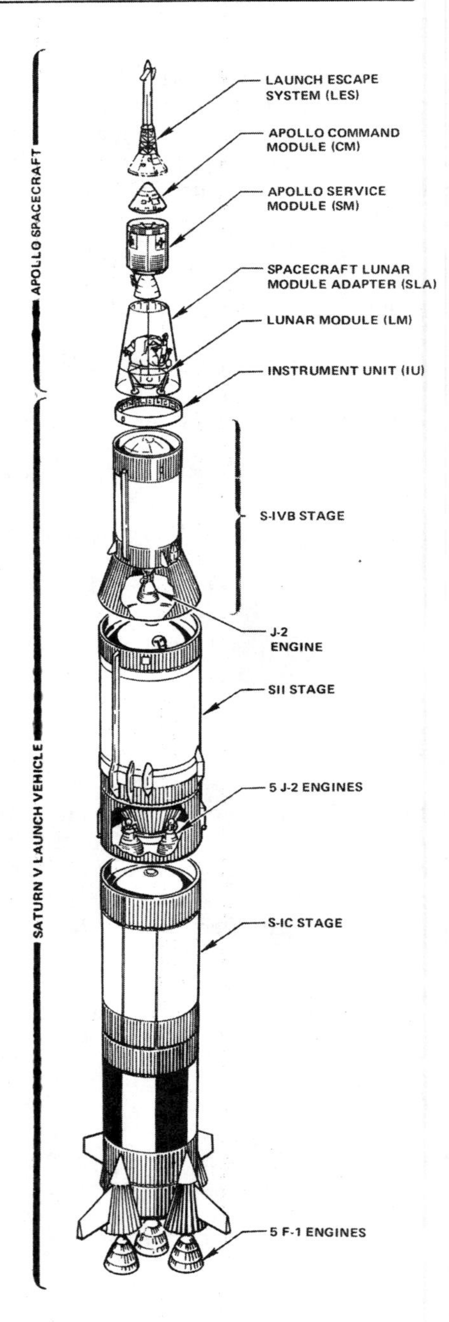

JSC-07904

NATIONAL AERONAUTICS AND SPACE ADMINISTRATION

APOLLO 17 MISSION REPORT

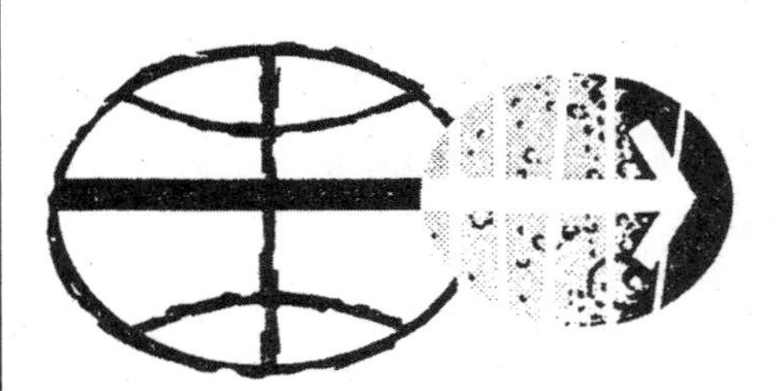

LYNDON B. JOHNSON SPACE CENTER
HOUSTON, TEXAS
MARCH 1973

APOLLO SPACECRAFT FLIGHT HISTORY

Mission	Mission report number	Spacecraft	Description	Launch date	Launch site
PA-1	Postlaunch memorandum	BP-6	First pad abort	Nov 1, 1963	White Sands Missile Range, N. Mex.
A-001	MSC-A-R-64-1	BP-12	Transonic abort	May 13, 1964	White Sands Missile Range, N. Mex.
AS-101	MSC-A-R-64_2	BP-13	Nominal launch and exit environment	May 28, 1964	Cape Kennedy, Fla.
AS-102	MSC-A-R-64-3	BP-15	Nominal launch and exit environment	Sept. 18, 1964	Cape Kennedy, Fla.
A-002	MSC-A-R-65-1	BP-23	Maximum dynamic pressure abort	Dec. 8, 1964	White Sands Missile Range, N. Mex.
AS-103	MPR-SAT-FE-66-4 (MSFC)	BP-16	Micrometeoroid experiment	Feb. 16, 1965	Cape Kennedy, Fla.
A-003	MSC-A-3-65-2	BP-22	Low-altitude abort (planned high-altitude abort)	May 19, 1965	White Sands Missile Range, N. Mex.
AS-104	Not published	BP-26	Micrometeoroid experiment and service module reaction control system launch environment	May 25, 1965	Cape Kennedy, Fla.
PA-2	MSC-A-R-65-3	BP-23A	Second pad abort.	June 29, 1965	White Sands Missile Range, N. Mex.
AS-105	Not published	BP-9A	Micrometeoroid experiment and service module reaction control system launch environment	July 30, 1965	Cape Kennedy, Fla.
A-004	MSC-A-R-66-3	SC-002	Power-on tumbling boundary abort	Jan. 20, 1966	White Sands Missile Range, N. Mex.
AS-201	MSC-A-R-66-4	SC-009	Supercircular entry with high heat rate	Feb. 26, 1966	Cape Kennedy, Fla.
AS-202	MSC-A-R-66-5	SC-011	Supercircular entry with high heat load	Aug. 25, 1966	Cape Kennedy, Fla.

Mission	Mission report number	Spacecraft	Description	Launch date	Launch site
Apollo 4	MSC-FA-R-68-1	SC-017 LTA-10R	Supercircular entry at lunar return velocity	Nov. 9 1967	Kennedy Space Center Fla.
Apollo 5	MSC-PA-R-68-7	LM-1	First lunar module flight	Jan. 22, 1968	Cape Kennedy Fla.
Apollo 6	MSC-PA-2-68-9	SC-020 LTA-2R	Verification of closed-loop emergency detection. system	April 4, 1968	Kennedy Space Center Fla.
Apollo 7	MSC-PA-R-58-15	CSM 301	First manned flight; earth-orbital	Oct. 11, 1968	Cape Kennedy Fla.
Apollo 8	MSC-PA-R-69-1	CSM 103	First manned lunar orbital flight; first manned Saturn V launch	Dec. 21, 1968	Kennedy Space Center Fla.
Apollo 9	MSC-PA-R-69-2	CSM 104 LM-3	First manned lunar module flight; earth orbit rendezvous; extra- vehicular activity	March 3. 1969	Kennedy Space Center Fla.
Apollo 10	MSC-00126	CSM 106 LM-4	First lunar orbit rendezvous low pass over lunar surface	May 18 1969	Kennedy Space Center Fla.
Apollo 11	MSC-00171	CSM 107 LM-5	First lunar landing	July 16, 1969	Kennedy Space Center Fla.
Apollo 12	MS C-01855	CSM 108 LM-6	second lunar landing	Nov. 14, 1969	Kennedy Space Center Fla.
Apollo 13	MSC-02680	CSM 109 LM-7	Aborted during trans-lunar flight because of cryogenic oxygen less	April 21, 1970	Kennedy Space Center Fla.
Apollo 14	MSC-04112	CSM 110 LM-8	Third lunar landing	Jan. 31. 1971	Kennedy Space Center Fla.
Apollo 15	MSC-05161	CSM 112 LM-10	Fourth lunar landing and first extended science capability mission	July 26, 1971	Kennedy Space Center Fla.
Apollo 16	MSC-07230	CSM 113 LM-11	Fifth lunar landing and second extended science capability mission	April 16. 1972	Kennedy Space Center Fla.
Apollo 17	JSC-07904	CSM-114 LM-12	Sixth lunar landing third extended science capability mission. Final mission of the Apollo Program.	December 7 1972	Kennedy Space Center Fla.

WE STOOD ON THE "SHOULDERS OF GIANTS"
..... IN REACHING BEYOND
OUR GRASP. Gene Cernan

APOLLO 17 MISSION REPORT

PREPARED BY Mission Evaluation Team

APPROVED BY

Owen G. Morris Manager, Apollo Spacecraft Program

NATIONAL AERONAUTICS AND SPACE ADMINISTRATION
LYNDON B. JOHNSON SPACE CENTER HOUSTON, TEXAS
March 1973

APOLLO 17 LANDING REGION

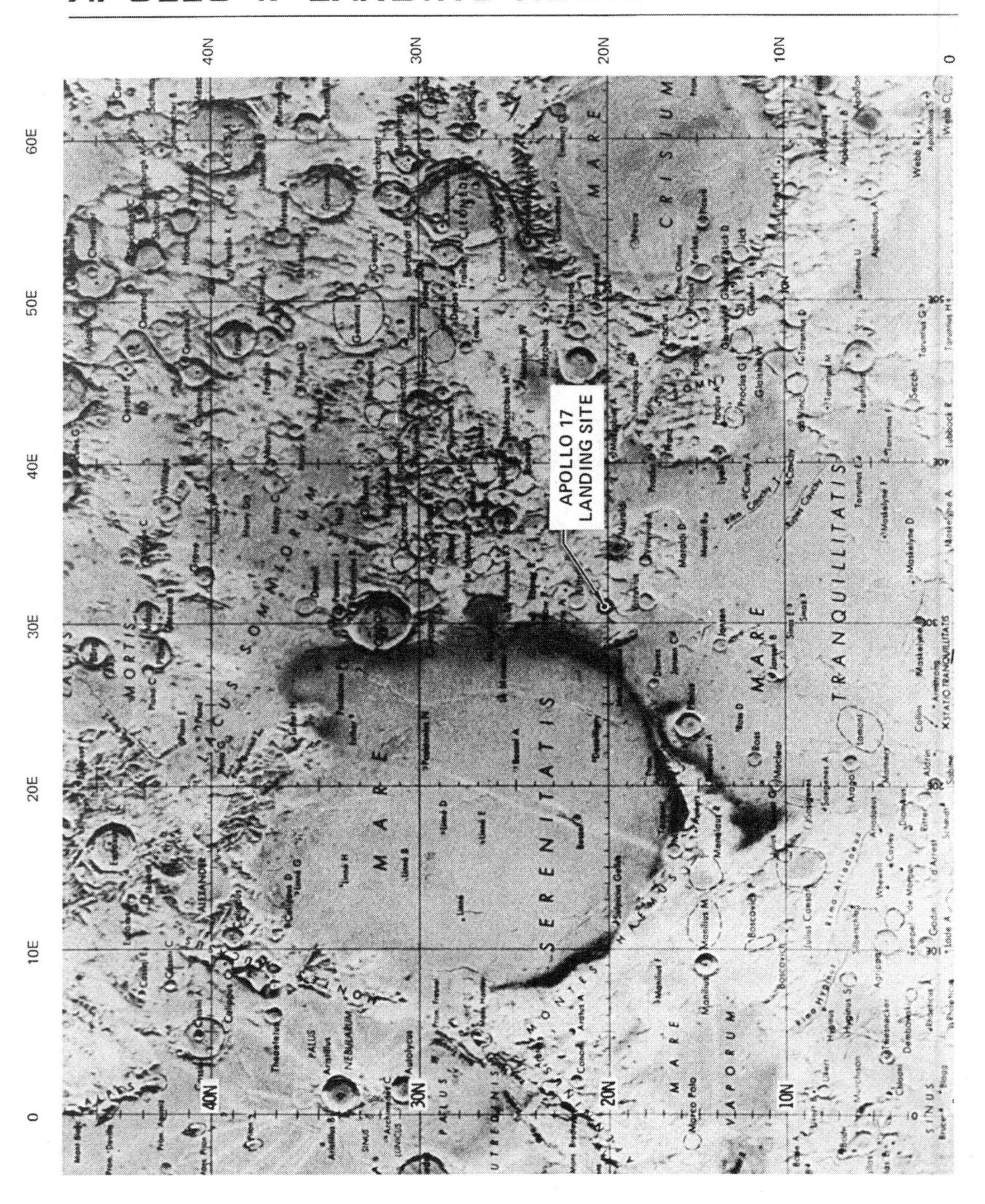

1.0 SUMMARY

Apollo 17, the final Apollo mission, was launched at 05:33:00 G.m.t. (12:33:00 a.m. e.s.t.), December 7, 1972, from Launch Complex 39 at the Kennedy Space Center. The spacecraft was manned by Captain Eugene A. Cernan, Commander; Commander Ronald E. Evans, Command Module Pilot; and Dr. Harrison H. Schmitt, Lunar Module Pilot.

The launch countdown had proceeded smoothly until T minus 30 seconds, at which time a failure in the automatic countdown sequencer occurred and caused a launch delay of 2 hours and 40 minutes. This was the only launch delay in the Apollo program that was caused by hardware failure. As a result, the launch azimuth was adjusted and an earth parking orbit of 92.5 miles by 91.2-miles was achieved. The vehicle remained in earth orbit for approximately 3 hours before the translunar injection maneuver was initiated. The translunar coast time was shortened to compensate for the launch delay. Transposition, docking, and lunar module ejection were normal. The S-IVB stage was maneuvered for lunar impact, which occurred about 84 miles from the pre-planned point. The impact was recorded by the Apollo 12, 14, 15, and 16 passive seismometers.

The crew performed a heat flow and convection demonstration and an Apollo light flash experiment during the translunar coast period. One midcourse correction was performed to achieve the desired altitude of closest approach to the lunar surface. The scientific instrument module door was jettisoned approximately 4 1/2 hours prior to lunar orbit insertion. The Apollo 17 spacecraft initiated the lunar orbit insertion maneuver and entered into a 170-mile by 52.6-mile orbit. Approximately 4 1/2 hours later, the command and service module performed the first of two descent orbit insertion maneuvers lowering the orbit to 59 by 14.5-miles. The command and service module and lunar module stayed in this orbit about 17 1/4 hours before undocking and separating. After undocking, the command and service module orbit was circularized to 70 miles by 54 miles and the lunar module lowered its orbit to 59.6-miles by 6.2-miles by performing the second descent orbit insertion maneuver. From this orbit, the lunar module initiated its powered descent and landed at 20 degrees 9 minutes 55 seconds north latitude, 30 degrees 45 minutes 57 seconds east longitude at 110:21:58.

The first extravehicular activity began at about 114:22. The offloading of the lunar roving vehicle and unstowage of equipment proceeded normally. The lunar surface experiment package was deployed approximately 185 meters west northwest of the lunar module. The Commander drove the rover to the experiments package deployment site and drilled the heat flow and deep core holes and emplaced the neutron probe experiment. Two geologic units were sampled, two explosive packages were deployed and seven traverse gravimeter measurements were taken during the extravehicular activity. About 31 pounds of samples were collected during the 7 hour and 12 minute extravehicular activity.

The second extravehicular activity began at about 137:55. The traverse was conducted with real-time modifications to station stop times because of geological interests. Orange soil was found and has been the subject of considerable geological discussion. Five surface samples and a double-core sample were taken at this site. Three explosive packages were deployed, seven traverse gravimeter measurements were taken, and all observations were photographed. The crew traveled 7370 meters away from the lunar module, and this is the greatest radial distance any crew has traveled away from the lunar module on the lunar surface. About 75 pounds of samples were gathered during the 7 hours 37 minutes of extravehicular activity.

The third extravehicular activity began at about 160:53. Specific sampling objectives were accomplished at stations 6 and 7 among some 3 to 4-meter boulders. Nine traverse gravimeter measurements were made. The surface electrical properties experiment was terminated because the receiver temperature was approaching the upper limits of the data tape and the recorder was removed at station 9.

At the completion of! the traverse, the crew selected a breccia rock, which was dedicated to nations represented by students visiting the Mission Control Center. A plaque on the landing gear of the lunar module commemorating all Apollo lunar landings was then unveiled. Samples amounting to about 137 pounds were obtained during the 7-hour and 15-minute third extravehicular activity for a total of approximately 243 pounds for the mission. The lunar roving vehicle was driven about 36 kilometers during the three extravehicular activities. The total time of the three extravehicular activities was 22 hours and 04 minutes.

Numerous orbital science activities were conducted in lunar orbit while the lunar surface was being explored. In addition to the panoramic camera, the mapping camera, and the laser altimeter, three new scientific instrument module experiments were included in the Apollo 17 complement

of orbital science equipment. An ultraviolet spectrometer measured lunar atmospheric density and composition, an infrared radiometer mapped the thermal characteristics of the moon, and a lunar sounder acquired data on subsurface structure. The orbital science experiments and cameras have provided a large amount of data for evaluating and analyzing the lunar surface and the lunar environment.

The command and service module orbit did not decay as predicted while 'the lunar module was on the lunar surface. Consequently, a small orbital trim maneuver was performed to lower the orbit, and in addition, a planned plane change maneuver was made in preparation for rendezvous.

Lunar ascent was initiated after 74 hours 59 minutes and 39 seconds on the lunar surface, and was followed by a normal rendezvous and docking. Samples and equipment were transferred from the ascent stage to the command module, and the ascent stage was jettisoned for the deorbit firing. The ascent stage impacted the lunar surface at 19 degrees 57 minutes 58 seconds and 30 degrees 29 minutes 23 seconds about a mile from the planned target. An additional day was spent in lunar orbit performing scientific experiments, after which transearth injection was initiated.

A 1-hour and 6-minute transearth extravehicular activity was conducted by the Command Module Pilot to retrieve the film cassettes from the scientific instrument module bay. The crew performed the Apollo light flash experiment and operated the infrared radiometer and ultraviolet spectrometer during the transearth phase. One midcourse correction was performed during this phase.

Entry and landing were normal. The command module landed in the Pacific Ocean west of Hawaii, about 1 mile from the planned location. The Apollo 17 mission lasted 301 hours, 51 minutes, and 59 seconds. The Apollo 17 mission thus brought to a close the Apollo Program, one of the most ambitious and successful endeavors of man.

2.0 INTRODUCTION

The Apollo 17 mission was the final mission in the Apollo program. The mission accomplished the sixth lunar landing and also completed the series of three orbital-science-oriented missions.

The Lunar Module Pilot was the first Scientist-Astronaut assigned to an American manned spaceflight mission. His academic background includes a Doctorate in Geology, and he has participated in many unique geological activities. He was selected as a Scientist-Astronaut in June, 1965, and this was followed by a year of flight training. His first mission assignment was as the backup Lunar Module Pilot for Apollo 15o In 1972, he was assigned as the prime Lunar Module Pilot for the Apollo 17 mission.

The vehicle configuration was similar to those of Apollo 15 and 16. There were significant differences in the science payload for Apollo 17. Spacecraft hardware differences and experiment equipment are described in Appendix A.

The mission achieved a landing in the Taurus-Littrow region of the moon and returned samples of the pre-Imbrium highlands and young craters. An assessment of the mission objectives is presented in section 13.

This report primarily provides information of the operational and engineering aspects of the mission. Preliminary scientific results and launch vehicle performance are reported in references 1 and 2, respectively. A complete analysis of all applicable data is not possible within the time frame of the preparation of this report. Therefore, report supplements will be published as necessary. Appendix E lists the reports and gives their status, either published or in preparation.

Standard English units of measurement are used in those sections of the report pertaining to spacecraft systems and trajectories. The International System of Units (SI) is used in sections pertaining to science activities. Unless otherwise specified, time is expressed as elapsed time from range zero (established as the integral second before lift-off), and does not reflect the time update shown in table 3-I. Mileage is given in nautical miles and weight is referenced to earth gravity.

3.0 TRAJECTORY

The basic trajectory profile for this mission was similar to that planned for the Apollo 16 mission. The major differences, aside from those required to reach the Taurus-Littrow landing site, were those required because of a night launch, translunar injection being initiated over the Atlantic Ocean rather than the Pacific Ocean, descent orbit insertion being performed in two maneuvers rather than one, and the elimination of the orbit shaping maneuver and the satellite jettisoning event. The sequence and definition of events for the Apollo 17 mission are shown in tables 3-I and 3-II. Tables 3-III and 3-IV contain the listing and definition of trajectory parameters, and table 3-V contains a summary of the maneuvers.

3.1 LAUNCH AND TRANSLUNAR TRAJECTORIES

The launch trajectory is presented in reference 3. The launch azimuth was updated from 72 degrees east of north to 91 degrees 30 minutes east of north. The translunar injection differed from the original plan because of a 2-hour 40-minute launch delay. This delay resulted in the translunar coast time being shortened (accomplished automatically by the launch vehicle guidance system), so that the arrival time at the moon would remain the same as that planned prelaunch. This constant time of arrival plan simplified the crew training by providing them with only one lunar lighting condition and one set of lunar groundtracks with which they had to become familiar, resulting in a single set of conditions on which the crew could concentrate their training.

One translunar midcourse correction of 10.5 ft/sec was required and performed at the second option point. The scientific instrument module door was jettisoned about 4 1/2 hours prior to lunar orbit insertion.

3.2 S-IVB STAGE

Separation from the S-IVB stage and the S-IVB evasive maneuver were completed normally. The S-IVB stage was targeted for lunar impact by two firings of the auxiliary propulsion system. Lunar impact occurred approximately 87 hours into the mission at 4 degrees 12 minutes south latitude and 12 degrees 18 minutes west longitude, about 84 miles from the planned target point. The impact was recorded by the passive seismometers at the four lunar surface experiment stations. Figure 3-1 shows the location of the S-IVB impact on the lunar surface.

TABLE 3-1.- SEQUENCE OF EVENTS

Events	Elapsed time Hr:min:sec
Lift-off (Range zero = 342:05:33:00 G.m.t.)	00:00:00.6
Earth orbit insertion	00:11:53
Translunar injection maneuver	03:12:37
S-IVB/command and service module separation	03:42:29
Translunar docking	03:56:45
Spacecraft ejection	04:45:00
First midcourse correction	35:30:00
Mission control center time update (+2:40:00)	65:00:00
Scientific instrument module door jettison	81:32:40
Lunar orbit insertion	86:14:23
S-IVB lunar impact	86:59:41
Descent orbit insertion	90:31:37
Lunar module undocking and separation	107:47:56
Circularization maneuver	109:17:29
Lunar module descent orbit insertion	109:22:42
Powered descent initiation	110:09:53
Lunar landing	110:21:57
Start first extravehicular activity	114:21:49
Apollo lunar surface experiment package first data	117:21:00
End first extravehicular activity	121:33:42
Start second extravehicular activity	137:55:06
End second extravehicular activity	145:32:02
Start third extravehicular activity	160:52:48
End third extravehicular activity	168:07:56
Orbital trim maneuver	178:54:05
Plane change	179:53:54
Lunar ascent	185:21:37
Lunar module vernier adjustment maneuver	185:32:12
Terminal phase initiation	186:15:58
Docking	187:37:15
Lunar module jettison	191:18:31
Separation maneuver	191:23:31
Lunar module deorbit firing	192:58:14
Lunar module impact	193:17:21
Transearth injection	234:02:09
Start transearth extravehicular activity	254:54:40
End transearth extravehicular activity	256:00:24
Second midcourse correction	298:38:01
Command module/service module separation	301:23:49
Entry interface (400 000 feet)	301:38:38
Begin blackout	301:38:55
End blackout	301:42:15
Forward heat shield jettison	301:46:20
Drogue deployment	301:46:22
Main parachute deployment	301:47:13
Landing	301:51:59

TABLE 3-II.- DEFINITION OF EVENTS

Events	Definition
Range zero	Final integral second before lift-off
Lift-off	Time of instrumentation unit umbilical disconnect as indicated by launch vehicle telemetry
Earth orbit insertion	S-IVB engine cutoff time plus 10 seconds as indicated by launch vehicle telemetry
Translunar injection maneuver	Starts when tank discharge valve opens, allowing fuel to be pumped to the S-IVB engine
S-IVB/command module separation, translunar docking, spacecraft ejection, scientific instrument module door jettison, lunar module undocking and separation, docking, lunar module jettison, and lunar landing	The time of the event based on analysis of timing data on air-to-ground voice transcriptions
Spacecraft maneuver initiation	Engine on time as indicated by onboard and/or ground computers
S-IVB lunar impact	Time based upon loss of signal from telemetry
Beginning of extravehicular activity	The time cabin pressure reaches 3 rsia dying depressurization indicated by telemetry data
End of extravehicular activity	The time cabin pressure reaches 3 psia during repressurization indicated by telemetry data
Apollo lunar surface experiment package first data	The receipt of first data considered valid from the Apollo lunar surface experiments package telemetry
Command module/service module separation	The time of separation indicated by loss of telemetry data from service module
Entry Interface	The time the command module reaches s geodetic altitude of 400 000 feet indicated by' ground radar tracking data
Begin blackout	The time of S-band communication is lost during entry
End blackout	The time of acquisition of S-band communications following blackout
Forward heat shield jettison, drogue deployment, and main parachute deployment	Time of first telemetry indication of system actuation by the relay
Earth landing	The time the spacecraft was observed to touch the water
Time update	A given increment of time change made to onboard timers that sets timers to flight plan time.

TABLE 3-III.- TRAJECTORY PARAMETERS[a]

Event	Reference body	Time, hr:min :sec	Latitude, deg:min	Longitude, deg:min	Altitude, mile	Space-fixed conditions		
						Velocity ft/sec	Flight-path angle, deg	Heading angle, deg E of N
ranslunar P ase								
Translunar injection cutoff	Earth	03:18:28	5.14 N	53.86 W	162.4	35 589.6	6.947	118.040
Command and service module/lunar module ejection from S-IVB	Earth	04:45:00	27.91 S	37.68 E	13 393.6	16 012.8	61.80	83.485
irst id ourse orre tion								
Ignition	Earth	35:30:00	17.04 S	22.82 W	128 217.7	4058.1	76.40	66.71
Cutoff	Earth	35:30:02	17.04 s	22.82 W	128 246.9	4066.8	76.48	66.84
Scientific instrument module door jettison	Moon	81:32:40	0.49 N	69.50 W	11 370.6	3774.6	-79.90	258.16
unar or it ase								
unar or it insertion								
Ignition	Moon	86:14:23	12.33 S	177.38 E	76.8	8110.2	-9.90	273.70
Cutoff	Moon	86:20:56	6.81 S	151.84 E	51.2	5512.1	0.43	288.89
irst des ent or it insertion								
Ignition	Moon	90:31:37	12.40 S	164.16 F	51.1	5512.7	-0.39	286.50
Cutoff	Moon	90:31:59	11.06 S	163.0. E	50.9	5322.1	-0.89	286.80
Command and service module/lunar module separation	Moon	107:47:56	5.02 S	135.91 E	47.2	5342.8	-1.26	289.41
o and and ser i e odule ir ulari ation								
Ignition	Moon	109:17:29	20.03 S	149.17 W	58.6	5279.9	0.45	270.13
Cutoff	Moon	109:17:33	20.02 S	149.30 W	58.8	5349.9	0.47	270.17
e ond des ent or it insertion								
Ignition	Moon	109:22:42	19.22 S	165-T8 W	59.6	5274.5	0.04	2T5.74
Cutoff	Moon	109:23:04	19.12 S	166.77 W	59.6	5267.0	0.02	276.06
Powered descent initiation	Moon	210:09:53	19.13 N	48.75 3	8.7	5550.3	-0.90	2T6.07
Orbital trim maneuver ignition	Moon	178:54:05	12.37 S	124.36 E	64.9	5315.1	0.08	285.74

TABLE 3-IV.- DEFINITION OF TRAJECTORY AND ORBITAL PARAMETERS

Trajectory parameters	Definition
Geodetic latitude	The spherical coordinate measured along a meridian on the earth from the equator to the point directly beneath the spacecraft, deg:min
Selenographic latitude	The definition is the same as that of the geodetic latitude except that the reference body is the moon rather than the earth, deg:min
Longitude	The spherical coordinate, as measured in the equatorial plane, between the plane of the reference body's prime meridian and the plane of the spacecraft meridian, deg
Altitude	The distance measured between the spacecraft end the reference radius of the earth along a line from the center of the earth to the spacecraft. When the reference body is the moon, it is the distance measured from the spacecraft along the local vertical to the surface of a sphere having a radius equal to the distance from the center of the moon to the lending site, ft or miles
Space-fixed velocity	Magnitude of the inertial velocity vector referenced to the body-centered, inertial reference coordinate system, ft/sec
Space-fixed flight-path angle	Flight-path angle measured positive upward from the bodycentered local horizontal plane to the inertial velocity vector, deg
Space-fixed heading angle	Angle of the projection of the inertial velocity vector onto the body-centered local horizontal plane, measured positive eastward from north, deg
Apogee	The point of maximum orbital altitude of the spacecraft above the center of the earth, miles
Perigee	The point of minimum orbital altitude of the spacecraft above the center of the earth, miles
Apocynthion	The point of maximum orbital altitude above the moon as measured from the radius of the lunar lending site, miles
Pericynthion	The point of minimum orbital altitude above the moon as measured from the radius of the lunar landing site, miles
Period	Time required for spacecraft to complete 360 degrees of orbit rotation, min
Inclination	The true angle between the spacecraft orbit place and the reference body's equatorial plane, deg
Longitude of the ascending node	The longitude at which the orbit plane crosses the reference body's equatorial plane going from the Southern to the Northern Hemisphere, deg

TABLE 3-V.- MANEUVER SUMMARY

(a) Translunar

Maneuver	System	Ignition time, hr:min :sec	Firing time, sec	Velocity change, ft/sec	Resultant pericynthion conditions				
					Altitude miles	Velocity, ft/sec	Latitude, deg:min	Longitude, Deg:min	Arrival time, hr:min:sec
Translunar injection	S-IVB	3:12:37	351.0	10 376.0	-	8393	10:21 S	173:36 E	83:40:52
First midcourse correction	Service propulsion	35:30:00	1.7	10.5	52.1	8203	9:46 S	159:48 E	83:38:14

(b) Lunar orbit

Maneuver	System	Ignition time, hr:min:sec	Firing time, sec	Velocity change, ft/sec	Resultant orbit	
					Apocynthion, miles	Pericynthion, miles
Lunar orbit insertion	Service propulsion	86:14:23	393.2	2998	170	52.6
First descent orbit insertion	Service propulsion	90:31:37	22.3	197	59	14.5
Command end service module separation	Reaction control	107:47:56	3.4	I.0	61.5	11.5
Lunar orbit circularization	Service propulsion	109:17:29	3.5	70.5	70	54
Second descent orbit insertion	Lunar module reaction control	109:22:42	21.5	7.5	59.6	6.2
Powered descent initiation	Descent propulsion	310:09:53	725	6698	-	-
Orbital trim	Reaction control	178:54:05	31.3	9.2	67.3	62.5
Lunar orbital plane change	Service propulsion	179:53:54	20.1	366	62.6	62.5
Ascent	Ascent propulsion	185:21:37	441	6075.7	48.5	9.1
Vernier adjustment maneuver	Reaction control	185:32:12	10	10	48.5	9.4
Terminal phase initiation	Ascent propulsion	186:15:58	3.2	53.8	64.7	48.5
Separation maneuver	Reaction control	191:23:31	12	2	63.9	61.2
Lunar module deorbit maneuver	Reaction control	192:58:14	1166	286	-	-

(c) Transearth

Event	System	Ignition time, hr:min:sec	Firing time, sec	Velocity change ft/sec	Resultant entry interface condition				
					Flight Path angle, deg	Velocity, ft/sec	Latitude, deg:min	Longitude, deg:min	Arrival time, hr:min:sec
Transearth injection	Service propulsion	234:02:09	143.7	3046.3	-6.31	366 090	2:09 N	173:43 W	301:38:13
Second midcourse	Reaction control	298:38:01	9	2.I	-6.73	36 096	0:44 N	173:20 W	301:38:32

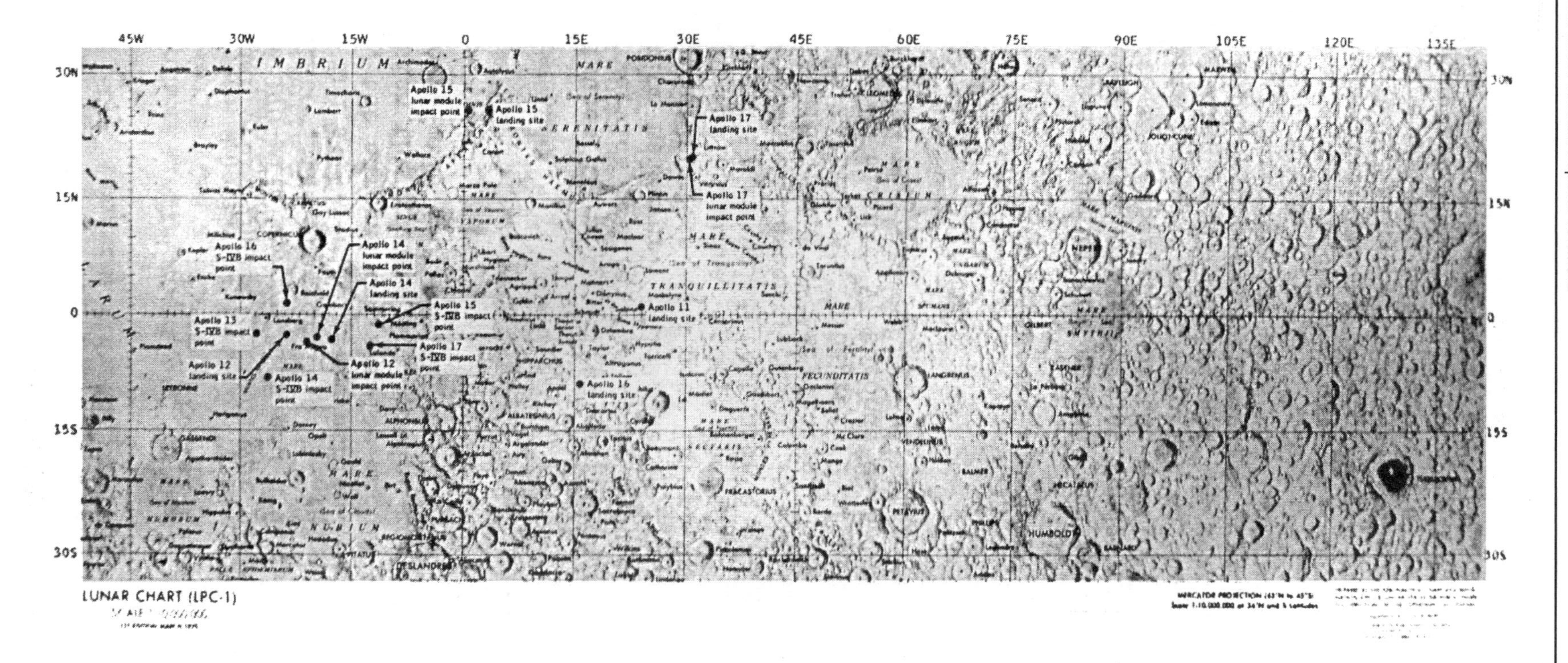

Figure 3-1.- Apollo landing sites and impact locations on the lunar surface.

3.3 LUNAR ORBIT

3.3.1 Orbital Phase

The lunar orbit insertion maneuver placed the spacecraft into an orbit having a 170-mile apocynthion and a 52.6-mile pericynthion. About four hours later, the spacecraft orbit was lowered to one having a 59-mile apocynthion and a 14.5-mile pericynthion. After spending 17 hours in this lower orbit, the command and service module separated from the lunar module after which the command and service module orbit was circularized into one having a 70-mile apocynthion and a 54-mile pericynthion.

3.3.2 Descent

Five minutes after the circularization maneuver was initiated by the command and service module, the lunar module performed the second descent orbit insertion maneuver. This lowered its pericynthion to within 6.2 miles of the lunar surface. An hour later, the lunar module powered descent was initiated and the lunar module landed on the moon at 110 hours 21 minutes 58 seconds. A manual target update of 3400 feet was incorporated early in the powered descent. Later in the descent maneuver, the Commander made eight landing point redesignations. These redesignations resulted in the spacecraft landing at 20 degrees 9 minutes 55 seconds north latitude and 30 degrees 45 minutes 57 seconds east longitude on the 1:25 000-scale Lunar Topographic Photo-map of Taurus Littrow, First Edition, September, 1972.

3.3.3 Ascent and Rendezvous

The planned decay of the command and service module altitude to match the lunar module trajectory at rendezvous was not realized. This was similar to the experience of the Apollo 15 mission. Because of this, an orbital trim maneuver was performed to change the command and service module apocynthion to 67.3 miles and the pericynthion to 62.5 miles. An hour later, a plane change maneuver was performed to provide the proper orbital plane for rendezvous with the lunar module.

The lunar module ascended from the lunar surface at 185 hours 21 minutes 37 seconds after having been on the lunar surface for almost 75 hours. Approximately 7 1/2 minutes later, the ascent stage was inserted into lunar orbit. The achieved orbit required a vernier adjustment maneuver of 10 ft/sec to return the orbit to the planned conditions for rendezvous. The rendezvous was then completed normally, and the two vehicles were docked at 187 hours 37 minutes 15 seconds.

3.3.4 Lunar Module Deorbit Maneuver

The lunar module was jettisoned four hours after docking. The lunar module deorbit maneuver began about an hour and a half after jettisoning and impact occurred at 19 degrees 57 minutes 58 seconds north latitude, and 30 degrees 29 minutes 23 seconds east longitude, about 9.9 kilometers from the Apollo 17 landing site, and about 1.75 kilometers from the planned impact point (figs. 3-1 and 4-1).

3.4 TRANSEARTH AND ENTRY TRAJECTORY

The command and service module remained in lunar orbit approximately 43 hours after the lunar module was jettisoned. The transearth injection maneuver was initiated at 234 hours 2 minutes 9 seconds. The maneuver was so accurate that only one midcourse correction was required during transearth coast, and that was at three hours prior to entry with a differential velocity of 2.1 ft/sec.

The command and service modules were separated 15 minutes before entry into the earth's atmosphere. The command module entered the atmosphere 1200 miles from the landing point and the landing occurred

1.3 miles short of they targeted point. The earth landing coordinates, as determined from the spacecraft computer, were 17 degrees 52 minutes 48 seconds south latitude and 166 degrees 6 minutes 36 seconds west longitude.

4.0 LUNAR SURFACE SCIENCE

The Apollo lunar surface experiments package for this mission consisted of the heat flow experiment, the lunar seismic profiling experiment, the lunar atmospheric composition experiment, the lunar ejecta and meteorites experiment, and the lunar surface gravimeter experiment. Other lunar surface experiments included the traverse gravimeter experiment, the surface electrical properties experiment, the lunar neutron probe experiment, the cosmic ray detector experiment, the lunar geological investigation, and the soil mechanics experiment.

Descriptions of the experiment equipment or references to documents in which the descriptions may be found are contained in Appendix A. A comprehensive discussion of the preliminary scientific results of the mission are contained in reference 1.

4.1 SUMMARY OF LUNAR SURFACE ACTIVITIES

The landing point was in a cratered valley between two massifs. Figure 4-1 is a panoramic camera photograph of the Taurus-Littrow landing site. The variety of topographic features at the Taurus-Littrow landing site provided a valuable asset in the exploration of the lunar surface. The crew completed three periods of extravehicular activity during the 75 hours on the surface. The events of each of the three periods are summarized in table 4-1 and the routes traversed are shown in figure 4-2. The arrangement of the experiment equipment is shown in figure 4-3. More detailed descriptions of the lunar surface activities are provided in sections 4.12 and 10.8.

4.2 APOLLO LUNAR SURFACE EXPERIMENTS PACKAGE CENTRAL STATION

The site selected for deployment of the Apollo lunar surface experiments package was located approximately 185 meters west northwest (bearing of 287°) of the lunar module (fig. 4-3). During preparations for the traverse, the Lunar Module Pilot had difficulty removing the dome from the fuel cask where the fuel capsule is stowed. Insertion of the dome removal tool and dome rotation to the unlocked position went smoothly; however, the tool extraction pull resulted in a separation of the tool from the dome. Using the chisel end of the geological hammer, the dome was pried from the cask. (Section 15.4.4 contains a discussion of this anomaly.) Removal of the fuel capsule from the cask and installation in the generator was completed normally. During the traverse to the deployment site, one of the two central station leveling blocks was knocked off; however, this did not adversely affect the deployment.

TABLE 4-I.- LUNAR SURFACE EXTRAVEHICULAR ACTIVITY EVENTS[a]

Elapsed time, hr:min:sec	Event
144:15:58	Departed for the lunar module with a short stop to deploy seismic profiling experiment explosive charge 8 documented with photographs, and a stop at the lunar surface experiments site to allow the Lunar Module Pilot to relevel the lunar surface gravimeter experiment.
144:32:24	Arrived at the lunar module and started extravehicular activity closeout.
145:19:24	Traverse gravimeter experiment reading obtained.
145:32:02	Lunar module cabin repressurized.
Third Extravehicular Activity	
160:52:48	Beginning of third extravehicular activity.
161:02:40	Traverse gravimeter experiment reading obtained.
161:16:15	Lunar roving vehicle loaded for traverse, and performed panoramic and 500-mm photography.
161:19:45	Traverse gravimeter experiment reading obtained.
161:20:17	Cosmic ray experiment retrieved.
161:36:31	Departed for surface electrical properties experiment site.
161:39:07	Arrived at surface electrical properties experiment site. Activated the experiment, gathered samples, and performed documentary photography.
161:42:36	Departed for station 6 with two short stops to gather enroute samples.
162:11:24	Arrived at station 6. Traverse gravimeter reading obtained, gathered samples including a single core tube sample and a rake sample, and performed documentary, panoramic, and 500-mm photography.
163:22:10	Departed for station 7.
163:29:05	Arrived at station 7. Gathered samples and performed documentary and panoramic photography.
163:51:09	Departed for station 8 with one short stop to gather enroute samples.
164:07:40	Arrived at station 8. Two traverse gravimeter experiment readings obtained, gathered samples including rake and trench samples, and performed documentary and panoramic photography.
164:55:33	Departed for station 9.
165:13:10	Arrived at station 9. Seismic profiling experiment explosive charge 5 deployed, two traverse gravimeter experiment readings obtained, gathered samples including a trench sample and a double core tube sample, and performed documentary, panoramic, and 500-mm photography. Removed data storage electronics assembly from surface electrical properties receiver.
166:09:25	Departed for the lunar module with two short stops, one to gather enroute samples and the other to deploy seismic profiling experiment explosive charge 2 and perform documentary and panoramic photography.
166:37:51	Arrived at lunar module and started extravehicular activity closeout.

Elapsed time, hr:min:sec	Event
166:55:09	Traverse gravimeter experiment reading obtained.
167:11:11	Final traverse gravimeter experiment reading obtained.
167:33:58	Apollo lunar surface experiments package photography completed.
167:36:43	Lunar neutron probe experiment retrieved.
167:39:57	Lunar roving vehicle positioned to monitor lunar module ascent.
167:44:41	Seismic profiling experiment explosive charge 3 deployed.
168:07:56	Lunar module cabin repressurized.
Second Extravehicular Activity	
137:55:06	Beginning of second extravehicular activity.
138:04:08	Traverse gravimeter experiment reading obtained.
138:39:00	Lunar roving vehicle loaded for traverse and a traverse gravimeter experiment reading obtained.
138:44:02	Departed for surface electrical properties experiment site.
138:47:05	Arrived at surface electrical properties experiment site. Activated experiment, gathered samples, and performed panoramic photography.
138:51:43	Departed for station 2 with four short stops; one to deploy seismic profiling experiment explosive charge 4, and three to gather enroute samples.
140:01:30	Arrived at station 2. Traverse gravimeter experiment reading obtained, gathered samples including a rake sample, and performed documentary and panoramic photography.
141:07:25	Departed for station 3 with one stop to obtain a traverse gravimeter experiment reading, gather samples, and perform panoramic and 500-mm photography.
141:48:38	Arrived at station 3. Traverse gravimeter experiment reading obtained, gathered samples including a double core-tube sample and a rake sample, and performed panoramic and 500-mm photography.
142:25:56	Departed for station 4 with two short stops to gather enroute samples.
142:42:57	Arrived at station 4. Traverse gravimeter experiment reading obtained, gathered samples including a trench sample and a double core-tube sample, and performed documentary and panoramic photography.
143:19:03	Departed for station 5 with one stop to deploy seismic profiling experiment explosive charge 1, gather samples, and perform panoramic photography.
143:45:15	Arrived at station 5. Traverse gravimeter experiment reading obtained, gathered samples, and performed documentary and panoramic photography.

Elapsed time, hr:min:sec	Event
First Extravehicular Activity	
114:21:49	Beginning of first extravehicular activity.
114:51:10	Lunar roving vehicle offloaded.
115:13:50	Lunar roving vehicle deployed, test drive performed and documented with photography, gathered samples and performed 500-mm and panoramic photography.
115:40:58	United States flag deployed and documented with photographs and stereo photography.
115:50:51	Traverse gravimeter experiment reading obtained.
115:54:40	Cosmic ray experiment deployed.
115:58:30	Apollo lunar surface experiment package offloaded.
116:06:01	Traverse gravimeter experiment reading obtained.
116:11:54	Traverse gravimeter experiment reading obtained.
116:46:17	Traverse gravimeter experiment reading obtained.
118:07:43	Apollo lunar surface experiments package deployment completed and documented with photographs and panoramic photography.
118:35:27	Deep core sample obtained and lunar neutron probe experiment deployed.
118:43:08	Traverse gravimeter experiment reading obtained.
119:11:02	Departed for station 1.
119:24:02	Arrived at station 1 and deployed seismic profiling experiment explosive charge 6, obtained traverse gravimeter experiment reading, and documented rake samples and performed panoramic photography.
119:56:47	Departed for surface electrical properties experiment site with a stop to deploy seismic profiling experiment explosive charge 7, and perform panoramic photography.
120:11:02	Arrived at surface electrical properties experiment site. Deployed antennas and the transmitter, gathered samples, and performed documentary and panoramic photography. Traverse gravimeter experiment reading obtained.
120:33:39	Departed for the lunar module.
120:36:15	Arrived at lunar module and started extravehicular activities closeout.
121:16:37	Traverse gravimeter experiment reading obtained.
121:21:11	Traverse gravimeter experiment reading obtained.
121:33:42	Lunar module cabin repressurized.

[a] All times are completion times unless otherwise noted.

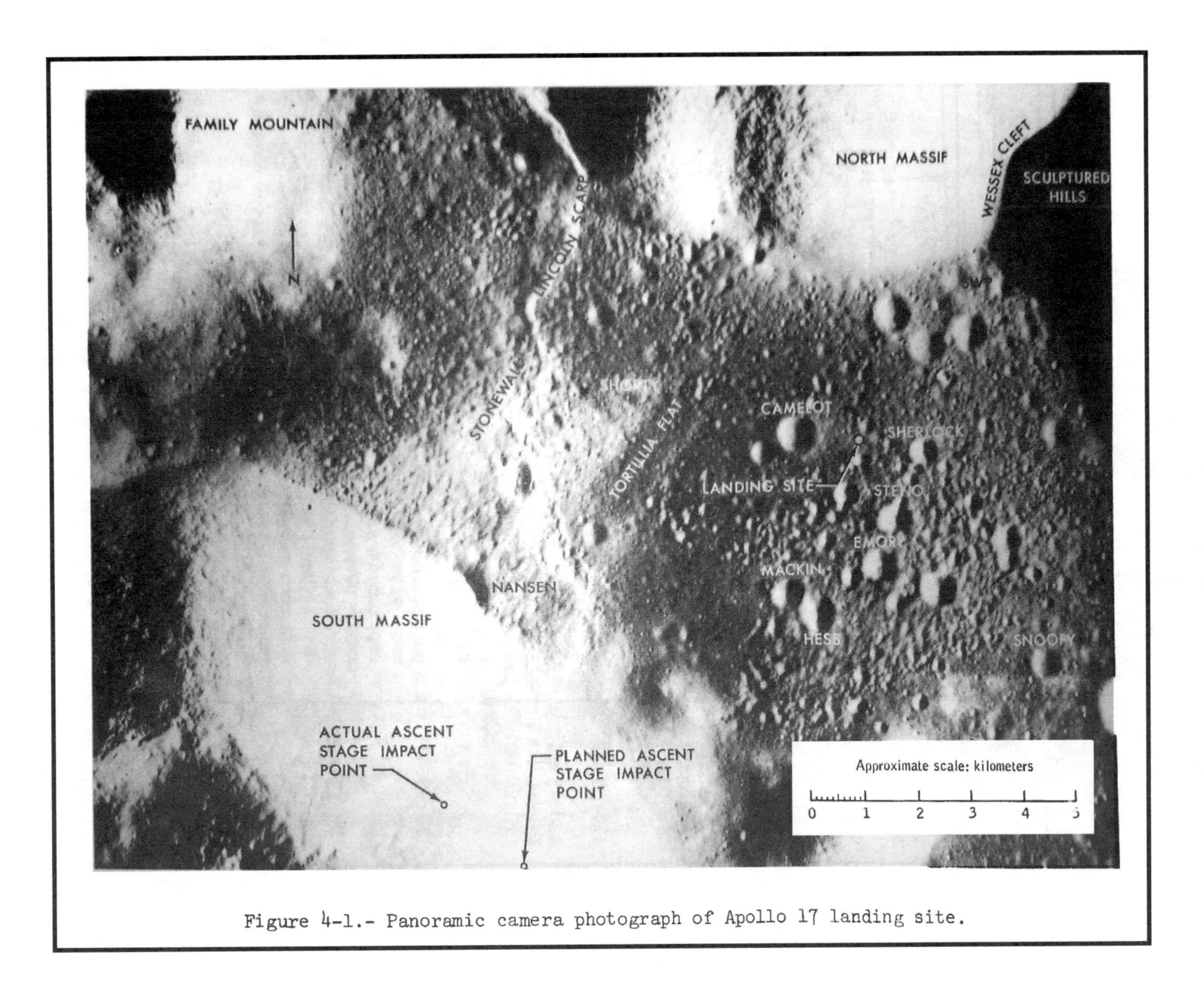

Figure 4-1.- Panoramic camera photograph of Apollo 17 landing site.

Figure 4-2.- Extravehicular activity traverses.

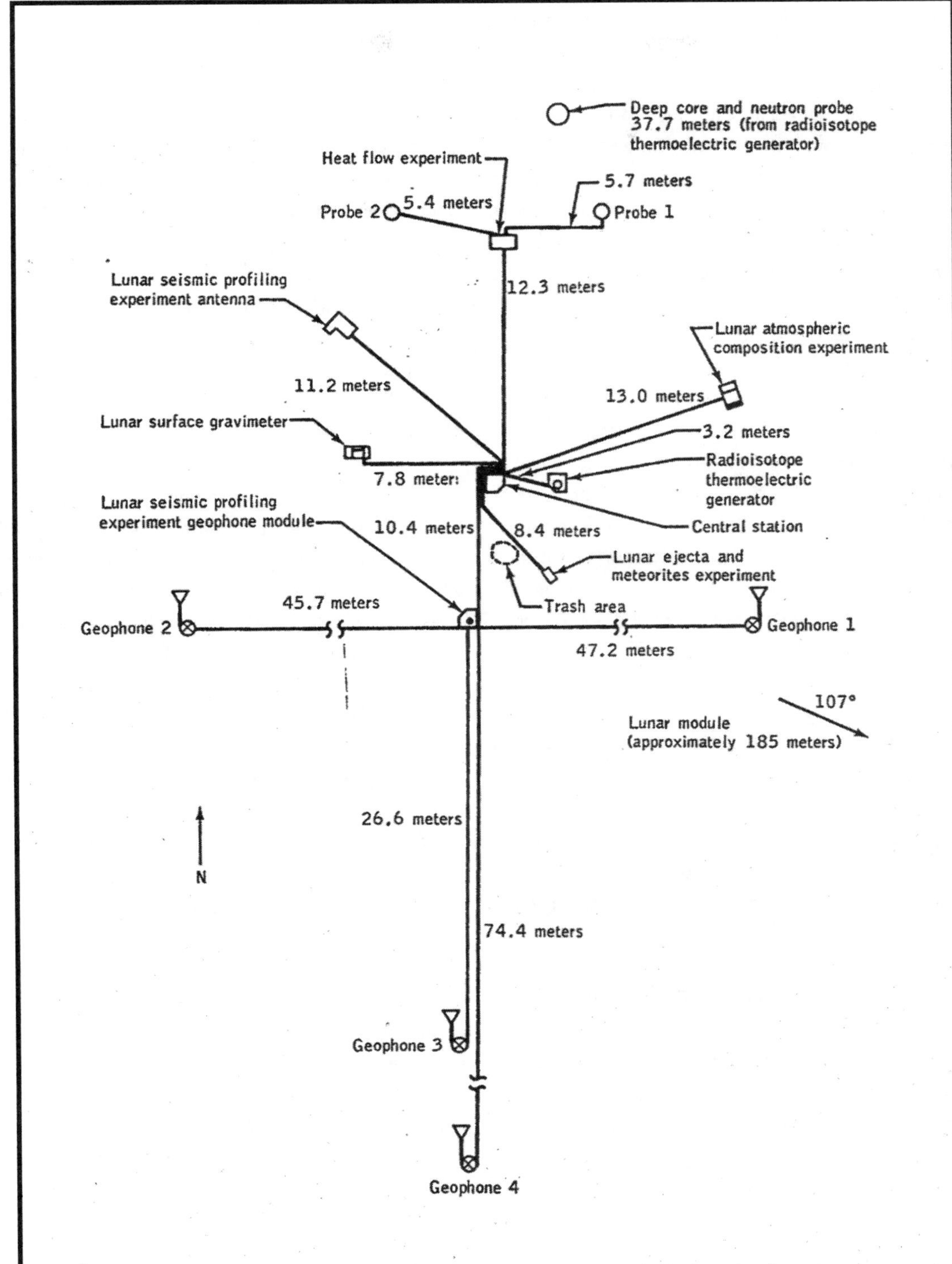

Figure 4-3.- Apollo lunar surface experiments package and neutron probe deployment.

During the antenna alignment, centering of the east/west bubble level was difficult, and near the end of the third extravehicular activity, the antenna level settings were rechecked. The east/west bubble was against the east edge and the north/south bubble was centered, and no changes were made. This amount of misalignment causes a somewhat greater variation in signal strength during the lunar libration cycle, but it will not impact system commands or the transmission and reception of telemetry data.

Initial data were received at 0254 G.m.t. on December 12 and the received signal strength, radioisotope thermoelectric generator power, reserve power, and temperature status were all near the pre-mission predictions. The power output stabilized at 75.8 watts during the first lunar day and increased to 77.2 watts during the first lunar night. The automatic power management circuit is maintaining the average thermal plate temperature of the central station between 270.1° K and 324.8° K.

The telemetry signal power level, as received at various Spaceflight Tracking and Data Network sites, varied ±1.5 dBm in a sinusoidal manner around the normal level and this had no affect on the data. This variation was probably caused by a multipath phenomena produced by an antenna side-lobe reflection from the South Massif.

4.3 HEAT FLOW EXPERIMENT

Two heat flow experiment (S-037) bore stems were drilled into the lunar surface to the planned depth of 254 centimeters and the probes were inserted during the first extravehicular activity (fig. 4-4).

The total power-on time required to drill bore stem 1 into the soil was 3 minutes 46 seconds. The penetration rate was variable with a particularly low penetration rate (40 centimeters per minute) occurring at a depth of about 80 to 100 centimeters. Between 120 and 200 centimeters depth, the penetration rate was 84 centimeters per minute, and in the final 50 centimeters, the rate slowed to about 60 centimeters per minute. Based on crew comments and televised visible sudden torques on the drill handles, some rock fragments were probably encountered during the drilling operations. The first meter of probe hole 2 was drilled more rapidly than that of probe 1 and below 200 centimeters, a resistant layer was encountered which slowed progress to about 60 centimeters per minute. Rock fragments were frequently encountered during the drilling of probe hole 2 which required a total power-on time of approximately 3 minutes.

The heat flow experiment was turned on at 0302 G.m.t. on December 12 and valid temperature data were received from all sensors. The operation of the experiment has been satisfactory. Probe 2 was operating before being inserted into the bore stem and a bore stem-and-probe temperature of about 300° K was indicated immediately after insertion. From December 12 to January 3, the experiment was operated in the normal gradient mode, which samples each sensor every 7.2 minutes. Between January 3 and January 24, eight low-conductivity experiments were conducted with a heater power of 0.002 watt. One high-conductivity experiment was performed on January 26.

The reference thermometer attached to the experiment electronics package radiator plate indicates that the package reaches a maximum temperature of 328° K at lunar noon, and remains at 290° K throughout the lunar night.

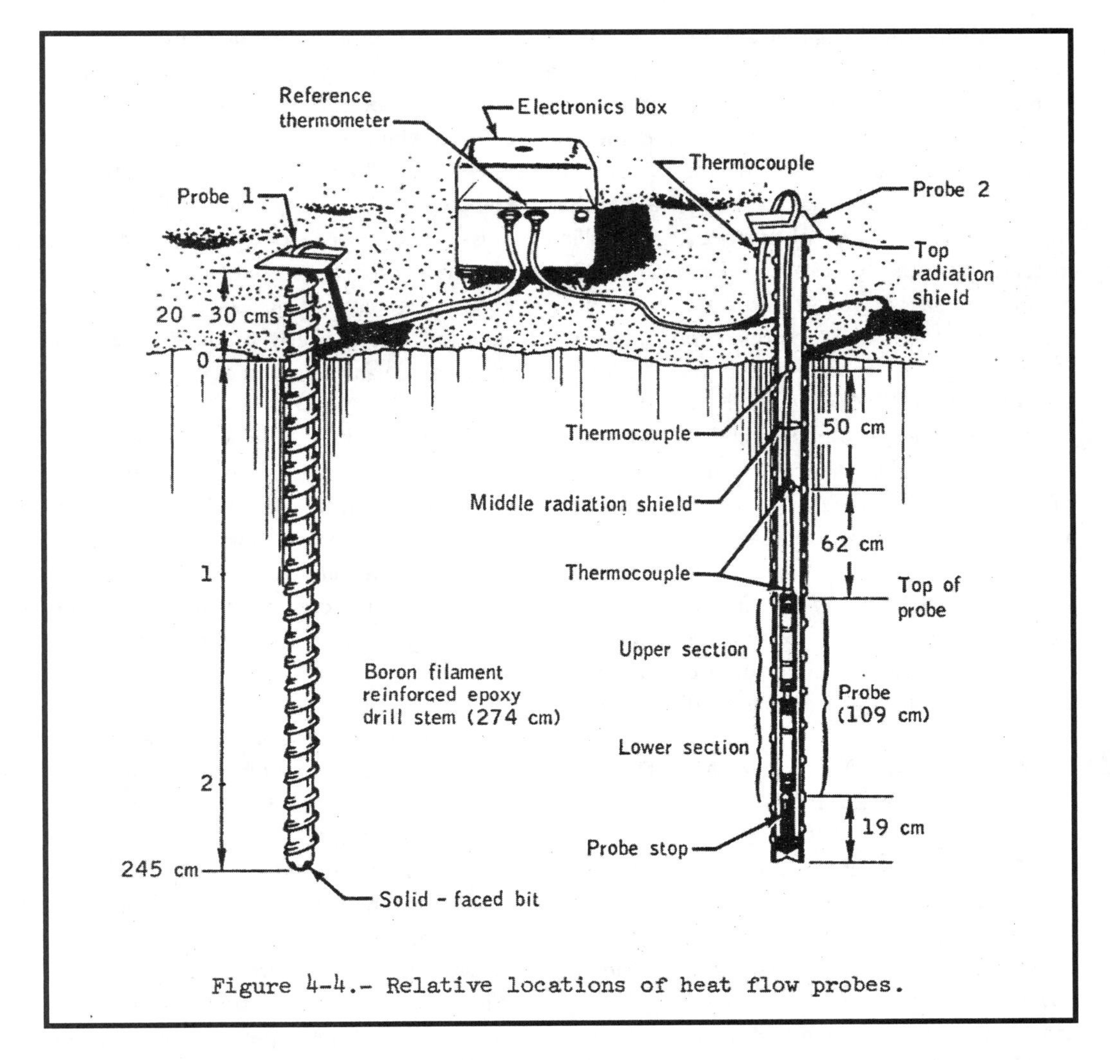

Figure 4-4.- Relative locations of heat flow probes.

4.4 LUNAR SEISMIC PROFILING EXPERIMENT

The lunar seismic profiling experiment (S-203) geophone array was deployed at the lunar surface experiments site (fig. 4-3). The experiment was commanded on to verify instrument operation at 0358 G.m.t. on December 12.

The explosive packages were deployed as shown in figure 4-2 during the three extravehicular activity periods. All of the explosive charges were detonated by command and each of the geophones responded to the detonations. The detonation of explosive package 7 was observed through the television camera.

The central station was commanded to the high-bit-rate mode at 2229 G.m.t. on December 14 to record the impulse produced by the lunar module ascent. A strong seismic signal response from the geophone array was recorded.

The lunar seismic profiling experiment was again commanded to the high-bit-rate mode at 0636 G.m.t. on December 15 to record the lunar module ascent stage impact. The impact occurred at 19 degrees 59 minutes 24 seconds north. and 30 degrees 30 minutes 36 seconds east at 0650 G.m.t. on December 15. The impact point was on the south slope of South Massif, about 8.4 kilometers southwest of the Apollo 17 lending site.

The recording of the seismic signals produced by the detonation of the eight explosive packages, together with the signal from the lunar module ascent stage impact, have enhanced the knowledge of the lunar structure. At the Taurus-Littrow site, the lunar near-surface material has an average seismic velocity of 250 meters per second to a depth of 248 meters. There is an indication of increased apparent velocities within the lunar near-surface material, suggesting the presence of possible interstratified material, perhaps thin lava flow. Beneath the 250 meters per second material, the seismic velocity increases to 1200 meters per second, characteristic of a competent lava flow. The thickness of the 1200 meters per second material is about 925 meters. Underlying the 1200 meters per second layer is material of an undetermined thickness which possesses a seismic velocity of about 4000 meters per second.

When the Apollo 17 data are combined with data from the earlier missions, it should be possible to determine the structure of the lunar crust to a depth of approximately 10 kilometers.

4.5 LUNAR ATMOSPHERIC COMPOSITION EXPERIMENT

The lunar atmospheric composition experiment (S-205) deployment is shown in figure 4-3. All activities associated with the deployment were completed as planned. The dust cover was opened on December 18 at approximately 0420 G.m.t. after the last lunar seismic profiling experiment explosive package was detonated. After allowing the radiator temperature to decrease from a peak of 340.8° K (before cover removal), to 327.5° K, nine hours of ion source outgassing was accomplished the following day with a temperature of 523° K having been reached.

The first activation of the experiment occurred on December 27 at approximately 1800 G.m.t. The instrument responded well to commands and was operating normally except for a background count ramp in the low- and mid-mass channels. The presence of this interference will not cause the loss of data, but it will, increase the difficulty in reducing the data from this portion of the spectrum. This anomaly is discussed in section 15.4.5.

Operation throughout the first lunation was characterized by good performance of the instrument except for two occasions during lunar sunrise when the instrument switched into the high-voltage-lock mode, which stopped the stepping of the sweep voltage. Two logic system noise bursts, which occurred just after sunrise, may have caused the sweep high voltage to be commanded into lock. On both occasions, the situation was properly rectified by commanding the sweep high voltage back on.

Many residual peaks were observed in the spectrum, but only the helium peak is clearly native to the moon. The remainder are at least partially due to outgassing of the instrument, the lunar module, and the Apollo lunar surface experiments package, as evidenced by the peaks decreasing in amplitude throughout the lunar night.

Sunrise brought a large increase in all peaks in the spectrum except helium which decreased as would be expected if it were a native non-condensible gas. Operation was curtailed 24 hours after sunrise because of the very high gas densities in the ion source chamber (approximately 109 molecules per cubic centimeter) from material outgassing. A 15-minute period of operation was performed at lunar noon. Full-time operation was begun again a few hours before sunset on January 23. There was a marked decrease in gas densities about the time of sunset.

Daytime operation will be limited to those times when the outgassing levels are tolerable. As indicated by the data from previous missions, the instrument, site, and other artifacts will eventually be sufficiently free of outgassing contaminants to enable the lunar atmospheric composition experiment to produce high quality data.

4.6 LUNAR EJECTA AND METEORITES EXPERIMENT

The lunar ejects. and meteorites experiment (S-202) deployment is shown in figure 4-3. The radiator mirror was uncovered at 0957 G.m.t. on December 21, after detonation of the last explosive package and when the internal structural temperature had decreased to 344° K. The experiment was operated for about 15 hours on December 22 but was turned off because of rising temperatures. The experiment was turned on again on December 23 to record the background noise rate for the instrument. The sensor covers were not removed until December 28, after completion of continuous operation which encompassed about 60 hours of lunar day and about 50 hours of lunar night. The instrument data indicate that the background noise rate was essentially zero.

The instrument is, in the full operating mode from sun angles of 20° before sunset through 20° after sunrise because of a thermal control problem during the lunar day. Section 15.4.3 contains a discussion of this anomaly. Fortunately, the instrument measurements of ejects are best performed during the lunar night when the primary particle impact rates are near zero. The instrument will remain off during most of the lunar day until the science objectives which can be achieved during lunar night are fulfilled or until satisfactory thermal control can be attained.

The measurements indicate that the detected number of lunar ejects, particles compare within an order of magnitude to the number of primary particles. The measurements are verifying that the bulk of cosmic dust material comes from the general direction of the sun which agrees with the results obtained from similar instruments carried on Pioneer 8 and 9.

4.7 LUNAR SURFACE GRAVIMETER EXPERIMENT

The lunar surface gravimeter experiment (S-207) (fig. 4-3) was activated at 0523 G.m.t. on December 12, and following activation, the data (science and engineering) indicated normal operations. However, the addition of all masses apparently set the sensor beam against the lower stop and removal of one mass apparently moved the beam against the upper stop. Operation of the adjustment screws failed to null the beam and the second adding of all masses failed to move the beam to the lower stop. This anomaly is discussed in section 15.4.1.

On January 3, 1973, the experiment was re-configured, and the sensor beam was centered by adjusting the mass change mechanism to obtain long-term seismic and free-mode science data. In this revised configuration, tidal data are nit being obtained, but the experiment is collecting long-term seismic and free mode information.

The lunar surface gravimeter experiment sensor's initial on-scale temperature of 321.5° K occurred at about 2054 G.m.t., December 14, some 63 hours after initial turn-on. The experiments sensor temperature now remains stabilized at 322.323° K. The instruments subsystem components continue to operate normally providing engineering status data.

4.8 TRAVERSE GRAVIMETER EXPERIMENT

The traverse gravimeter experiment (S-19,) made an earth-moon gravity tie and obtained the value of gravity at various stops along the traverses, relative to the value if gravity at the landing site. Using absolute gravity measurements in the Earth, a preliminary value if 162 694 (±5) milligals was obtained at the Taurus-Littrow landing site. The number of stations (12) at which discrete gravity measurements were made (fig. 4-2 and table 4-I) was about as planned with an extra measurement being made at station 2A and no measurements taken at station because if time constraints.

The hardware performed satisfactorily except for differences between measurements made in the lunar roving vehicle and measurements made with the experiment resting on the surface at the same sites. The values measured in the surface were always lower. The difference in the gravity measurement at the landing point was 4.6 milligals, at station 8 it was 6.9 milligals, and at station 9 it was 6.2 milligals. The reason for this discrepancy is not known. In initial postflight calculations, an empirical correction of -6 milligal was established for all values measured when the experiment was mounted on the lunar roving vehicle.

A preliminary analysis if the data has been made by projecting it to a northwest-southeast profile and by making two dimensional approximations for all the reductions and interpretations. Free-air and Bouguer corrections were applied to the data. The resultant Bouguer corrections shows values at stations 2A and 8, near the South and North Massifs, respectively, which are more than 20 milligals lower than the value at the landing site. The variation of Bouguer values in the central part if the valley are relatively small, although the value at station 4 is a few milligals higher and the value at station 5 is a few milligals lower than the value at the landing site. A preliminary interpretation of the gross features of the gravity profile is a model with basalt flows having a positive density contrast of 0.8 gram per cubic centimeter and a thickness if 1 kilometer buried under the valley floor.

4.9 SURFACE ELECTRICAL PROPERTIES EXPERIMENT

The surface electrical properties experiment (S-204) was deployed as shown in figure 4-2, and was utilized during portions of both the second and third extravehicular activity periods.
During the transmitter deployment, a problem was encountered in keeping the solar panel open because of memory in the solar panel wiring harness. The crew resolved the problem by taping the panel fully open. Also, during the deployment of the transmitter antennas, the two sets of dipoles were reversed from the planned orientation. The effect of the reversal was corrected in the data reduction process with no loss of data.

A thermal control problem with the surface electrical properties receiver caused the premature termination of the experiment. This problem is discussed in detail in section 15.4.2. Despite these problems, when the surface electrical properties experiment recorder was returned to earth, 1 hour and 42 minutes of data had been obtained.

4.10 LUNAR NEUTRON PROBE EXPERIMENT

The lunar neutron probe experiment (S-229) was deployed as shown in figure 4-3 and emplaced in the deep drill core hole during the first extravehicular activity period. The 2-meter probe was retrieved and deactivated at the end of the third extravehicular activity period, accruing 49 hours of exposure. The site was approximately 38 meters north of the Apollo lunar surface experiments package radioisotope thermoelectric generator (fig. 4-3) Corrections required because of the proximity of the generator, which is a strong source of neutrons, should be small; although experimentation is necessary to determine the size of the correction. The probe was returned, disassembled, and the targets and detectors were in excellent condition. The background due to the direct interaction of the fast neutrons from the radioisotope thermoelectric generator with the plastic was measured, and appears to be negligible.

Only the mica detectors have been completely examined. Analysis of the remaining detectors and data is continuing and will be completed after postflight calibration of the probe. Although the calibration data have not been completely processed, the track densities are in the expected range. The neutron capture rates appear to be within a factor of two of those estimated from the theoretical calculations.

4.11 COSMIC RAY DETECTOR EXPERIMENT

The cosmic ray detector experiment (S-152) was deployed by pulling the slide cover open and hanging the cover in the shade while the box was in the sun on the lunar module. In both cases, the nuclear particle detectors faced outward from the lunar module. The detectors were exposed to the lunar environment for approximately 45.5 hours. No degradation of any of the detector surfaces was found. Microscopic examination of the detector surfaces showed very little dust. The maximum temperature of approximately 400° K (as shown by temperature labels) was well below the critical limit.

The tracks of heavy solar wind ions are clearly visible, as are the tracks of intermediate-energy heavy particles. The flux of the latter is surprisingly high (approximately 6 x 103 tracks per square centimeter)

and indicate that the sun emits an appreciable flux of particles in the range of 10 keV per nucleon to 10 meV per nucleon, even at times of quiet sun. These particles were found in both the shade and sun detectors and thus are not directly associated with the solar wind.
The tracks have been seen in the mica, glass, and plastic detectors. The energy spectrum is similar to that of solar flares. This intermediate energy component is possibly associated with a visually active sun spot area that was present during the entire Apollo 17 mission. This experiment is the first flown that could have detected the presence of the intermediate energy particles. The presence of the intermediate-energy particles will limit the degree to which the radon atmosphere can be established at the Apollo 17 site, but this constitutes no degradation of the basic experiment.

4.12 LUNAR GEOLOGY

4.12.1 Sample Collection

Areas sampled for the lunar geology investigation (S-059) during the extravehicular activities included all of the mapped units at the site (fig. 4-2). The variation and location of the 110.405 kilograms of rocks and soils collected are presented in the following table.

Most of the rocks collected were described as crystalline, with basalts and blue-gray breccias being the dominant rock types. The breccias exhibit a very complex multi-cycle history, and breccias apparently occur as inclusions in the anorthositic gabbro. A variety of compositions was obvious among the small clasts in the breccias.

Only station 10 was deleted from the preplanned stations. The crew drove past the station 10 area and verbally reported the rock types. The station 4 sampling tasks on the bright mantle were deleted because of the short time at the station in preference to sampling the first priority target - Shorty Crater. Sampling at all other stations was as planned. A 3.2-meter core was obtained at the Apollo lunar surface experiments package site. Single cores were collected at the lunar module and at station 6.

Type of material	Locations where found [a]
Dark mantle	Stations 1A, 5, 8, 9, landing site, and lunar roving vehicle stops 1, 3, 7, 8, 9, and 11.
Subfloor	Stations 1A, 5, 9, landing site, and lunar roving vehicle stops 3 and 9.
Bright mantle	Stations 2, 3, lunar roving vehicle stops 2, 4, 5, 6, and possible station 4.
South Massif	Station 2.
North Massif	Stations 6, 7, and possibly at lunar roving vehicle stop 10.
Young crater ejecta	Stations 4 and 9.

[a] Figure 4-2 shows the locations of all stations and sampling stops.

Double cores were collected at stations 3, 4, and 9 (the lower stem from station 3 was placed in the core sample vacuum container). The special environmental sample container was used for the fuel products contamination sample at the lunar module. Rake samples were collected at stations 1A, 2, 3, 6, and 8. Sampling of permanently shadowed areas were attempted at station 2 and 6, but were probably successful only at station 6. Samples from the top and bottom of a boulder and the soil under it were collected at station 8. Documented rock and soil samples were collected at all stations and most lunar roving vehicle sampling stops. Panoramic photographs taken from the vicinities of stations 4, 5, and 6 are shown in figures 4-5 through 4-7.

4.12.2 Summary of Geology

To provide a context for the individual sample location descriptions, a brief description of the geologic characteristics of each sampling station is included. The stations and lunar roving vehicle sample stops are described in the sequence in which they occurred on the traverses. The interpretations in these summaries are tentative and some will almost certainly be modified after more information on the samples is available.

4.12.2.1 Station at landing site area.

The landing site station is located in three general areas ranging from 200 meters east to 200 meters west of the lunar module and covering approximately 0.5 square kilometer. The station is located near the center of Taurus-Littrow valley in an area where fine- to coarse-grained subfloor basalts are overlain by a regolith that may contain or be overlain by dark mantling material.

The valley floor at the station is smooth, locally flat, and only gently rolling. The abundance of surface rocks is higher than the regional average for the valley floor and ranges from 2 to 7 percent at the lunar module and experiments package sites to less than 1 to 2 percent at the surface electrical properties transmitter/antenna site. Blocks range up to a maximum of 4 meters in the experiments package area, and rock burial and filleting in the station area is pronounced on some meter-sized blocks.

Rock types at the station are extremely limited and consist predominantly of coarse-grained subfloor basalt with local fine-grained variations. Clods of soil breccia associated with impact events are also present. Rock sample types collected at the three areas of the station include: Lunar module site (1 glass-coated breccia; 2 coarse-grained basalts; 1 fine-grained basalt); Experiments package site (1 soil breccia; 1 coarse-grained basalt; 1 fine-grained basalt); Surface electrical properties antenna site (possibly 3 fine-grained basalts). There is evidence that the fine-grained basalts are more common at the surface electrical properties antenna area east of the lunar module site. Other samples include a deep core sample (3.2 meters) and a single core tube.

Soils are commonly medium dark gray and more cohesive with depth to a depth of 25 to 35 centimeters. The soil consists of layers of different drilling resistance judging from the alternating zones of easy to difficult drilling that were encountered in all holes, with the most resistant zone being at about 2 meters. There are no conspicuous surface lineaments in the station area. "Raindrop" texture is common over the area but most pronounced at the surface electrical properties antenna site. Exotic feldspar-rich components from the North and South Massif units do not appear to be abundant in the regolith at the station.

The surface in the station area contains many subtle 20-centimeter to 2-meter craters, the latter probably representing the current steady state limit of crater size. The nearest large craters to the station are Rudolph (80-meter diameter), 70 meters to the north of the experiments package, and highly subdued Poppy (100-meter diameter), 70 meters south of the lunar module site. Most of the observable surface blocks are associated with larger crater ejecta such as Camelot (600-meter diameter and to the west a distance of about 1-crater diameter).

Figure 4-5. Panorama looking east from Station 5 at Camelot Crater

Figure 4-6. Panorama of exploration area from vicinity of Shorty Crater

Figure 4-6. Concluded

Figure 4-7. Panorama looking south from station 6 above Henry crater

Soils are commonly medium dark gray and more cohesive with depth to a depth of 25 to 35 centimeters. The soil consists of layers of different drilling resistance judging from the alternating zones of easy to difficult drilling that were encountered in all holes, with the most resistant zone being at about 2 meters. There are no conspicuous surface lineaments in the station area. "Raindrop" texture is common over the area but most pronounced at the surface electrical properties antenna site. Exotic feldspar-rich components from the North and South Massif units do not appear to be abundant in the regolith at the station.

The surface in the station area contains many subtle 20-centimeter to 2-meter craters, the latter probably representing the current steady state limit of crater size. The nearest large craters to the station are Rudolph (80-meter diameter), 70 meters to the north of the experiments package, and highly subdued Poppy (100-meter diameter), 70 meters south of the lunar module site. Most of the observable surface blocks are associated with larger crater ejecta such as Camelot (600-meter diameter and to the west a distance of about 1-crater diameter).

The samples collected at the station represent varieties of coarse to fine-grained basalts that make up the sub-regolith valley floor. A distinct layer of dark mantle material overlying regolith was recognized neither on photographs, from the mission nor by the crew during the mission, suggesting that the upper part of the dark mantle, if it exists as a separate entity, has been gardened into a normal regolith overlying the subfloor basalts. The deep core sample taken at the experiments package site should provide the best stratigraphic data on the postulated dark mantle material. However, gardening of dark mantle and normal regolith may have destroyed original dark mantle depositional characteristics.

4.12.2.2 Station 1A

Station 1A is located about 1 kilometer south-southeast of the surface electrical properties antenna site and about 150 meters from the northwest rim of Steno Crater. Steno is not visible in any of the station photographs, but its north rim is apparently the blocky area that rises to the south in the photographs. The station is on the subdued ejecta blanket of Steno Crater that has been covered by dark mantle.

The station area is gently rolling, a reflection of the influence of the ejects, blanket of Steno Crater, a 70-meter very subdued crater to the east, and other hummocks that are remnants of older crater forms. The station is characterized by a scattering of rocks that range up to 1/2 to 1 meter. Some distant boulders on the rim of Steno range up to several meters. Boulders are concentrated around smaller craters only near the 10-meter station crater. Other concentrations do not seem to be related to crater rims. Fillets are not well developed; burial of blocks ranges from perched to almost totally buried.

The major sampling areas are a 10-meter blocky rim crater and a rake sample from a relatively flat area free of blocks 15 meters east of the blocky crater. The sample areas represent the extremes of rock concentrations. Sampling of the 10-meter blocky crater yielded only subfloor basalt fragments from the two large boulders that were sampled. Sampling of the intercrater area yielded some subfloor basalt samples plus some fragments that no doubt represent exotic material.

The fine-grained soil seems to be the same gray tone everywhere in the station area. There are no obvious lighter-toned zones, and no concentration of raindrop impressions or lineaments.

Craters in the station area range from several centimeters up to tens of meters. Most are moderately subdued to very subdued. Ejects is readily apparent only around the 10-meter blocky crater that is the prime sampling area.

The fragment samples are important because they are examples of subfloor basalt, derived from the crater floor or possibly re-excavated Steno ejecta. The soil samples are significant as examples of material that probably represents dark mantle.

4.12.2.3 First lunar roving vehicle sample stop.

The first lunar roving vehicle sample stop is located along the route from the lunar module area to station 2, between Horatio and Bronte Craters. Like the lunar module area and station 1, the sample stop area is in the dark mantle unit shown on premission photogeologic maps; the surface appearance, however, is quite different.

Both the lunar module area and station 1 are within a cluster of large (0.5- to 0.7-kilometer diameter) craters, and the landscape is characterized by boulders and smaller blocks probably ejected from those craters. The visible younger craters more than a few meters in diameter are also blocky, probably because of re-excavated blocks from the ejects blankets of the large craters. The block-studded dark mantle of the lunar module area extends to the vicinity of Horatio Crater.

West of Horatio, the dark mantle surface appears strikingly smooth; less than one percent of the area is covered by rock fragments, and these are noticeably smaller than those typical of the lunar module area. Craters more than one meter across are widely separated, and some craters as large as 20 or 30 meters in diameter are block free. Craters with ejected fragments of soil breccia are also present.

This difference in the nature of the crater ejecta and the general block population indicates that the dark mantle west of Horatio, where this sample was obtained, is likely to be less mixed with material derived from the subfloor unit. The first lunar roving vehicle sample stop area was selected by the crew on a ray from one of the rare, blocky rimmed craters in this generally block-free area. The crater is 10 to 15 meters in diameter, and the sample was taken from about three-fourths of a crater diameter out from the rim. The rocks are irregular and jagged, but some have rounded tops as if easily eroded. All the blocks from the crater are less than 1 meter in diameter with the largest resting on the crater rim. Burial seems insignificant, fillets are not present, and dust is not seen upon the rocks. The soil of the area appears typical of the dark mantle region. Other craters in the area are either block-free or have ejected what seem to be clods of soil breccia. The rock fragment sampled from the crater ejecta is a piece of hard rock, probably subfloor basalt.

4.12.2.4 Second lunar roving vehicle sample stop.

The second lunar roving vehicle sample stop was made on the tongue of light mantle about 3.6 kilometers west of the lunar module. The sample was scooped from the bright rim of a crater. The location of the sampled area in the photographs taken along the traverse route is very tentative.

Rock fragments up to several centimeters across are sparse, but the driving photographs, which were taken down-sun, are not adequate to determine the population of rock fragments in detail. The crew reported that the population of fragments larger than 1 centimeter across is less than 1 percent. They also reported a few larger blocks that look like subfloor material but noted that it was difficult to be sure. They reported that the rims of the craters in the light mantle are somewhat brighter than those in the dark mantle, and that the albedo of the surface appeared somewhat brighter looking down-sun. Otherwise, the surface appears similar to that of the dark mantle. Craters up to 5 meters in diameter are fairly common in the sample area. Most are subdued or have only slightly raised rims. Small cloddy craters are scattered along the traverse route. The material sampled probably represents light mantle material that was ejected from the crater.

4.12.2.5 Third lunar roving vehicle sample stop.

The third lunar roving vehicle sample stop was made at a site in the dark mantle between the main body of light mantle and a finger of light mantle that lies to the southeast. The area resembles the dark mantle surface at the first lunar roving vehicle sample stop rather than the dark mantle surface at the lunar module and station 1. It appears in the surface photography as an exceptionally smooth-surfaced flat area with a pock-marked appearance caused by widely spaced craters that are nearly free of blocks on their rims; the soil between craters has a "raindrop" appearance. The block population covers less than 1 percent of the area, and all blocks are less than 1 meter across. Clods of soil breccia are present on the rims of a few craters less than 3 meters across.

Samples include a rock fragment and a "couple of teaspoons" of soil. The rock is from a cluster of rocks on the surface that are not clearly related to any crater. The rocks in the area all appear similar and resemble subfloor basalt. The soil sample is a scoop of soil from the same area, and can be considered typical of the surface material in a dark mantle area free of large blocks. There is no evidence of the presence of light mantle debris in the area of the third lunar roving vehicle sample stop.

4.12.2.6 Station 2.

Station 2 is located at the foot of South Massif where it intersects the southeast margin of Nansen Crater, and near the contact between the light mantle and materials of the South Massif.

Rock chips were collected from three boulders, and soil was collected from near and under the boulders, in a small area low on. the slopes of the South Massif. A rake sample was collected from the same area and another from an area about 40 meters northeast of the base of the massif, in the light mantle.

Rocks visible in the photographs in and around the sampled area vary in size from less than 1 centimeter across up to boulders more than 2 meters across. Rocks are more abundant near the base of the slope than at intermediate elevations or out on the light mantle. Most of the rocks larger than 25 centimeters are rounded, and smaller rocks range from angular to rounded.

Rocks of all sizes vary from almost totally buried to virtually no burial. Fillets are poorly developed except on the uphill sides of rocks that are on slopes greater than 10 degrees. The downhill sides of such rocks commonly have no fillets.

Except for rocks in the rake sample, the rock samples that were collected from the lower slopes of South Massif were chipped from three boulders, two of which are about 2 meters across and the third about 2/3 meter across. The boulders are complex crystalline breccias which probably either rolled or were ejected from craters higher on South Massif. Soil was collected from beneath overhangs on the two larger boulders, and the soil beneath the smaller boulder was collected after rolling the boulder. Soil was collected from the fillet of the southernmost boulder, and a rake sample was taken from an area about 5 meters east of this boulder. Another rake sample was collected from the light mantle about 50 meters north of the break in slope at the base of the South Massif, in an area of sparse rock fragments.

Soil in the sample area is light to medium gray. The surface is saturated by craterlets up to about 5 centimeters in diameter. The soil appears from boot prints to have a rather low cohesiveness.

Craters 3 to 5 meters in diameter are common, but most are either subdued or have only slightly raised rims. No blocky craters are visible in the area of station 2, and there is no evidence that any of the fresher craters penetrate to consolidated rock. A few small, cloddy craters are visible, but none are in the immediate sampling areas.

The crew reported seeing lobes of unconsolidated material extending from the slopes of South Massif onto the light mantle. The lobes are especially prominent on the south wall of Nansen Crater and encroach onto the north wall, suggesting down-slope mass wasting. Visible lobes do not extend as far as the rake sample area of the light mantle.

The boulders are probably representative of material from high on the slope of South Massif; the fines in the vicinity of the boulders are probably derived largely from South Massif, but with some contamination from crater ejecta off the light and dark mantles. The rake sample on the light mantle probably contains light mantle and some material ejected off South Massif and from the dark mantle.

4.12.2.7 Station 2A (fourth lunar roving vehicle sample stop).

Station 2A is located about 500 meters

northeast of Nansen Crater on the light mantle. It was originally planned as a lunar rover sample stop, but during the mission, it was decided to take a traverse gravimeter reading at this stop. While off the rover, the crew collected four samples.

Rock fragments up to several centimeters in diameter cover less than one percent of the surface. Many appear to be only slightly buried, and fillets are poorly developed. The crew described the surface material as blue-gray soil to a depth of about 5 to 10 centimeters, and a lighter-colored soil below. The surface is saturated with craters up to 5 centimeters in diameters.

Larger craters up to about 5 meters in diameter are sparse to fairly common. Most of the craters are subdued or have only slightly raised rims. None of the craters have blocky ejecta, but a few appear to be cloddy. Samples collected include a fragment, probably breccia; a clod which disintegrated to soil by the time it reached the Lunar Receiving Laboratory; and a sample of the upper blue-gray material, which may have some contamination by ejects. from the dark mantle. A sample of light material from the bottom of a 15-centimeter trench was also collected, and it is likely that this is the most representative sample of light mantle fines that was returned by the mission.

4.12.2.8 Station 3.

Station 3 is located on the light mantle near the base of the scarp, approximately 50 meters east of the rim of Lara Crater. All of the samples except the drive tube were collected from the raised rim of a crater 10 meters in diameter. Time at the station was limited and was not sufficient for detailed, systematic documentation of the samples. Some of the individual rock fragments sampled cannot be recognized on pre-sampling photographs, but the locations of all the samples are known with reasonable certainty. The drive tube was taken about 20 meters south-southeast of the rim of the 10-meter-diameter crater.

Rock fragments 1 to 25 centimeters across are fairly common but cover less than 1 percent of the surface, and several boulders up to 1 meter across are visible in the panorama. Most of the rocks are rounded, except for the largest boulders, which are generally angular to subrounded. The largest rocks are generally near craters several meters in diameter. Rock fragments were collected from the rim of a crater 10 meters in diameter and include several samples of breccia similar to samples collected at station 2. Fillets are poorly developed, but a few of the rocks are apparently more than one-half buried.

The soil at the surface was described as medium-gray, but light-gray material was kicked up from just below the surface material at the crater rim. A trench 20-centimeters deep showed that medium-gray surface material about one-half centimeter thick overlies a light-gray layer 3 centimeters thick, which, in turn, overlies marbled or mottled light and medium-gray material. The soil appears to be loosely compacted and saturated with craterlets up to about 5 centimeters in diameter.

Craters larger than 4 centimeters and up to about 2 meters in diameter are fairly common. Several of these have fresh cloddy ejecta, but all of the small craters in the immediate vicinity of the sampled area appear to be subdued shallow pans. Craters ranging from 2 meters up to about 15 meters in diameter are also fairly common, and several have raised and somewhat blocky rims. One such crater is the 10-meter diameter crater at the sampling site. These craters appear to have penetrated only unconsolidated and somewhat rocky material of the light mantle. Lara Crater, 500 meters in diameter, and the largest crater in the area is probably covered by light mantle material, as suggested by premission mapping. It is probable that the samples collected are related only to the light mantle and not to materials that were ejected from depth from Lara Crater.

4.12.2.9 Fifth lunar roving vehicle sample stop.

The fifth lunar roving vehicle sample stop is located about 800 meters northeast of station 3 in the blocky ejecta of a crater in the light mantle. Rock fragments range from 1 centimeter to 50 centimeters in diameter, and cover 15 to 20 percent of the ejects blanket surface. The fragments are dominantly angular and partially buried. No fillets are visible in the photographs.

The crew commented that the ejecta is much different than they had previously seen around other craters. They also suggested that the crater may penetrate through the light mantle into bedrock. The sample is a fragment of ejecta from this crater and may not be representative of light mantle material.

4.12.2.10 Sixth lunar roving vehicle sample stop.

The sixth lunar roving vehicle sample stop is located about 1.3 kilometers northeast of station 3, on the light mantle. The sequence of driving photographs suggests that the sample was taken somewhere in the vicinity of two small cloddy craters.

Rock fragments 1 to 5 centimeters across are sparse and cover less than 1 percent of the surface. A few boulders 1 to 2 meters across are scattered over the surface. Many of the rocks are either sitting on the surface or only slightly buried. Fillets appear to be poorly developed. The surface is saturated with craters up to 5 centimeters across. Craters 5 centimeters to 2 meters across are common and most are shallow subdued pans. Craters 2 to 10 meters across are sparse and are mostly subdued or have slightly raised rims.

The sample collected is soil from the surface of the light mantle; it is probably representative of light mantle fines except for contamination by ejects. from dark mantle materials.

4.12.2.11 Station 4.

Station 4 is located on the south rim crest of Shorty, a 110-meter crater near the north edge of the light mantle (fig. 4-5) Shorty resembles Van Serg Crater and is similar to other craters that have been interpreted as young impact craters. The floor is hummocky, with a low central mound and with marginal hummocks that resemble slumps forming discontinuous benches along the lower parts of the crater wall. The rim is distinctly raised and is sharp in orbital views. The dark ejects, blanket is easily distinguished from the reflective surface of the surrounding light mantle, which it overlies. However, the low albedo of the ejecta is similar to that of the dark mantle elsewhere on the plains surface.

The central mound is blocky and extremely jagged, and the hummocks or benches that encircle the floor are also blocky. Although some portions of the walls are blocky, the walls, the rim, and the outer flank of Shorty Crater consist largely of dark material that is much finer grained than the floor. On the crater rim, fragments ranging up to about 15 centimeters in diameter typically cover less than 3 percent of the surface. Scattered coarser fragments, ranging up to at least 5 meters in diameter, are present.

Sampling was carried out in a low place on the rim crest of Shorty just south of a 5-meter boulder of fractured basalt. Debris that may have been shed from the boulder lies on the nearby surface, and blocks are abundant on this part of the inner crater wall. All of the rocks examined are basalt. They are commonly intensely fractured and some show irregular knobby surfaces that resemble the surfaces of terrestrial flow breccias. Rocks range from angular to subrounded; some are partially buried; some are filleted, including the up-slope sides of a few of the larger boulders on the inner crater wall.

The crater rim and flanks are pitted by scattered, small (up to several maters) craters whose rims range from sharp to subdued. Typically their ejects are no blockier, except for clods, than the adjacent surfaces.

A trench dug in the rim crest exposed compact reddish soil buried beneath a 1/2-centimeter-thick gray soil layer typical of the general soil surface at the station. The reddish soil occurs in a meter-wide zone that trends parallel to the crater rim crest for about 2 meters. Color zoning within the colored soil occurs as 10-centimeters-wide yellowish bands that form the southwest and northeast margins of the deposit. They are in steep sharp contact with gray soil adjacent to the colored soil, and grade inward to ;the more reddish soil that makes up the major part of the zone. A drive tube placed in the axial portion of the colored zone bottomed in black fine-grained material that reminded the crew of magnetite. Similar reddish material has been excavated by a small, fresh crater high on the northwest interior wall of Shorty and perhaps also on the rim crest a short distance southeast of the lunar rover.

Although a volcanic origin has been considered for Shorty Crater, no compelling data to support the volcanic hypothesis have been recognized. Most probably Shorty is an impact crater. Its blocky floor may represent the top of the subfloor basalt, which is buried by 10 to 15 meters of poorly consolidated regolith, dark mantle, and light mantle. The predominantly fine-grained wall, rim, and flank materials would then be ejecta derived largely from materials above the subfloor, and the basalt blocks would be ejects. derived from the subfloor. Regardless of its origin, the crater is clearly younger than the light mantle.

The origin of the red soil is currently enigmatic. It and the underlying black soil may represent a single clod of ejecta excavated by the Shorty impact from similar materials previously deposited in the target area by impact or volcanism. However, the symmetrical color zonation of the red' soil and parallelism of the zone's steep sharp boundaries with both the internal color banding and the axis of the rim crest are improbable features for a clod of ejecta. Perhaps volcanic origin along a fissure at the crater rim crest should be considered.

The 1/2-centimeter-thick gray soil that mantles the red soil unit should be present at the top of the drive tube sample. It could have been formed by either volcanism, impact processes, or regolith formation and deserves special attention in analysis. However, Shorty Crater is younger than any widespread dark mantle deposit near station 4.

4.12.2.12 Seventh lunar roving vehicle sample stop.

The seventh lunar roving vehicle sample' stop is located at the apex of the "v"-shaped Victory Crater, very near the contact of dark mantle with the finger of light mantle. The area was mapped as dark mantle near the light mantle contact, but the presence of light mantle material was not detected by the crew at the sample site. Victory was judged by the crew to be a series of craters with a definite rim, and probably is an impact feature with normal ejects. The ejects blanket and the rim appeared to be blanketed by dark mantle, and the surface on the rim looked like it had a "normal block population", although the inner walls of the crater were locally very blocky.

The blocks on the rim near the sample site are all less than one meter in diameter and cover two to three percent of the area. Burial of the rocks seems to be moderate to slight and fillets are only rarely present. The rocks are subangular to rounded, and some have planar sides. The soil appears normal, and lunar rover tracks in the soil are very sharply defined. Scattered subdued craters 1 to 5 meters in diameter are present near the ample site, but none seem to have a direct relation to the sample.

The sample is soil that should be representative of the surface material in the area. This soil sample is probably mostly dark mantle, but could contain a considerable amount of soil fragments derived from subfloor basalt.

4.12.2.13 Eighth lunar roving vehicle sample stop.

The eighth lunar roving vehicle sample stop is located in an area mapped as dark mantle slightly more than one crater diameter from Camelot and Horatio Craters. The area can be considered as' dark mantle away from the ejects. of the cluster of large craters. The surface appearance is similar to that at the first, third, and seventh lunar roving vehicle sampling stop areas, with a fragment coverage of less than !percent of the area. The photographs taken from the rover show that this very low abundance of fragments continues from Victory to the rim of Camelot. The narrow tongue of light mantle crossed by the traverse is not obvious in the surface photographs and was not noticed by the crew.

The sample collected is a scoop of soil from an inter-crater area. The sample site is within a few meters of a crater one meter in diameter which has abundant rim fragments that appear to be clods of soil breccia. All fragments near the sample site are less than 10 centimeters across, and the soil in the area appears typical of soils in dark mantle areas. Nearby craters from 1 to 5 meters in diameter may be members of two small, overlapping clusters. Several subdued, nearly rimless craters are present, as well as at least 3 craters with distinct ejects blankets covered with clods. The sample is probably typical of surface material in areas of dark mantle at some distance from contributions from other formations, and thus should be comparable to soil from the third lunar roving vehicle sample stop. These two samples are good subjects for study of variation in the dark mantle map unit.

4.12.2.14 Station 5.

Station 5 is located within a block field on the southwest rim of the large (600-meter) crater Camelot (fig. 4-4). The blocks, which are partly buried by dark mantle material, are exposed near and along the low, rounded rim crest of the crater and extend downward into the crater walls where, as in other craters, outcrops are most abundant. Blocks are absent on the crater floor. Outward from the rim crest, the block population decreases rapidly within a few meters and the terrain becomes smooth and undulating, but pitted by small craters up to several meters across.

Within the block field, individual rocks, varying from cobble to boulder-size, are subrounded to subangular, moderately to deeply buried, and cover about 30 percent of the surface. Except for one locality where soil occurs on the surface of a large flat rock, the tops of boulders impressed the crew as having been cleaned by the zap-pitting process. Filleting in this area appears to be minimal.

All of the rocks described by the crew constitute subfloor material having a very uniform appearance. Planar to subplanar concentrations of vesicles, linear arrangements of crystals, and possible gray zones of finer material cause some variations in rock textures and structures. Descriptions by the crew indicate that the rocks are subophitic pyroxene-bearing basalts with shiny ilmenite platelets in the vugs and vesicles. The soil-like mantle seems to consist of cohesive particles of uniform small size. A "raindrop" pattern is ubiquitous on the mantle. At Camelot, the mantle appears thinner than at the crater Horatio.

Shallow depressions and subdued craters up to several meters in diameter are superposed on the rim and floor of Camelot. These craters are younger than Camelot; however, within the observed block field only one younger crater of moderate size (4 to 5 meters) was noted by the crew.

The composition, textures, and uniform lithology of the ejects blocks around the crater Camelot indicate that subfloor basalts covered this part of the valley floor to a depth of at least 100 meters prior to the formation of the crater. Subsequent to the formation of Camelot, deposits of dark mantle material partly buried rocks around the rim and in the floor of Camelot and partly filled and subdued many younger craters in the area. The samples of basalt from blocks at station 5 should provide a sample of the subfloor unit to a depth of 30 meters.

4.12.2.15 Ninth lunar roving vehicle sample stop.

The ninth lunar roving vehicle sample stop is about halfway between the lunar module and station 6 in an area mapped as dark mantle about 2/3 of a crater diameter south of Henry Crater. Between the lunar module/surface electrical properties area and a point about 200 meters south of the ninth lunar roving vehicle sampling stop, the surface is characterized by 3- to 5-percent coverage of rock fragments and some large boulders. At this ninth lunar roving vehicle stop, the fragment coverage is less than 1 percent. To the north of this stop, the fragment coverage increases as the rim of Henry is approached, but never reaches more than a few percent except near young craters. The sample stop is in the part of the traverse from the lunar module to station 6 with the fewest rock fragments. The fragments in the area include some that seem to be blocks excavated by a 10-meter crater and some that appear to be blocks and clods excavated by several 1- to 2-meter craters. The sparseness of blocks ejected by craters suggests that the blocks were derived from the regolith rather than directly from the subfloor unit. The blocks are angular to subrounded, less than 30 centimeters across, and slightly to mostly buried. Only a few of the rocks have fillets.

The sample is a soil sample from a small crater with abundant clods of soil breccia on its rim. It should be reasonably representative of surface material in the nearby area of dark mantle. Some mixture of material from the re-excavated ejects. from Henry Crater is to be expected. However, the sample should be of considerable value in determining geographic variation in dark mantle material.

4.12.2.16 Tenth lunar roving vehicle sample stop.

The tenth lunar roving vehicle sample stop is adjacent to Turning Point rock, 2.8 kilometers north of the surface electrical properties transmitter. The area was mapped as dark mantle near the gradational contact with North Massif material. The sample site is on a gentle slope above a moderately distinct break in slope with the valley floor and below an equally distinct break in slope with the North Massif. The material underlying the area between the breaks in slope may be debris from the North Massif. Turning Point rock rests within a halo of smaller boulders and rock fragments that are probably derived from it. The rocks are rounded, partly buried, and filleted on the up-slope side. Turning Point rock probably reached its present position as one large boulder that has since been fragmented.

The sample was taken about 4 meters north of Turning Point rock, and it consists of at least three rock fragments and some soil. Most likely, the fragments in the sample are derived from Turning Point rock, but they could be debris derived separately from North Massif. Although the soil may contain fine debris from the rock, it most likely is a sample of the soil from North Massif that has moved down-slope and banked against the rock.

4.12.2.17 Station 6.

Station 6 is on the south slope of the North Massif, approximately 250 meters north of the break in slope between the valley floor and the massif. The area slopes approximately 11 degrees to the south and is covered by many large blocks and smaller fragments which have come from higher on the North Massif (fig. 4-6). Twenty meters from the rover are five large boulders aligned down-slope from the end of a single boulder track as if a single boulder broke into five parts. Four were sampled.

The boulder track made by the station 6 boulders can be traced one third of the way up the North Massif. This is the lowest level on the mountain where high concentrations of boulders appear. The sharpness of the boulder "track of the station 6 boulders indicates that the boulder may have been in its present position only a short length of time. The shadowed soil will probably reflect this. One of the boulders has a prominent north overhang, beneath which a shadowed soil was collected along with two soil samples collected outside the shadow. A soil sample was collected from the surface of one of the boulders. A rake sample and a single drive tube were taken within a few meters of the boulders. Two undocumented grab samples were collected probably near the rover, and another sample was collected down-slope from the station area.

Less than 1 percent of the surface in the station 6 area is covered by fragments. There appears to be a bimodal distribution of fragments in the area, though not as striking as the station 7 area. As at station 7, there appear to be relatively few fragments in the size range of 3 to 15 centimeters. Fragments less than 1/2 meter are scattered randomly over the surface, larger ones are generally in clusters. Most blocks are subrounded to rounded. A few angular blocks are scattered over the surface. Fillets are well developed on the up-slope side of some of the large boulders. Fillets on blocks less than 20 to 30 centimeters are poorly developed or absent.

The medium-gray soil in the station 6 area is moderately firm away from crater rims. The rover wheels made shallow impressions and threw very little spray, 10 to 15 meters away from craters. Close to a 10-meter crater in the area, boot prints and rover wheel penetration were deeper, and spray generated by them was considerably more. The drive tube collected near the rover was pushed by hand to a depth of 10 centimeters.

Craters, randomly scattered over the surface, range in size from the limit of resolution to 10 meters. Craters larger than 1 meter are sparse. The rims are smooth and unblocky, and usually more raised on the down-slope side. A few small fresh craters have blocky rims, with the ejects deposited preferentially down-slope.

The cluster of five boulders was probably at one time a single boulder.' There are at least two major rock types represented: a highly vesicular light-gray breccia, and a darker blue-gray breccia. The lower three boulders are highly vesicular, light gray breccias. The contact between the two breccias is in the second boulder from the top. Near the contact are inclusions of blue-gray breccia in the vesicular breccia. The blue-gray breccia is partially recrystallized, and has inclusions of friable breccia. The contact between the inclusions and blue-gray breccia is sharp. If all five boulders were at one time part of the same rock, there are at least four stages of brecciation. The similarity of the station-6 boulders to the station-7 boulder suggests strongly that the North Massif is composed primarily of multi-cycle breccias. Other samples of the North Massif are the rake sample, drive tube, and three grab samples, all of which are probably composed primarily of eroded material from the massif with possible minor contribution of subfloor material.

4.12.2.18 Station 7.

Station 7 is located at the base of the North Massif, just above the break in slope between the valley floor and the massif. The slope toward the valley is about 9 degrees. The station is located on North Massif material and the site contains boulders from slopes high on the massif. Several rock chips were collected at the station from such a 3-meter boulder of complex, multi-stage breccias. The 3-meter breccia boulder is the largest one in the station area, but other large boulders can be seen in the panorama. The large rock appears similar in color and weathering characteristics to the other boulders in the area. Several additional fragments were selected from the regolith surface.

Less than one percent of the surface is covered by blocks. There appears to be a bimodal distribution: Fragments from the limit of resolution to 2 or 3 centimeters are abundant, and blocks 30 centimeters and larger are common, but there is a scarcity of blocks in the size range of 3 to 30 centimeters. Blocks smaller than 30 centimeters are scattered randomly over the surface; those larger than 30 centimeters are in clusters. Most blocks, in all size ranges, are rounded, but a small percentage are angular. One 2- to 3-meter boulder down-slope from the sampling area is strikingly more angular than most other rocks. The rocks range from being deeply buried to perched. The sampled boulder overhangs the surface on the east, west, and, presumably, south sides. A few half-meter boulders in the same area are almost totally buried, and most fragments less than 1 meter are at least partially buried. Filleting is well developed on blocks larger than 1/2 meter, and poorly developed or absent on smaller blocks. Fillets are restricted to the uphill side of the blocks. One meter-sized boulder has a fillet extending three-fourths of the distance up the up-slope face. The boulder is a vesicular hornfelsic breccia (described by the crew as vesicular anorthosite) in contact with a dense blue-gray breccia. In the dense blue breccia is a 1 1/2-meter crushed light-colored inclusion, which is intruded by dikelets of breccia. One sample is a piece of blue-gray breccia in fairly sharp contact with a tan breccia which intrudes it. A large separate sample appears to be similar to the blue-gray breccia in the boulder. Several small chips collected on the surface are probably similar to the boulder, and also representative of North Massif material, although there may be a minor contribution from subfloor basalts.

The medium-gray soil in the station 7 area is relatively firmly compacted. Boot prints and rover tracks penetrate between 1 and 2 centimeters. Very little spray was generated from boot scuffs or the lunar rover wheels. The soil is quite cohesive (as shown by vertical walls of boot prints and the intact state of tread imprints).

Craters in the station 7 area range from the limit of resolution to 3 or 4 meters in diameter. Craters 1/2 meter or less are common. Most craters have somewhat subdued non-blocky rims, and ejecta is not visible. A few small (less than 1/2-meter) craters have raised blocky rims. The ejects. around these craters is deposited preferentially down-slope. On a meter-diameter crater down-slope from the rover, the ejecta is piled against a rock down-slope and forms only a slight raised rim on the uphill side.

The large boulder at station 7 probably originated at least one-third of the way up the North Massif because this is the lowest level on the massif where clusters of boulders appear. The rock types seen in this boulder are similar to those in large boulders at station 6, making the possibility of its being an exotic improbable, and suggesting that samples of the boulders are representative of the massif. These samples indicate that the North Massif is composed of multi-cycle breccias.

4.12.2.19 Eleventh lunar roving vehicle sample stop.

The eleventh lunar roving vehicle sample stop is located between stations 7 and 8 on the southeast rim of SWP Crater. Before the mission, the area was mapped as dark mantle blanketing SWP Crater. The sample site was chosen by the crew in the ejects. blanket of a fresh crater estimated to be 30 to 40 meters in diameter. The rocks in the sample area are part of the ejects blanket and were identified by the crew as clods of soil breccia that break easily and are "chewed up" by the rover wheels. At the rim, and just within the rim, the clods cover as much as 70 percent of the surface and at the sample site cover as much as 50 percent of the surface. The clods are very angular with some rounding of the tops. All are football size or smaller. The soil between the clods is the same color and probably the same composition as the clods. Within the ejects. blanket of the 30- to 40-meter crater, there are no visible younger craters.

The area surrounding the 30- to 40-meter crater appears typical of dark mantle surfaces that have a rock fragment population of less than one percent. The 30- to 40-meter crater should have penetrated some 6 to 8 meters of the soil (possibly dark mantle), which is a considerable greater thickness than the soil at station 1A. Possibly the crater at the eleventh lunar roving vehicle sample stop penetrated the same type of material as did Van Serg Crater (station 9) and if so, that material is a distinct unit that is not present at station 1A. However, with our present knowledge, the dark mantle at the eleventh lunar roving vehicle sample stop, station 9, and station 1 is considered to be the same material but with a large range in thickness. The sample taken at the eleventh lunar roving vehicle sample stop will be significant in determining the near surface stratigraphy. It is representative of the ejecta blanket of the 30- to 40-meter crater, and its position on the ejecta blanket suggests that it came from a depth of 2 to 4 meters below the surface.

4.12.2.20 Station 8.

Station 8 lies near the base of the Sculptured Hills south of Wessex Cleft and about 4 kilometers northeast of the lunar module. The terrain is undulating and forms a moderately inclined transition zone between the hills and the valley floor to the southwest; slopes increase noticeably towards the hills within the sample area. Eleven samples were collected, representing two, possible three, major rock types, and including soils, a rake sample, and a suite of four samples from a trench.

Small cohesive clods and pebble-size coherent rock fragments are common throughout the station area, but larger rocks and boulders are rare. Most of the rocks are subrounded to subangular, are well exposed to partly buried, have poorly developed fillets, and are only thinly covered by dust. The population of block and fragments does not increase greatly around the rims of any craters in the area photographed.

With the exception of one coarse-grained gabbroic boulder, all of the rocks larger than 20 centimeters which were examined had the appearance of subfloor basalts. The gabbroic rock was estimated by the crew to be made up of about equal amounts of blue-gray plagioclase (possibly maskelynite) and a light yellow-tan mineral, probably orthopyroxene; its average grain size appeared to be about 3 to 5 millimeters. The surface of this boulder was coated by glass. Both the top and bottom of the boulder were sampled. A white friable rock sampled by the crew in a small pit crater within the wall of a larger crater has probably been highly shocked by at least two episodes of fairly recent cratering.

The soil in this area, at least to the 20 to 25 centimeter depth of the trench samples, consists of fine-grained, cohesive particles. It has the dark appearance characteristic of the mantle throughout the valley floor. Surficial directional patterns related to structure were not observed, but many tracks made by the downhill movement of clods of all sizes were noted on the steeper slopes.

No large craters are present in the immediate area; those up to several meters in diameter are common, however, and have a continuum of morphologies from fresh-appearing, topographically sharp features to highly subdued depressions. None of the craters have either prominently raised or blocky rims.

At this stage of our knowledge, probably the most important results obtained from the field observations of the materials at station 8 is the apparent complete dissimilarity in lithology of the rocks on the surface at station 8 compared with those at the base of the massifs. If the gabbroic rock or the basalts are representative of materials forming the Sculptured Hills, the implications favored as to the composition and origin of the hills will be quite different from the premission interpretation that the hills are Serenitatis basin ejecta. However, only a small number of rocks were available for examination, and these may not be representative. The absence of boulder tracks, together with the glass coating on the gabbroic boulder, suggests that the rocks may be exotic blocks ejected from distant impact craters. It is possible that the rake sample and the sample from the crater wall may provide better clues to the composition and origin of the Sculptured Hills. Dark mantle material in this area is believed to be relatively thin, thus rake and soil samples may contain a mixture of dark mantle, subfloor ejecta, and debris mass wasted from the Sculptured Hills.

4.12.2.21 Station 9.

Station 9 is located on the southeast rim and nearby outer flank of Van Serg Crater. The crater, 90 meters in diameter, has a blocky central mound about 30 meters across, discontinuous benches on the inner walls, and a raised blocky rim with a distinct crest out from which slopes the blocky ejects, blanket. In both orbital and lunar surface photographs, Van Serg resembles other craters that have been interpreted as typical young impact craters. Its ejecta blanket is distinct in lunar surface views because of its blockiness, which is greater than that of the adjacent plains. The ejecta blanket can be recognized, at least in part, in premission orbital photographs as s distinct topographic feature, but it is inseparable from the adjacent plains on the basis of albedo.

Exploration at station 9 was concentrated in two areas: (1) the southeast rim crest of the crater, and (2) the surface of the ejecta blanket about 70 meters out from the crater rim to the southeast, where the lunar rover was parked. In both sample areas the predominant fragment size ranges up to about 30 centimeters, with a few boulders as large as 1 to 2 meters in diameter. At the rim crest fragments larger than 2 centimeters cover about 10 percent of the surface, but they cover no more than 3 percent of the surface in the sampling area near the rover.

The predominant rock type at station 9 is soft or friable dark-matrix breccia. White clasts up to about 2 centimeters in diameter are visible in some rocks on the crater rim, and light-colored clasts possibly as large as 1/2 meter in diameter were seen in rocks of the central mound. Some rocks are strikingly slabby. Closely spaced platy fractures occur in some, and a few show distinct alternating dark and light bands. Some frothy glass agglutinate was also sampled. In spite of their apparent softness, the rocks are typically angular. Many are partially buried, but there is little or no development of fillets even on the steep inner walls of Van Serg Crater.

Soil at the surface is uniformly fine and gray with no visible linear patterns. The uppermost one or two centimeters is loose and soft, but compacts easily to preserve boot prints. The trench near the lunar rover exposed about 10 centimeters of a white or light gray soil unit below a 7-centimeter upper dark unit.

Craters younger than Van Serg are extremely rare in the station area. A few small (about 1-meter) fresh craters are present. A large subdued depression southwest of the rover may be an old crater now mantled by Van Serg ejecta. Frequency and angularity of blocks, paucity of craters, general absence of fillets, and uneroded nature of crater rim and central mound attest to the extreme youth of Van Serg Crater. Such crater forms elsewhere have been attributed to impact, and no direct evidence of volcanic origin has been recognized. An impact origin seems likely.

The ejecta, unexpectedly, is dominated by soft dark matrix breccia instead of subfloor basalt. The rocks may be soil breccias, indurated and ejected in the Van Serg impact. If so, a fragmental unit that may be as much as 15 to 20 meters thick must overlie the subfloor basalt in the Van Serg area. Development of a regolith this thick in situ is difficult to reconcile with the apparent youth of the valley floor. Hence, the Van Serg rocks may represent a young mantle of transported fragmental material or a mature regolith that was developed on the surface of the subfloor basalt and was buried by the dark mantle prior to the Van Serg impact.

Soil samples probably consist of fine-grained Van Serg ejecta. The upper 7 centimeters of the trench may have been darkened by soil modifying processes, which would also have affected other soil samples and the upper part of the drive tube. Alternatively, the upper 7 centimeter zone of the trench may represent a young dark mantling unit deposited after the Van Serg event. Deposition of such a young thin unit would account for the similar albedos of the Van Serg ejects blanket and the general plains surface.

4.12.2.22 Twelfth lunar roving vehicle sample stop.

The twelfth lunar roving vehicle sample stop is located north of Sherlock at about 1/3 of a crater diameter out from the rim in an area mapped as dark mantle. The samples consist of a fragment of basalt and a separate soil sample. They were collected within a meter of a 1-meter boulder and a 1-meter crater that has clods of soil breccia.

The rock sample is probably a subfloor basalt derived from a depth of 50 to 90 meters. The soil sample probably is representative of the surface material in the area, which is mapped as dark mantle, although it may contain a small fraction of material derived from the nearby boulder. Surface photographs show an abundance (5 percent) of rock fragments, including several boulders (1 meter across), only the tops of which are visible. The rocks are very likely ejects of subfloor material from Sherlock Crater. Their size suggests that not more than a few meters of dark mantle have been deposited or formed since the Sherlock Crater impact. The slightly protruding boulders are smooth and rounded; most of the smaller fragments are rounded, but a few are subangular and have planar sides.

Younger craters have clearly re-excavated a few of the smaller blocks. Very small craters (1-meter diameter) with many fragments on their rims are also present and the rim fragments seem to be clods of soil breccia. The soil in the area appears normal in color and compaction.

4.13 SOIL MECHANICS EXPERIMENT

The soil mechanics experiment (S-200) is being carried out using data from other lunar surface activities. At the time of this report, only the following few observations have been made:

a. The lunar drill deep core hole remained open, as predicted for insertion of the lunar neutron flux probe.

b. Resistances to drilling and core tube driving were within established ranges and indicated no unusual mechanical properties.

c. Considerable local variation in soil mechanical properties is indicated, as has been observed at previous landing sites.

d. Surface texture, dust generation, cohesiveness, and average footprint depth indicates soil properties at the surface comparable to those at other Apollo landing sites.

4.14 PHOTOGRAPHY

A total of 2218 photographs were taken on the lunar surface to provide documentation of the landing site, through 360° panoramas at various stops, Apollo lunar surface experiments package placement data, and sample documentation as listed in table 4-1. The crew also obtained several photographs of scientific interest to document real-time observations.

5.0 INFLIGHT SCIENCE AND PHOTOGRAPHY

The inflight science experiments, performed during lunar orbit and transearth flight, were the. S-band transponder, lunar sounder, infrared scanning radiometer, ultraviolet spectrometer, gamma ray spectrometer, and the Apollo window meteoroid experiment. The photography equipment consisted of a panoramic camera, a mapping camera, and a laser altimeter, all of which had been flown on the two previous J-series missions. Other cameras used were the electric, 35-mm, and data acquisition.

5.1 S-BAND TRANSPONDER EXPERIMENT

The S-band transponder experiment (S-164) systems performed satisfactorily. There were no significant malfunctions during the data collection period and all planned data were obtained. The quality of the data was good; however, as in past missions, spacecraft maneuvers and ventings degraded the data for short periods of time.

The gravity variations recorded in the Mare Serenitatis region are one of the largest measured on the moon. Another large high-gravity variation was recorded in the Mare Crisium region. The landing site indicates a low gravity measurement that was consistent with the surface gravimeter measurements. Additional low measurements were recorded that are probably the effect of the large craters Eratosthenes and Copernicus, (fig. 3-1 near 15° north latitude and 15° west longitude) which lie adjacent to the ground track.

There are definite gravity variations in the Mare Procellarum region, but correlation with topographic features is not very evident. There seems to be some correlation with the infrared and ultraviolet earth-based photography of this area that is interpreted to be areas of old and recent lava flows. The more recent lava flows are the areas of gravity highs. The continuing detailed analysis of the Serenitatis mascon will be greatly enhanced with the new Apollo 17 profile. Figure 5-1 shows the Apollo 15 and 17 gravity profiles for the Serenitatis region. After analysis of the Apollo 15 data, the best estimate of the mascon was a near-surface symmetric disk that matched the profile in figure 5-1 fairly well (i.e. better than 95 percent of the variation removed). The dotted curve in figure 5-l is the predicted Apollo 17 profile from the best model, but it was poor, because of being erroneously low by 37 percent. Based on Apollo 17 data, new models will be formulated and many models now under consideration may be deleted.

Figure 5-1. Serenitatis gravity profile.

5.2 LUNAR SOUNDER

All planned operations of the lunar sounder experiment (S-209) were accomplished. Eleven hours of film data (active mode) were recorded, 7 hours of which were in high frequency band, and 4 hours in the very high frequency band. In addition, 30 hours of telemetry data were collected in the passive (receiver only) mode. Six of the 30 hours were collected during lunar orbit and 24 hours were collected during the transearth coast period.

The images produced by optical processing indicated that the energy in high frequency channel 2 was down approximately 10 to 20 dB relative to high frequency channel 1. The predicted signal differences were 7 to 10 dB. The greater difference will limit the analysis of the 15 mHz data. The high frequency channel 1 and very high frequency images are of excellent quality.

The very high frequency profile is

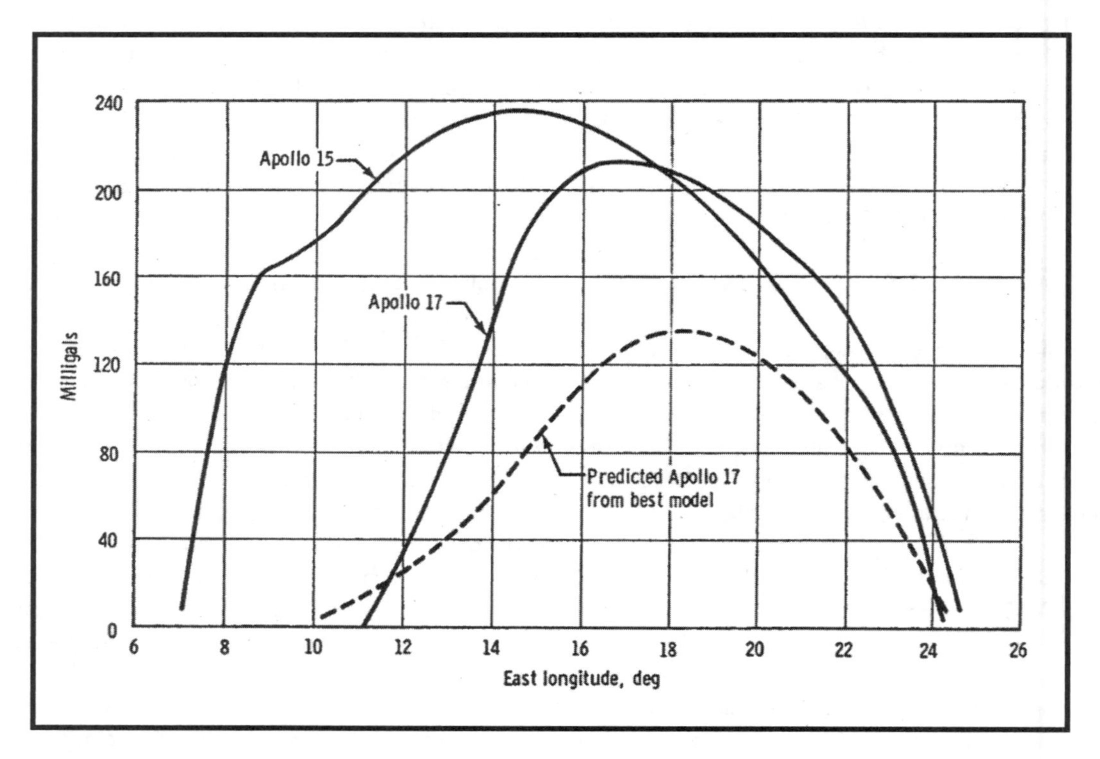

satisfactory for addressing local selenomorphological problems. Tentative subsurface returns have been identified in both the high frequency I and very high frequency channels. Evaluation is continuing, however, to substantiate that these features are not sidelobes of the main surface return or off-vertical surface features.

In the passive mode, the telemetry data exhibits a high terrestrial noise component on the front side of the moon. The lunar backside offered the opportunity to measure the true galactic noise background. However, the noise from the guidance and navigation system was incorrectly predicted and a reassessment of the guidance and navigation noise level is required before an accurate assessment of the galactic noise can be made. The 24 hours of data collected during transearth coast will be used to analyze terrestrial noise sources and improve the calibration of the antenna pattern.

Minor problems were encountered with operation of the equipment and film processing. The film recorded occasional static discharge marks as a result of static electricity generated when the film edge rubbed against the roller flange. Also, streaks and flow patterns were noted and apparently were acquired during the film development process. The nature and spacing of the discharge and development marks had little effect on the data.

Random dark bands' were observed on the film but coherent data existed between the bands. The bands are believed to have resulted from arcing in the high voltage power supply for the recorder cathode ray display tube. This problem was observed at turn-on during preflight vacuum testing of the Apollo 17 recorder. Two bands were noted during the first eight hours of operation and there was negligible data loss. The rate of occurrence increased to five or six discharges per minute during the final three hours of operation. Each band obscured from one fourth of a second to one second of data.

Two problems were encountered in extending and retracting the high frequency antennas:

The HF antenna boom I retract limit sensing switch failed to operate, resulting in telemetry and command module display data that indicated the boom had not fully retracted. Drive motor current was then used to determine the antenna retract status. Section 15.1.3 discusses this problem.

Extension of the HF antenna boom 2 required more time than expected. Motor current data indicates friction was higher than predicted. Section 15.1.4 discusses this problem.

5.3 ULTRAVIOLET SPECTROMETER

The ultraviolet spectrometer (S-169) was operated as planned during the mission. Data were obtained both while the command and service module was in lunar orbit and during transearth coast with a total of 164 hours and 45 minutes of operating time. Data were obtained in the ultraviolet spectrum region (1180 to 1680 angstroms). Observations were accomplished of the lunar surface, lunar atmosphere, zodiacal light, and galactic and extragalactic sources.

Instrument performance was normal except for a background count rate that was higher than expected. The higher count rate, which has been tentatively attributed to cosmic radiation, raises the lower limit for detection of weak signals, but does not affect the precision with which stronger light signals can be measured. The high count rate also does not affect the absolute sensitivity of the spectrometer.

On the last day of operation of the ultraviolet spectrometer, two telemetered temperature measurements ceased to operate. Section 15.5.4 contains a discussion of this problem. This anomaly did not in any manner affect the ability of the spectrometer to complete its mission. A portion of the spectral data has been evaluated, and some scientific observations are presented in the following paragraphs.

The total lunar atmosphere appears to be much less dense than the maximum particle density of 10 7 per cubic centimeter reported by the Apollo 14 cold cathode pressure gage. The Apollo 17 mission ultraviolet spectrometer was sensitive to all possible atmospheric constituents except argon, helium, and neon. The upper limit to the total particle density of all constituents observed on the Apollo 17 mission is 105 per cubic centimeter.

Only atomic hydrogen was positively identified in the lunar atmosphere. The particle density, was less than 50 per cubic centimeter, which is about 1 percent of the density expected, if solar-wind protons undergo a charge exchange at the surface and thermally escape.

Lunar albedo measurements confirm laboratory measurements performed with lunar samples from previous missions. The flight data show the same angular scattering function observed in the visible region. Albedo variations and their relationship to the lunar geography are being analyzed. New information about the solar atmosphere (including zodiacal light) and the earth's atmosphere was obtained during the mission. Additionally, very good data were obtained during transearth coast on stellar and extragalactic sources, and a general ultraviolet sky survey was conducted.

5.4 INFRARED SCANNING RADIOMETER

The infrared scanning radiometer (S-171) performed normally during all lunar orbit phases of the mission and during transearth coast. In more than 90 hours of operation, approximately 100 million lunar temperature measurements were made and all test objectives were met. The temperatures in the infrared scanning radiometer remained well within design limits.

The infrared scanning radiometer output was non-zero when the instrument viewed space during the final portion of each scan. The non-zero value varied over a range of approximately 7 to 15 telemetry bits as a function of orbital position. Furthermore, during transearth operations, a non-zero value appeared near the mid-position of each scan and remained there until the mirror entered the housing at the end of a scan. The cause of the non-zero condition is attributed to some object aboard the spacecraft and within the outskirts of the infrared scanning radiometer field-of-view. The effect of this object on the data can possibly be removed with only a small degradation of the signal-to-noise ratio in the affected data.

The infrared scanning radiometer low-gain channel saturated when the instrument measured subsolar point temperatures. The effect was not anticipated since the infrared scanning radiometer was calibrated to temperatures higher than the predicted subsolar point value, however, no test objective was compromised by the saturation.

Oceanus Procellarum has a considerable variation in temperature, because its surface features range from large craters to small features which are below the infrared scanning radiometer resolution (less than 2 kilometers). The remainder of the ground track has far fewer thermal variations, and the night-time scans on the backside show only a few variations.

5.5 PANORAMIC CAMERA

The panoramic camera operation was normal except for two problems. On revolution 15, the velocity /altitude sensor data were erratic (see sec. 15.5.1), and the camera was placed in the velocity/altitude override position for all subsequent photography.

On revolution 74, telemetry indicated a stereo drive motor failure 8 minutes prior to completion of the last section of coverage. The loss of forward motion compensation degraded the resolution of the remaining frames. This anomaly is discussed in section 15.5o3. However, post-transearth injection photography was according to the flight plan. Telemetry data in revolution 49 indicated film usage was slightly high, and to compensate for this, approximately 15 frames were deleted from the end of revolution 62 at 90° east. At acquisition of signal on revolution 62, the spacecraft was in a pitch-up attitude; consequently, photographs from this pass are west-looking obliques rather than convergent stereo as planned.

Figure 5-2 shows the cumulative panoramic camera coverage through revolution 74. The coverage for revolution 62 (133° east to 93° east) is the normal ground coverage rather than the actual western oblique coverage. Examination of a quick-look film positive indicated satisfactory quality. However, because the camera was in velocity/ altitude override, stereo overlap is not always 100-percent and models as low as 74 percent may be expected.

The 5 oblique frames from revolution 36 cover the Apollonius Highlands and Mare Fecunditatis north of Langrenus. This area includes the Luna 16 landing site. The oblique photography obtained on revolution 62 produced some unexpected results. The sequence starts over Tsiolkovsky and the pitch increases until the entire horizon is included. This spectacular photograph is the only one in existence that distinctly shows the mountain range that encircles Mare Smythii (fig. 5-3).

5.6 MAPPING CAMERA

The mapping camera and its associated stellar camera performed satisfactorily throughout the mission. In accordance with the flight plan, the camera was operated in both the extended and the retracted positions during revolution-1-2. Camera deployment was normal but the retraction was sluggish (sec. 15.5.5). Deployment for the next camera operation on revolution 13-14 required 3 minutes, 19 seconds as compared with the expected 1 minute and 20 seconds. To reduce the possibility of total failure of the deployment mechanism, the camera was left extended until after operation on revolution 38, when it was necessary to retract for the plane change maneuver and rendezvous.

Operation on revolution 49 was in the retracted position with a consequent loss of stellar photography on that revolution. The camera was extended for operation on revolution 62. The attitude of the spacecraft on part of revolution 62 resulted in the photographs being west-looking obliques rather than the planned stereo verticals.

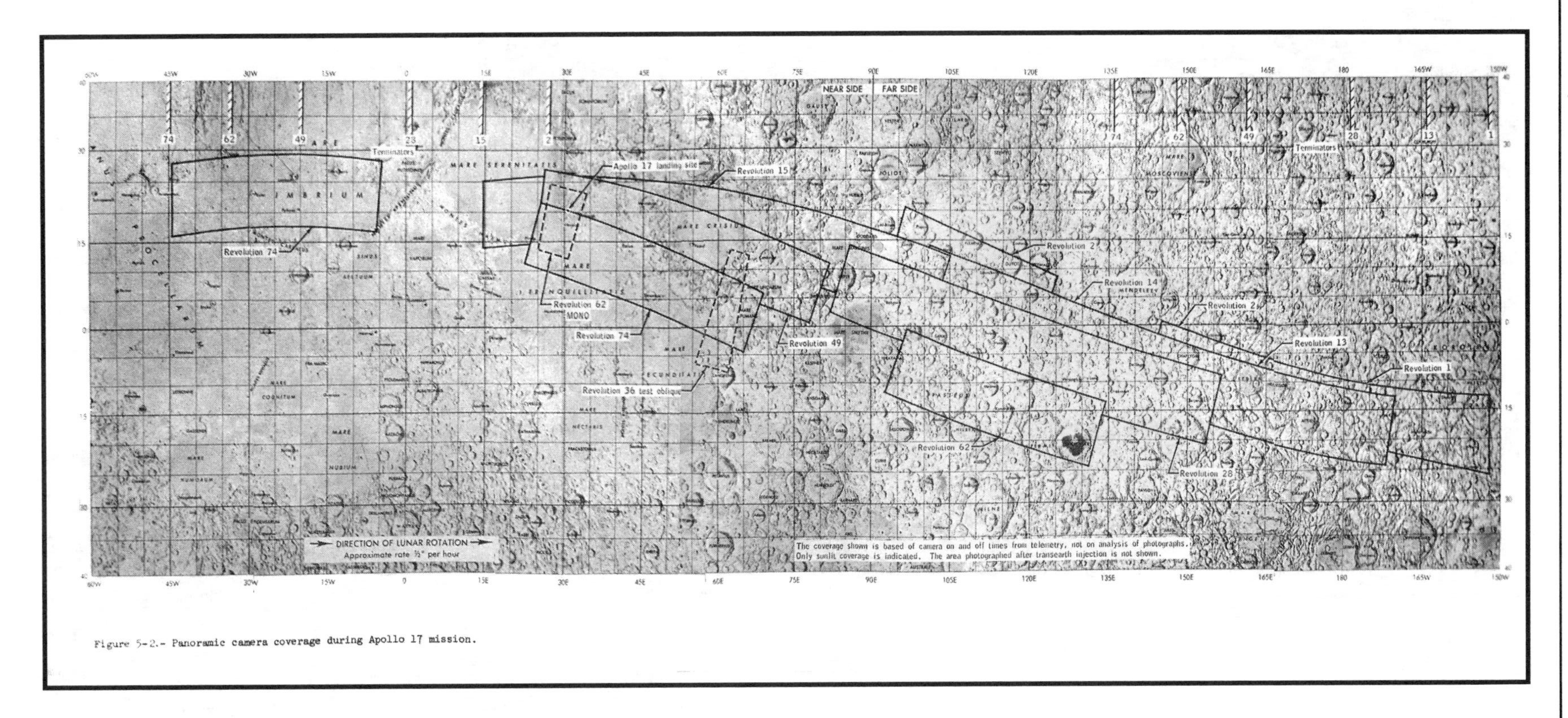

Figure 5-2.- Panoramic camera coverage during Apollo 17 mission.

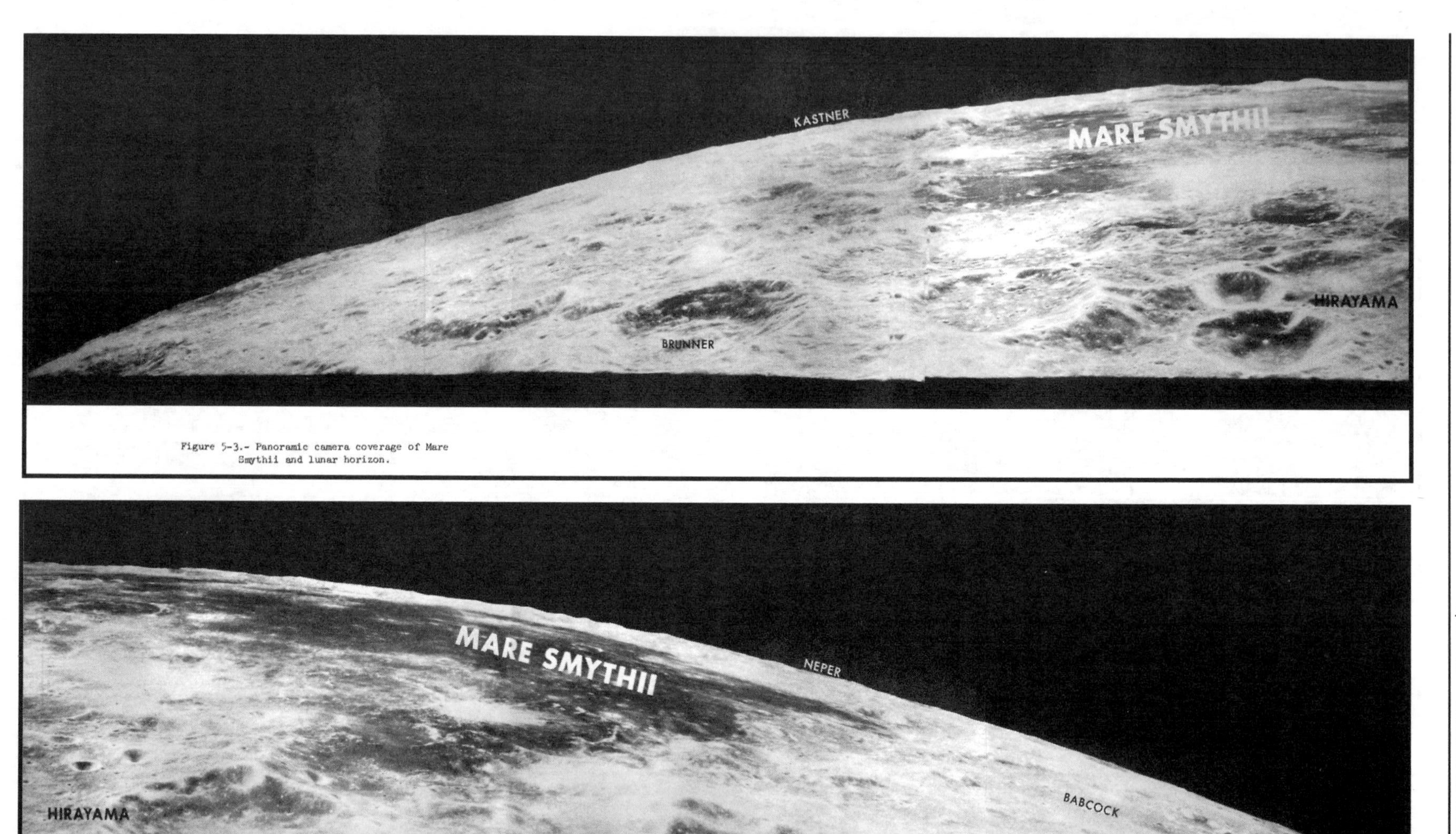

Figure 5-3.- Panoramic camera coverage of Mare Smythii and lunar horizon.

Figure 5-3.- Concluded.

On revolution 65, the camera was not turned on until after acquisition of signal and approximately 15 minutes of planned north oblique photography was lost. The delayed turn-on was the result of the camera being too cold following a complete power-down for lunar sounder operations. The remaining operation, including post-transearth injection photography was normal. Figure 5-4 shows cumulative mapping camera coverage through revolution 74.

Throughout the mission, the exposure pulses were missing at the lower light levels, thus indicating a possible overexposure condition (see sec. 15.5.2); however, a check during film processing indicated no adverse affects.

When the camera was operated in the retracted position, the camera covers appear in the field of view. Photography on revolution 14-15 was planned to be vertical, however, the spacecraft was in a north oblique attitude at the beginning of the pass. This resulted in the bonus acquisition of the only reasonably good Apollo photographs of the crater Icarus.

All of the objectives of mapping camera photography were successfully accomplished and in general, the quality of the photographs is excellent. The portion which was taken under high sun conditions are somewhat over exposed and as might be expected this results in some loss of definition. Newton's rings are present on the imagery, more so than on previous photography. Film duplicates exhibit electrostatic discharges on some of the frames; however, none of these effects should degrade the capability to reduce the photographic data.

5.7 LASER ALTIMETER

The laser altimeter performed satisfactorily throughout the mission with a total of 3769 measurements. Approximately 4 minutes of data were lost on revolution 24 when the power was inadvertently turned off. About 38 minutes of dark-side altimetry was deleted from revolution 62 to permit a maneuver away from the scientific instrument module bay attitude for operation of the ultraviolet spectrometer.

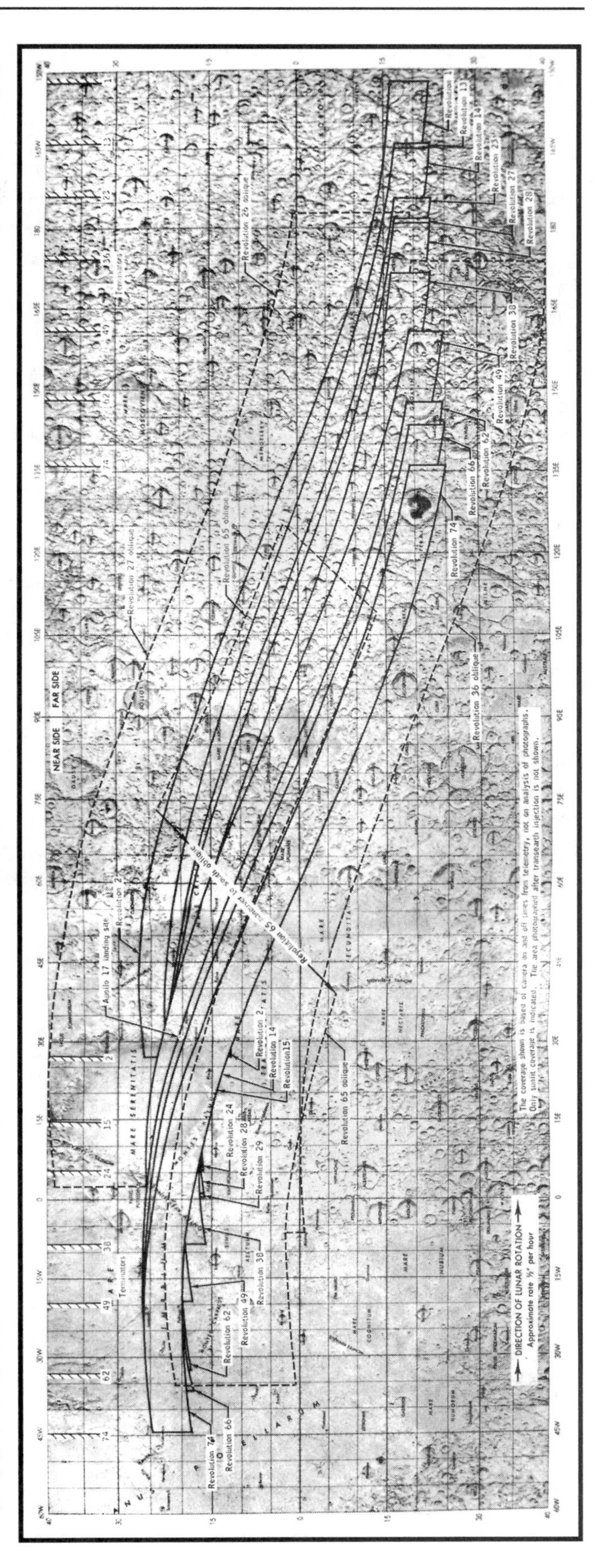

The altimeter was operated continuously for an additional 10 hours during the pre-transearth injection sleep period.

5.8 OPERATIONAL AND COMMAND MODULE SCIENCE PHOTOGRAPHY

Crew camera and photographic operations, from a data and equipment standpoint, were good. The photographic data requirements were fulfilled with only minor exceptions which resulted from the rescheduling of flight plan activities to make up the time lost due to a late earth lift-off. The primary photographic data not obtained were the lunar surface polarimetric photography sequence, and one solar corona photographic sequence from the command module.

Valuable unscheduled photographs were obtained on the lunar surface, in lunar orbit, and during the translunar and transearth coast periods, all of which were accompanied by vivid descriptions. The orange soil documentation and the earthshine photographs were of particular interest. An interesting observation was that the crew could see more details in the earthshine photographs on high-speed film, than they could detect visually from orbit.

5.8.1 Operation and Documentary

Documentary and operational photography was accomplished covering all aspects of the mission. Table 5-I lists the equipment and film types used, along with the photographic objectives for each system. The crew photographic complement differed from Apollo 16 in that the lunar surface 16-mm sequence system was deleted from the Apollo 17 mission.

5.8.2 Lunar Orbital Science

All the planned photography was taken. Near-terminator photography of mare regions produced valuable photography with the 70-mm camera system; however, the camera system was not fast enough for best data of far-side terminator targets using high-speed black-and-white film.

The 35-mm camera provided better earthshine photography than on previous missions because of the higher speed lens. Five planned targets: Eratosthenes, Copernicus, Reiner, Gamma Riccioli, and Orientale were photographed along with crew option photographs of Tsiolkovsky, Mare Imbrium, and the landing site using the blue, red, and polarizing filters. These data are a valuable addition to the

TABLE 5-I.- PHOTOGRAPHIC EQUIPMENT AND TARGETS

Equipment	Lens focal length	Film 'type	Task/target
Command module 70-mm electric data camera	80 250	S0368 2485	Undocking, lunar module ejection, lunar module inspection, rendezvous, and docking. Earth end noon orbital science. Solar corona. Orbital science, stereo strip. Contamination. Near terminator.
Lunar surface 70-mm electric data camera	60 500	S0368 3401	Geology sample documentation, surface panorama Apollo lunar surface experiments package deployment, geology documentation, soil mechanics, lunar module inspection, distant features.
Command module 35-mm camera	55	S0168 2485	Apollo light flash moving emulsion device position data. Zodiacal light, galactic libration point. Lunar surface in earth-shine, dim light phenomena, and far side terminator.
Command module 16-mm data acquisition camera	75 18	S0368 S0168 2485	Transposition, docking, undocking, rendezvous. Lunar-module inspection, scientific instrument module door .jettison, transearth extravehicular activity. Entry and parachutes. Heat flow demonstration. Comet. Contamination. Intravehicular activity operations. Lunar strip sextant photography.
Lunar module 16-mm data acquisition camera	10		Descent, surface activity, ascent, rendezvous..

catalog of surface features which are being studied.

5.8.3 Astronomic Photography

The one solar corona photographic sequence obtained consisted of 7 photographs with exposure times 10, 4, 1, 1/2, 1/8, 1/30, and 1/60-second starting 75 seconds before sunrise and ending 10 seconds prior to sunrise.

The sequence was made using the Hasselblad electric camera with the 80-mm lens set at f/2.8, and high-speed, black-and-white film.

The sunrise coronal sequence provides data on the east limb of the sun. Jupiter appears in all seven of the photographs, thereby permitting simplified indexing and tracing of the data. Two coronal streamers are evident in the photograph taken 6.4 seconds prior to sunrise; and while these streamers seem to appear in the photograph taken 30 seconds prior to sunrise, photographic techniques and microdensitometry will be required to determine their full extent. One streamer lies nearly along the ecliptic, and the excellent sketches made by the crew during the mission as well as their descriptions during the debriefings show and include streamers extending along the ecliptic out to Jupiter, some 24.4 degrees from the center of the sun. Microdensitometry traces will be made from the lunar limb out along the ecliptic on all photographs to determine the brightness of the corona as a function of distance from the center of the sun.

Zodiacal light photography data were obtained with greater success on this mission than on any previous attempt, due primarily to refined procedures and equipment. The zodiacal light extending eastward from the lunar-occulted sun was recorded in three separate series of photographs. Each of these pre-sunrise series included overlapping frames covering elongation angles from 80 degrees to 0.5 degree eastward along the ecliptic plane. The 35-mm camera system was used with the 55-mm lens at f/1.2, and front-end (color or polarizing) filters using high-speed, black-and white film. The camera was bracket-mounted behind the right-hand rendezvous window and was pointed almost parallel to the vehicle plus X axis when color filters were in use, and 30 degrees from the plus X axis towards the minus Z axis when the larger polarizing filter was in use. For the zodiacal light tasks, the camera's optical axis was initially pointed in the direction of the vehicle instantaneous velocity vector by properly orienting the spacecraft. This allowed the camera axis to be aimed close to the ecliptic plane with one corner of the frame viewing a portion of the lunar limb cutting across the ecliptic. Stellar images on the data frames provide good postflight references for pointing accuracies.

The first series of 11 data frames was taken in red filtered light on revolution 23. An exposure time of 90 seconds was used initially for positions farthest away from the sun, and the exposure time was decreased as the zodiacal light's brightness increased, so that the final exposure was 1/30 second within 0.5 degree of the sun.

A second series of photographs was made in blue filtered light on revolution 28. Since it used the same timing and duration of exposures as the previous series, a direct frame-for-frame comparison will show spectral content differences. Both series will also be compared directly to matching duration exposures of a calibrated stepwedge, illuminated by a solar equivalent light source. Calibration exposures were carried on the flight so that they experienced the same level of radiation fogging as the data frames, and their combined analysis will remove emulsion reciprocity failure effects from the final photometric results.

The polarized series was composed of eleven sets of photographs over corresponding regions of the sky. Each set was made up of two photographs with the filter rotated 90 degrees. Equivalent polarized calibrations of these exposures were also carried on the flight.

The most striking initial results come from the comparison of corresponding red and blue images. The inner zodiacal light within about 15 degrees of the sun shows a stronger red component in and close to the ecliptic plane, while the inner zodiacal light well out of the ecliptic plane as well as almost all of the outer zodiacal light produced a stronger blue component. While a similar visual comparison of equivalent polarized frames does not show any obvious variation in features, very good isophote maps can be made for more sensitive comparisons.

5.9 VISUAL OBSERVATIONS FROM ORBIT

Fourteen lunar surface targets were visually studied from orbit to complement photography and other remotely-sensed data. Crew members were aided in performing the task by onboard graphics, 10 power binoculars, and a color wheel. All the supporting material was found adequate except for the color wheel which apparently did not include a color range comparable to the actual lunar colors.

Some salient results of the observations are:

a. Descriptions of both the regional and local geological settings of the landing site will aid in the study of the returned samples and their environments.

b. Color determination of the overflown lunar mania will help in extrapolation of results of sample analyses.

c. Orange-colored units detected on eastern and western Mare Serenitatis are probably similar to that sampled at Shorty Crater.

d. Domical structures within the crater Aitken on the lunar far-side were characterized as resembling the many dacite domes in northern California and Oregon.

5.10 GAMMA RAY SPECTROMETER EXPERIMENT

The gamma ray spectrometer experiment (S-160) was passive during the mission and consisted of only the sodium iodide crystal of the hardware flown on Apollo 15 and 16. This was flown on Apollo 17 to obtain a measure of induced crystal activation from primary protons and secondary neutrons while in the spacecraft. The induced activity was believed to be a major source of interference in the galactic flux data taken on Apollo 15 and 16.

The active part of the experiment started after landing and recovery of the crystal from the spacecraft. The crystal, activated during flight, was counted in a low-background shield aboard the recovery ship. An identical crystal not flown aboard Apollo 17 was used as a control throughout the measurement program.

At the time of publication, qualitative identifications have been made of the following nuclear isotopes; Sodium 24, Iodine 123, Iodine 124, Iodine 125, Iodine 126, and possible Antimony 124 and Sodium 22. Results indicate that the induced activity can be attributed mainly to species with half lives of about half a day and longer. Significant intensities of decay products with shorter half lives have not been seen, although if such products were present they should have been observed.

The final results of this experiment will be reported in a supplement to this report.

5.11 APOLLO WINDOW METEOROID EXPERIMENT

This Apollo window meteoroid experiment (S-176) utilizes the command module side and hatch windows for detecting meteoroids having a mass of 10-9 grams or less. The windows were scanned at a magnification of 20X

(200X magnification for areas of interest) to map all visible defects. Possible meteoroid craters have been identified and a possible correlation has been made between the meteoroid cratering flux on the glass surfaces and the lunar rock cratering studies.

The Apollo 17 windows were optically examined with more than the usual anticipation because the crew noted an impact of about 1-mm diameter. No meteoroid craters larger than 0.1-mm diameter have been detected, but two large

TABLE 5-II.- METEOROID CRATERS AND RELATED INFORMATION

Mission	Window exposure, m^2 -sec	Number of impacts	Meteoroid flux, number/m^2 -sec	95 percent confidence limits, number/m^2-sec	Minimum meteoroid mass, g
Apollo 7 (Earth orbital without lunar module)	2.21×10^{5}	5	2.26×10^{-5}	5.29×10^{-5} 7.23×10^{-6}	1.3×10^{-10}
Apollo 8 (Lunar orbital without lunar module)	1.80×10^{5}	1	1.07×10^{-5}	5.96×10^{-5} 1.07×10^{-6}	7.9×10^{-11}
Apollo 9 (Earth orbital with lunar module)	1.87×10^{5}	1	5.36×10^{-6}	3.0×10^{-5} 5.36×10^{-7}	5.4×10^{-10}
Apollo 10 (Lunar orbital with lunar module)	1.99×10^{5}	0	---	1.86×10^{-5} --	1.6×10^{-10}
Apollo 12 (Lunar landing)	2.43×10^{5}	0	---	1.52×10^{5} --	1.6×10^{-11}
Apollo 13 (Circumlunar abort with lunar module)	1.42×10^{5}	1	1.36×10^{-5}	7.6×10^{5} 1.37×10^{-6}	5.9×10^{-9}
Apollo 14 (Lunar lending)	2.35×10^{5}	2	1.64×10^{-5}	5.9×10^{-5} 1.64×10^{-6}	1.6×10^{-11}
Apollo 15 (Lunar landing)	2.88×10^{5}	0	---	1.28×10^{5} --	6.7×10^{-11}
Apollo 16 (Lunar landing)	2.55×10^{5}	0	--	1.39×10^{5} --	6.7×10^{-11}
Apollo 17 (Lunar landing)	2.95×10^{5}	0	---	1.25×10^{-5}	5.37×10^{-10}

bubbles had formed just below the window surface. The bubble in window 3 had a 0.42-mm diameter, and the bubble in window 1 had a 0.75-mm diameter. A tabulation of meteoroid impacts and related data from the previous missions is contained in table 5-II.

6.0 MEDICAL EXPERIMENTS AND INFLIGHT DEMONSTRATIONS

6.1 BIOSTACK EXPERIMENT

The biostack experiment (M-211) was conducted to determine the biological effects of heavy ions from cosmic sources and the space flight environment on plant and animal biological systems in a dormant state.

The experiment was hermetically sealed and self-contained.

The biostack container like the one flown on Apollo 16, was returned to the principal investigator in Frankfurt/Main Germany and opened within 10 days after landing. The results will be published in a supplement to this report.

6.2 BIOLOGICAL COSMIC RADIATION EXPERIMENT

The biological cosmic radiation experiment was a passive experiment to determine if damage could be detected in the brain and eyes of pocket mice after exposure to heavy cosmic particles in the spacecraft cabin environment. A dosimeter recorded the level of radiation in the vicinity of the experiment and two temperature recorders indicated the maximum and minimum temperatures within the canister. The experiment package was removed from the spacecraft about 4 1/2 hours after landing. The internal pressure of 10.1 psia was immediately raised to 14.7 psia with a 1-to-1 mixture of oxygen and helium. The canister was then flown to the Lyndon B. Johnson Tropical Medical Center in American Samoa. The experiment was opened approximately 7 1/2 hours after landing and 4 mice were observed moving about in their tubular compartments. The fifth mouse was dead and its death was not caused by radiation. The physical condition of the living mice appeared to be excellent. The recorders indicated that the temperature within the canister ranged from 71.0° F to 85.7° F. The implanted scalp dosimeters indicated penetration by a significant number of cosmic particles. Performance of the potassium superoxide granules in providing life support oxygen was considered to be normal.

The living mice were euthanized and all five mice were then prepared for postmortem examination. Autopsies were performed during which various required tissue samples were obtained for subsequent evaluation. Analysis of the dosimeters will provide information for sectioning the brains and eyes to maximize the probability for locating lesions caused by the radiation particles. The results of the analysis will be reported in a supplement to this report.

6.3 VISUAL LIGHT FLASH PHENOMENON

The visual light flash phenomenon functional objectives were completed during the transearth and translunar coast phases. During the 64-minute translunar coast test period, the Command Module Pilot wore the Apollo light flash moving emulsion detector, the Commander wore an eye shield, and the Lunar Module Pilot served as the recorder for the events observed by the other crewmen. A total of 28 events, occurring at random intervals, were reported by the two crewmen. This total is significantly lower than the 70 events reported in a comparable observation period on Apollo 16. The Apollo light flash moving emulsion detector control plates (nuclear emulsion) have been processed and contained a low background of proton particle traces in the emulsions.

The transearth coast observation period commenced at 279:55 and continued for 55 minutes. All three crewmen wore eye shields for this observation period. No light flashes were experienced during this observation period; however, both the Command Module Pilot and Lunar Module Pilot reported later that they experienced light flashes during the subsequent sleep period. No explanation of the lack of light flashes during the observation period can be made at this time. A supplemental report will be published summarizing the results of this experiment on all mission flown.

6.4 HEAT FLOW AND CONVECTION INFLIGHT DEMONSTRATION

A heat flow and convection demonstration was conducted by the Command Module Pilot during translunar coast. The demonstration was a modified version of the Apollo 14 demonstration and contained three separate experimental tests. Data were obtained with the 16-mm data acquisition camera and from crew observations; both of which were of excellent quality. Film data indicates that all demonstrations were performed successfully.

A brief description as well as the preliminary results of the demonstration are contained in the following paragraphs, and the final results will be presented in a supplement to this report.

Flow Pattern Experiment

The flow pattern experiment was to investigate convection caused by surface tension gradients. The gradients result from heating a thin layer of liquid which generates cellular patterns known as Benard Cells.

The apparatus consisted of an open aluminum pan approximately 7 centimeters in diameter with electrical heaters attached to the bottom. The liquid was Krytox oil with approximately 0.2-percent fine aluminum powder added for visibility, and the solution was released into the pan by a valve and pump arrangement. Baffles of KEL-F material around the inside periphery of the pan maintained the liquid level at 2 and 4 millimeters in depth. The baffles were redesigned after the Apollo 14 mission to assure an even layer of oil across the bottom of the pan. On the Apollo 14 demonstration, the fluid tended to adhere to the walls of the pan. On the Apollo 17 demonstration, the test was conducted twice, once with a 2-millimeter fluid-depth, and once with a 4-millimeter depth.

The fluid contained bubbles which were not easily dissipated by stirring. At the 2-millimeter depth, onset of convection occurred within a few seconds of heat application; whereas, on earth, the average onset time was approximately five minutes. The fluid was contained by the baffles around the periphery and assumed a convex shape, similar to a perfect lens. The surface was observed to be free of ripples and distortion, and the center thickness was approximately twice the baffle height of 2 millimeters.

The Benard cells formed in the 2-millimeter depth were less orderly and symmetrical than the ground-based patterns and they reached a steady state in about seven minutes (fig. 6-1). Cells formed in 4-millimeter test were more regular and larger than those in the 2-millimeter test, but the cells did not reach a steady-state condition during the 10-minute heating period.

Radial Heating Cell

The radial heating cell was to investigate heat flow and convection in a confined gas at low g conditions.

The experiment consisted of a cylinder which contained argon, and was approximately 6 centimeters in diameter and 2 centimeters in length. The initial internal pressure was approximately one atmosphere. Heat was applied by a post heater mounted in the center of the cell. Temperature changes and distribution were monitored by liquid crystal strips which changed color as the temperature changed. Clear color changes indicate proper operation; however, the results of the demonstration will not be available until analysis of the data is complete.

Lineal Heating Cell

The lineal heating cell unit was to investigate heat flow and convection in a confined liquid at low g conditions. The demonstration consisted of a cylindrical glass container approximately 3 centimeters in diameter and 9 centimeters long, containing Krytox oil. A disc-shaped heater was located at one end of the cylinder and the temperature changes were monitored by liquid crystal strips. The cell also contained afew magnesium particles to aid visibility. Clear color changes indicated proper operation, but final results are dependent on further analysis.

6.5 ORTHOSTATIC COUNTERMEASURE GARMENT

A custom-fitted orthostatic countermeasure garment was donned 2 hours prior to entry by

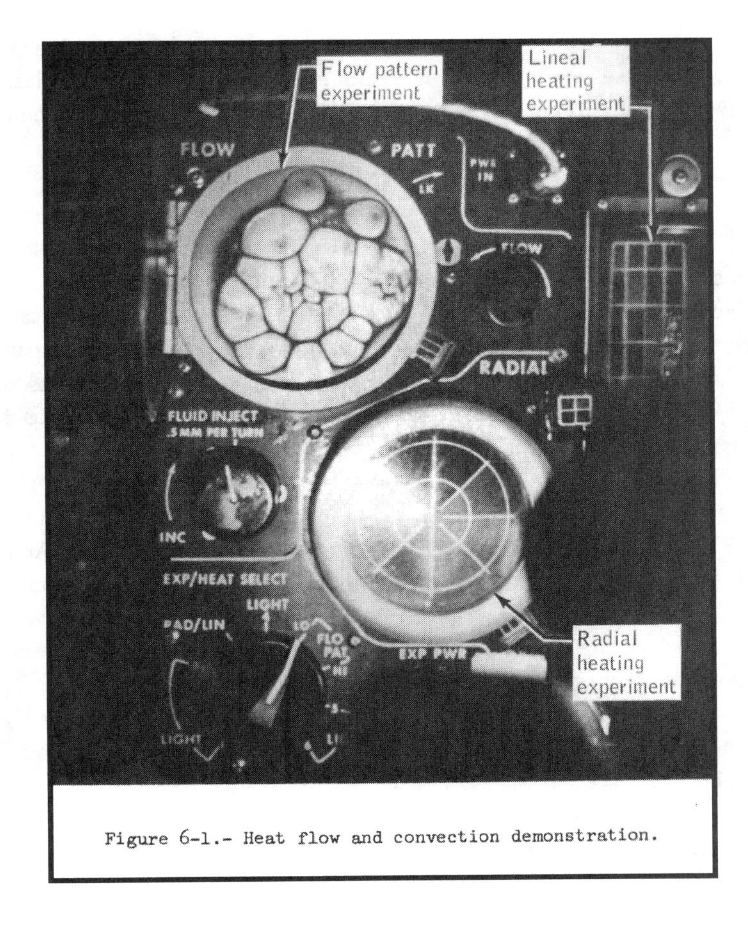

Figure 6-1.- Heat flow and convection demonstration.

the Command Module Pilot, and the garment was pressurized at landing, and remained pressurized until performance of the lower body negative pressure test 5 hours later. The garment, when pressurized, applied a pressure gradient from 50 mm Hg at the ankles to 0 at the waist. While wearing the garment in a pressurized condition, the Command Module Pilot's standing heart rate was 99 beats per minute and his pulse pressure was 46 mm Hg. Upon depressurization of the garment, the standing heart rate and pulse pressure changed to 104 beats per minute and 47 mm Hg, respectively. The Command Module Pilot's postflight resting heart rate (recumbent) was less than the preflight value and the lower body negative pressure test produced no significant change in heart rate or pulse pressure. Additionally, other related postflight findings such as a small decline in heart stroke volume, enlarged cardiac silhouette by X-ray, and the fact that the decrease in his leg volume was greatest of the three crewmen, suggest that the Command Module Pilot's cardiovascular orthostatic functions differed from the other two crewmen. The garment provided no apparent benefit to this crewman

7.0 COMMAND AND SERVICE MODULES

Performance of the command and service systems is discussed in this section. The sequential, pyrotechnic, thermal protection, and emergency detection systems operated as intended and are not discussed in this section. Discrepancies and anomalies in the command and service module systems are mentioned in this section and discussed in greater detail in section 15.1.

7.1 STRUCTURES AND MECHANICAL SYSTEMS

Command module window contamination, similar to that reported on previous flights, was observed shortly after earth orbit insertion and remained throughout the mission. A fine mist condensed on the inner surface of the heat shield panes of the two side windows and the hatch window. The outer surface of the heat shield pane on the left side window was also contaminated by the residue from the waste water dump.

During the transposition and docking sequence, the no. 9 docking latch switch did not open as evidenced by a barberpole indication and visual inspection of the latches. Examination of the latch verified that the latch hook was properly engaged and subsequent manual actuation of the latch resulted in proper latching and switch operation. Also, the handles for latches no. 7 and 10 did not lock automatically, thus requiring manual engagement. During preparation for the first lunar module ingress, the Command Module Pilot recocked no. 4 docking latch and all latches operated normally for the lunar orbit docking.

Following command and service modules/lunar module undocking, an extend/retract test was conducted for the lunar sounder HF antennas. The antennas were partially extended and then retracted; however, a retract indication was not received for antenna 1 (section 15.1.3). The antenna was then fully extended in preparation for the lunar sounder experiments. Subsequent antenna retraction verifications were based on retract power and time. Prior to the lunar orbit docking, the lunar module crew visually verified that the antenna was retracted.

Photographic data taken from the lunar module during rendezvous show the mapping camera reaction control system plume shield door in the open position with the mapping camera apparently retracted. A discussion of this anomaly is contained in section 15.1.7.

The earth landing system performance was normal. All three main parachutes and the forward heat shield were recovered. Post flight inspections of the main parachutes revealed no significant operational damage, and inspection of the forward heat shield revealed no unusual conditions.

7.2 ELECTRICAL POWER AND FUEL CELLS

Performance of the electrical power system

and fuel cells was normal. The entry, postlanding, and pyrotechnic batteries fulfilled all power requirements. Entry batteries A and B were each charged once before launch and, with the fuel cells, supplied the main buses through launch and during all service propulsion system maneuvers. Battery A was charged four times during the flight and battery B five times. The total ampere-hour status of entry batteries A, B, and C was maintained above a minimum capacity of 85 ampere-hours throughout the flight. This minimum, which was approximately 10 to 15 ampere-hours less than usual, occurred about nine hours into the mission and was due to the delayed earth lift-off which resulted in more battery usage.

The fuel cells were activated 58 hours prior to launch with fuel cell 2 on main bus A and fuel cells 1 and 3 oven circuited. Three and one-half hours before lift-off, fuel cell 1 was placed on bus A and fuel cells 2 and 3 were placed on bus B, a configuration which was not changed throughout the flight. The fuel cells supplied 686 kilowatt hours of energy at an average bus voltage and current of 29.0 volts and 76.5 amperes, respectively.

7.3 CRYOGENICS STORAGE SYSTEM

The cryogenic storage system supplied reactants to the fuel cells and metabolic oxygen to the environmental control system. The hydrogen tank 2 automatic pressure switch failed to operate properly after approximately 70 hours. After this time, pressure was manually controlled in hydrogen tanks 1 and 2. The oxygen and hydrogen systems performance was normal and as predicted except for this anomaly which is discussed in detail in section 15.1.8.

7.4 COMMUNICATIONS

Performance of the communications equipment was normal except for a 2-minute dropout of several pulse code modulation telemetry channels which occurred at about 191 hours. Details of this anomaly may be found in section 15.1.9.

7.5 INSTRUMENTATION AND DISPLAYS

The instrumentation and displays performed normally during the mission. One measurement, the fuel interface pressure, fluctuated about 17 psi twice, each time for about 10 hours. Oscillations of the same type were also experienced on Apollo 15 with the service module reaction control system quad A fuel manifold pressure. These oscillations result from minor instabilities in the signal conditioning amplifier as a result of temperature or other physical effects. These fluctuations did not prevent interpretation of the data.

On several occasions shortly after earth-orbit insertion, spurious master alarms occurred, as discussed in section 15.1.1. The condition was only a nuisance factor to the mission.

Initial checks after earth orbit insertion disclosed that the mission timer in the lower equipment bay was 15 seconds slow. The timer was reset and it operated properly for the remainder of the mission. This anomaly is discussed in section 15.1.2.

At about 71 hours, the caution and warning tone booster was inoperative. The tone booster is a photo-sensitive device which activates when a finite light intensity is generated by the master alarm light. Discussions with the crew indicate that the initial check of the tone booster was made using the lamp test mode. This mode activates about half of the warning lights including the master alarm light in a parallel electrical circuit, and this results in a lower intensity of the master alarm light. The subsequent test of the tone booster activated only the master alarm light; therefore, the light intensity was higher and the tone booster operated properly. The condition had been experienced during ground tests.

7.6 GUIDANCE, NAVIGATION, AND CONTROL

Performance of the guidance, navigation, and the primary and backup control systems was normal throughout the flight. The only anomaly was a large bias variation of the accelerometer in the entry monitor system (see section 15.1.5) The anomaly occurred late in flight, therefore, it had no effect on the mission and the system performed normally during entry.

The primary guidance system provided good boost trajectory monitoring during launch and during the translunar injection maneuver. The crew was unable to see any identifiable star groups during the boost phase.

This would preclude using out-of-the-window alignment techniques, if the primary and backup attitude reference systems were lost and it became necessary to abort.

At earth-orbit insertion, the differences between the primary guidance velocity vector and the Saturn guidance velocity vector were minus 3.0, minus 13.6, and minus 6.3 ft/sec in the primary guidance X, Y, and Z axes, respectively. The differences include an azimuth update of plus 0.9 degree in the Y axis made at 100 seconds and represent 0.6 and 0.7 sigma X and Z platform errors, respectively. The azimuth correction (Y axis) at 100 seconds was well within specification.

A history of primary guidance system inertial component errors is presented in table 7-I. The preflight performance values were obtained from system calibrations performed after the inertial measurement unit was installed in the command module. The flight performance values are from platform alignment data and accelerometer bias measurements made during the mission.

Table 7-II is a summary of inertial measurement unit realignments performed during the mission. Table 7-III summarizes the significant control parameters during service propulsion system maneuvers and the midcourse correction maneuvers.

The entry sequence, which begins with command module/service module separation, was normal through landing. The guidance system controlled the spacecraft attitude and lift vector during entry and guided the vehicle to landing coordinates of 17 degrees 52 minutes 48 seconds south latitude, 166 degrees 6 minutes 36 seconds west longitude, as determined from the spacecraft computer.

7.7 SERVICE PROPULSION SYSTEM

The service propulsion system performance was satisfactory during each of the six maneuvers with a total firing time of 584.7 seconds. The actual ignition times and firing durations are contained in table 7-III. All system pressures were normal during the service propulsion firings. The helium pressurization system functioned normally throughout the mission. All system temperatures were maintained within their redline limits without heater operation as in all previous Apollo flights. The nitrogen pressure system data showed normal usage for the six maneuvers. The propellant mass unbalance at the end of the transearth injection firing was approximately 30 lb decrease. The propellant utilization valve was in the decrease position for approximately 39 percent of the total service propulsion system firing time and this resulted in an overall propellant mixture ratio for the mission of 1.592. The predicted mixture ratio for the mission was 1.597 with the propellant utilization valve in the normal position. Adjusting for the service propulsion system firing time performed with the propellant utilization valve in the decrease position, the difference was within the predicted value.

TABLE 7-I.- INERTIAL COMPONENT HISTORY

	Preflight performance					Inflight performance		
Parameter	Number of samples	Sample mean	Standard deviation	Countdown value	Flight load	Number samples	Sample mean	Standard deviation
Accelerometers								
X - Scale factor error, ppm	6	-704	20	-715	-750	-	-	-
Bias, cm/sec^2	6	-0.08	0.06	-0.06	-0.19	112	0.04	0.12
Y - Scale factor error, ppm	6	-234	17	-233	-210	-	-	-
Bias, cm/sec^2	6	0.08	0.03	-0.04	-0.07	112	-0.02	0.03
Z - Scale factor error, ppm	6	824	22	-831	-810	-	-	-
Bias, cm/sec^2	6	0.71	0.03	0.69	0.71	112	0.74	0.03
Gyroscopes								
X - Null bias drift, meru	6	0.32	1.36	1.9	0.6	28	1.42	0.71
Acceleration drift, spin reference axis, meru/g	6	-6.22	0.33	-6.7	-6.0	-	-	-
Acceleration drift, input axis meru/g	6	5.72	2.97	4.1	7.0	-	-	-
Y - Null bias drift, meru	6	0.82	0.76	0.6	-0.1	28	0.17	0.53
Acceleration drift, spin reference axis, meru/g	6	5.10	0.62	4.7	6.0	-	-	-
Acceleration drift, input axis meru/g	6	8.98	3.05	10.3	13.0	-	-	-
Z - Null bias drift, meru	6	1.37	0.77	2.3	1.2	28	2.31	0.64
Acceleration drift, spin reference axis, meru/g	6	-3.50	0.54	-3.1	-4.0	-	-	-
Acceleration drift, input axis, meru/g	6	-6.95	4.69	1.2	-6.0			

TABLE 7-II.- PLATFORM ALIGNMENT SUMMARY

Time, hr:min	[a]Program option	Star used	Gyro torquing angle, deg			Star angle difference	Gyro drift, mere			Comments
			X	Y	Z	deg	X	Y	Z	
00:35	3	24 Gienah, 30 Menkent	0.080	0.029	0.018	0.01	-9.011	-3.27	2.03	Launch orientation
01:52	3	22 Regulus, 24 Gienah	-0.037	-0.007	-0.021	0.00	2.01	0.38	-1.14	
07:58	3	4 Achernar, 7 Menkar	-0.134	-0.018	0.175	0.01	1.46	0.20	1.90	
08:08	1	--	-	-	-	-	-	-	-	Passive thermal control orientation
08:11	3	4 Achernar, --	-0.05	-0.57	0.00	0.01	-	-	-	
16:55	3	32 Alphecca, 23 Denebola	-0.165	-0.134	0.153	0.00	1.28	1.04	1.18	
23:10	3	7 Menkar, 13 Capella	-0.151	0.019	0.140	0.02	1.60	-0.20	1.46	
35:04	3	32 Alphecca, 35 Rasalhague	0.089	-0.035	-0.023	0.00	-0.55	0.20	-0.13	
45:17	3	1 Alpheratz, 10 Mirfak	-0.022	0.017	-0.020	0.01	0.14	-0.11	-0.13	
58:19	3	1 Alpheratz, 36 Vega	0.021	-0.108	-0.005	0.01	-0.107	0.52	-0.27	
68:15	3	26 Spica, 32 Alphecca	0.114	-0.067	-0.033	0.00	-0.61	0.35	-0.18	
82:40	3	16 Procyon, 17 Regor	0.077	0.039	-0.002	0.00	-0.36	-0.18	0	
82:58	3	-	-0.022	-0.016	-0.054	0.01	-	-	-	Lunar orbit insertion orientation
84:50	3	13 Capella, 20 Dnoces	0.029	0.041	-0.041	0.01	-1 04	-1.48	1.48	
87:29	3	11 Aldebaran, 16 Procyon	-0.045	-0.039	-0.020	0.01	1.13	0.98	-0.50	
87:32	1	15 Sirus --	0.742	0.736	-0.735	0.00				Lending site orientation
89:30	3	15 Sirius, 22 Regulus	0.046	0.014	-0.069	0.00	-1.56	-0.48	-2.34	
93:31	3	20 Dnoces, 27 Alkaid	-0.036	0.052	0.030	0.00	0.60	-0.86	0.50	
105:40	3	14 Canopus, 25 Acrus	0.065	-0.076	-0.102	0.01	-0.53	0.63	-0.83	
108:34	3	30 Menkent, -	0.090	0.025	0.024	0.00	0	-0.11	0.11	
110:40	3	21 Alphard, 26 Acrux	0.013	0.015	-0.056	0.01	-0.42	-0.48	-1.78	
130:21	3	15 Sirius, 22 Regulus	0.143	-0.038	-0.006	0.01	-0.48	0.13	-0.02	
140:39	3	7 Menkar, 14 Canopus	0.046	-0.015	-0.006	0.01	-0.31	0.10	-0.04	
154:27	3	7 Menkar 14 Canopus	0.106	-0.012	-0.034	0.01.	-0.51	0.06	-0.16	
166:41	3	6 Acamar, 113 Deneb	0.053	-0.011	-0.039	0.01	-0.32	0.06	-0.21	
177:50	3	11 Aldebaran, 20 Dnoces	0.173	0.023	-0.045	0.00	-1.03	-0.14	-0.27	
177:55	1	20 Dnoces, 11 Aldebaran	-0.541	-0.147	-0.706	0.01	-	-	-	Plane change orientation
180:22	1	2 Diphda, 14 Canopus	-0.493	-0.798	-0.092	0.01	-	-	-	Lift-off orientation
183:55	3	22 Regulus, 24 Gienah	0.102	0.030	-0.040	0.00	-1.91	-0.56	-0.89	
207:43	3	12 Rigel, 21 Alphard.	0.163	-0.073	-0.019	0.00	-0.46	0.20	-0.05	
215:42	"	12 Rigel, 21 Alphard	0.046	-0.020	-0.009	0.01	-0.38	0.17	-0.07	
227:34	3	27 Alkaid, 31 Arcturus	0.092	0.043	-0.026	0.00	-0.52	-0.24	-0.15	
231:34	3	12 Rigel, 21 Alphard	0.065	-0.050	-0.039	0.01	-1.08	0.83	-0.65	
231:46	3	12 Rigel, 21 Alphard	-0.011	-0.007	-0.002	0.01	-	-	-	Transearth injection orientation
235:30	3	23 Denebola, 30 Menkent	0.105	-0.009	0.030	0.01	-1.88	0.16	0.54	
235:43	3	23 Denebola, 30 Menkent	-0.086	0.026	-0.045	0.00	-	-	-	passive thermal control orientation
249:31	3	26 Spica, 27 Alkaid	0.102	-0.011	-0.089	0.00	-0.49	0.05	-0.43	
262:19	3	1 Alpheratz, 36 Vega	0.121	0.009	-0.047	0.01	-0.63	-0.04	-0.23	
273:50	3	26 Spica, 27 Alkaid	0.086	-0.063	0.060	0.00	-0.54	0.40	0.38	
285:20	3	3 Navi, 36 Vega	0.079	0.106	-0.068	0.01	-0.46	-0.61	-0.39	
297:16	3	7 Menkar, 115 Fomalhaut	0.064	-0.059	-0.032	0.00	-0.36	0.33	-0.18	
297:29	3	7 Menkar, 45 Fomalhaut	-0.059	-0.003	-0.050	0.00	-	-	-	Entry orientation
300:03	3	24 Gienah, 33 Antares	0.084	0.040	-0.036	0.01	-2.20	-1.00	-0.90	

[a] The numbers used in this column represent the following: 1 - Preferred; 3 - REFSMAT.

TABLE 7-III MANEUVER SUMMARY

Parameter	[a] First midcourse correction	[a]Lunar orbit insertion	[a] Descent orbit insertion	[a]Lunar orbit circularization	[a] Orbital trim	[a]Lunar orbit plane change	[a] Transearth injection	Second midcourse correction
Time								
Ignition, hr:min:sec . .	35:29:59.91	86:14:22.60	90:31:37.43	109:17:28.92	178:54:05.45	179:53:53.83	234:02:09.18	298:38:01
Cutoff, hr:min:sec . . .	35:30:01.64	86:20:55.76	90:31:59.70	109:17:32.72	178:54:42.95	179:54:13.88	234:04:32.87	298:38:10
Duration, hr:min:sec . .	01.73	06:33.16	22.27	03.80	31:30	20.05	02:23.69	00:09
[b]Velocity gained, ft/sec (actual/desired)								
X-axis	9.1/9.8	-2982.4/2987.1	-112.8/-113.0	0.3/0.1	5.1/5.7	-266.8/-267.0	-3043.3/-3043.3	1.3/1.4
Y-axis	3.2/3.4	-48.3/-48.3	-161.3/-161.1	0.0/0.0	7.0/7.1	250.1/250.1	97.3/98.1	0.0/0.1
2-axis	-2.2/-2.4	-39.1/-39.1	-16.3/-16.4	68.9/70.5	0.4/0.2	-11.1/-11.2	86.8/87.0	1.6/1.6
[c] Velocity residuals, ft/sec (before/after trimming)								
X-axis	0.7/0.1	-0.3/No trim	0.1/No trim	1.3/0.0	-0.5/0.0	0.1/No trim	0.1/No trim	0.1/No trim
Y-axis	0.0/0.1	+0.1/No trim	0.2/No trim	0.2/0.0	0.0/0.1	0.1/No trim	0.8/No trim	0.0/No trim
g-axle	0.0/0.0	0.0/No trim	0.0/No trim	-0.6/-0.6	0.5/-0.1	0.1/No trim	0.2/No trim	-0.1./No trim

a Service propulsion system used for these maneuvers.
b Inertial coordinates before trimming.
c Body coordinates after trimming.

7.8 REACTION CONTROL SYSTEM

The performance of the command and service module reaction control systems were normal during all phases of the mission. During postflight purging of the oxidizer manifold, the plus pitch and plus yaw engines of the command module reaction control system I responded simultaneously to ground support equipment commands. The problem was located in the ground support equipment and that hardware will be modified for the Skylab program.

7.9 ENVIRONMENTAL CONTROL SYSTEM

The environmental control system provided an acceptable environment for the crew and spacecraft equipment. During the daily water chlorination operations, leakage was observed in and around the casing assembly of the injector mechanism: Despite the minor leakage, all chlorination procedures were accomplished as scheduled. This anomaly is discussed in section 15.1.6.

During four early lunar orbits, the water/glycol temperature control valve (mixing valve) failed to open properly as the radiator outlet temperature decreased. The mixed coolant temperature momentarily fell as much as 4° F below the specification control band of 42° to 48° F during mixing startup. No corrective action was taken and initiation of mixing was proper during all subsequent lunar orbits and during transearth coast. This anomaly is discussed in section 15.1.10.

The radiator flow proportioning valve automatically switched over to the backup control system for the primary coolant loop. About 4 hours later, the crew reset the flow control to the primary system and returned control to the automatic mode after which operation of the valve was satisfactory. Switching normally occurs for any one of the following conditions.

a. The indicated temperature difference between the two radiator panels is greater than 15° F and the flow control valve is not drawing fall current.

b. The indicated temperature difference between the two radiator panels is greater than 15° F and the valve current is of the wrong polarity.

c. A transient on the ac or do bus.

7.10 EXTRAVEHICULAR ACTIVITY EQUIPMENT

The extravehicular activity crew equipment performed successfully throughout the transearth extravehicular activity. Preparations for the extravehicular activity were delayed slightly when the Command Module Pilot's communication carrier did not receive a low pressure warning tone during the oxygen purge system checkout. The Command Module Pilot exchanged communication carriers with the Lunar Module Pilot and received the tone on a subsequent check, and the extravehicular activity proceeded as planned. Later, an inspection of the faulty communications carrier revealed two broken leads in the electrical pigtail. See section 15.3.1 for a discussion of this anomaly.

7.11 CONSUMABLES

The command and service module consumable usage during Apollo 17 was well within the redline limits.

7.10.1 Service Propulsion Propellant

Service propulsion propellant and helium loadings and consumption values are listed in the following table. The loadings were calculated from gaging system readings and measured densities prior to lift-off.

Condition	Propellant, lb		
	Fuel	Oxidizer	Total
Loaded	15 669.0	25 073.0	40 742.0
Consumed	14 917.0	23 754.0	38 671.0
Remaining after trans earth injection	752.0	1319.0	2071.0
Usable after trans earth injection	606.0	1024.0	1630.0

Condition	Helium, lb	
	Storage bottles	Propellant tanks
Loaded	87.6	5.4
Consumed	65.5	-
Remaining after transearth injection	22.1	70.9

7.11.2 Reaction Control System Propellant

Service Module.- The propellant utilization and loading data for the service module reaction control system were as shown in the following table. Consumption was calculated from telemetered helium tank pressure histories and was based on pressure, volume, and temperature relationships.

Condition	Propellant, lb		
	Fuel	Oxidizer	Total
Loaded Quad A	110	227	337
Quad B	110	225	335
Quad C	110	226	336
Quad D	110	225	335
Total	440	903	1343
[a] Usable loaded			1252
Consumed			654
Remaining at command module/service module separation			598

[a] Usable propellant is the amount loaded minus the amount trapped with corrections made for gaging system errors.
Command Module.- The loading of command module reaction control system propellant were as follows.

Condition		Propellant, lb	
	Fuel	Oxidizer	Total
Loaded			
System 1	39	78	117
System 2	38	78	116
Total	77	156	233
Usable loaded			
Consumed	25*	47*	72

*Based on amount of propellant off loaded.

7.11.3 Cryogenics

The total cryogenic hydrogen and oxygen quantities available at lift-off and consumed were as follows. Consumption values were based on quantity data transmitted by telemetry.

Condition	Hydrogen, lb		Oxygen, lb	
	Actual	Planned	Actual	Planned
Available at lift-off				
Tank 1	26.7		309.5	
Tank 2	26.9		313.8	
Tank 3	25.5		325.8	
Total	79.1	82.1	949.1	944.8
Consumed				
Tank 1	19.0		215.0	
Tank 2	19.9		197.5	
Tank 3	22.2		187.6	
Total	61.1	60.5	600.1	576.9
Remaining at command module/service module separation				
Tank 1	7.7		94.5	
Tank 2	7.0		116.3	
Tank 3	3.3		138.2	
Total	18.0	21.6	349.0	367.9

8.0 LUNAR MODULE PERFORMANCE

The lunar module systems performance is discussed in this section. All spacecraft displays, plus the thermal protective, pyrotechnic, and reaction control systems operated as intended and are not discussed in this section. The systems discrepancies and anomalies are mentioned in this section and discussed in greater detail in section 15.2.

8.1 STRUCTURES AND MECHANICAL SYSTEMS

All mechanical systems functioned properly. The loads at landing were low based on the rate of descent at footpad contact of about 3 ft/sec. The vehicle attitude on the lunar surface was approximately 5.3 degrees pitch up with a left roll of about 2.6 degrees.

8.2 ELECTRICAL POWER DISTRIBUTION AND BATTERIES

The electrical power system performed as expected. The descent batteries delivered 1585 ampere-hours out of a nominal total capacity of 2075 ampere-hours. At jettison, the ascent batteries, had delivered about 300 ampere-hours out of a normal capacity of 592 ampere-hours, and at impact, over 200 ampere-hours remained. The do bus voltage was maintained at about 28.8 volts, and the maximum observed current was 71 amperes during the powered descent.

8.3 COMMUNICATIONS

All functions of the communications system were acceptable during each phase of the mission. The S-band steerable antenna lost lock several times because of vehicle blockage or reaching the gimbal limits. No losses of automatic track, due to divergent antenna oscillations, were noted as experienced on several previous missions.

8.4 INSTRUMENTATION

The instrumentation system operated normally except for a battery 4 measurement shift which was noted at about 108 hours. This anomaly is discussed in section 15.2.1.

8.5 RADAR

Landing radar performance was normal during powered descent. Velocity acquisition was obtained at an estimated altitude of 42 000 feet, prior to changing the lunar module yaw attitude. from 70 to 20 degrees. Range acquisition was obtained during the yaw maneuver, at an altitude of approximately 39 000 feet. Antenna position and range scale change occurred at the predicted time and tracking was continuous to lunar touchdown. There was no lock-up on moving dust or debris near the lunar surface.

The rendezvous radar performance was normal for all mission phases including self-test, rendezvous radar/transponder checkout, and rendezvous tracking.

8.6 DESCENT PROPULSION SYSTEM

The descent propulsion system performed satisfactorily. The total time for the descent firing was approximately 727 seconds. The propellant quantity gaging system indicated about 1225 lb of usable propellant remaining at engine shutdown. This is equivalent to 117 seconds of hover time. The descent propulsion system pressures and temperatures were as expected during all phases of the mission. The supercritical helium system functioned normally. Performance of both the supercritical helium tank and the ambient start bottle is shown in table 8-I.

The gaging probe readings at engine shutdown are given in table 8-II. The low-level sensor was activated at lunar landing and was probably a result of propellant slosh, since all probes indicated propellant quantities above the sensor activation level.

8.7 ASCENT PROPULSION SYSTEM

The ascent propulsion system performance was satisfactory for the lunar ascent and a terminal phase initiation firings. The engine firing duration for the ascent maneuver was 441 seconds. The terminal phase initiation maneuver firing time was approximately 2.7 seconds, however, no data were received during the second firing. System pressures and temperatures were normal during all phases of the mission.

8.8 GUIDANCE, NAVIGATION, AND CONTROL

The performance of the primary guidance, navigation, and control system as well as the abort guidance system was normal throughout the mission. Two irregularities were noted during system activation.

The first irregularity was noted when power was applied to the computer. The crew did not observe the expected restart light, and 400 was displayed in register 2 of the display keyboard. The register should have been blank. Data indicate that 3 restarts had occurred instead of one and that two computer addresses contained useless data. All of these symptoms can be caused by very small voltage variations resulting from changing current demands as the computer is being activated. Subsequent computer self-tests were normal and the computer performance was normal for the

TABLE 8-I.- DESCENT PROPULSION SYSTEM
SUPERCRITICAL HELIUM SYSTEM PERFORMANCE

Parameter		Value
Supercritical helium bottle		
Pressure rise rate from lift-off to powered descent maneuver ignition	. . . .	6.8 psi/hr
Pressures during powered descent Ignition pressure	. . . .	1304 psia
Peak pressure	. . . .	1415 psia
Pressure at lunar touchdown	. . . .	482 psia
Ambient start bottle		
Pressure level from loading to use . . .	. . . .	1646 to 1625 psia
Pressure after squib valve opening . . .	. . . .	600 psia
Pressure at ignition for powered descent	. . . .	602 psia
Pressure at lunar landing	. . . .	602 Asia

TABLE 8-II.- PROPELLANT QUANTITY GAGING SYSTEM DATA AT ENGINE SHUTDOWN

	Tank 1	Tank 2	Total
	Oxidizer		
Quantity, percent	6.6	7.1	--
Weight, lb	379	407	796
	Fuel		
Quantity, percent	6.4	6.2	--
Weight, lb	230	223	453

remainder of the mission.

The second irregularity occurred when the engine pitch and roll gimbal trim settings were displayed.The numbers were correct, but the axes were reversed from those in the onboard checklist. Due to an error in the documentation supplied to Kennedy Space Center, the axes were reversed when the data were loaded into the computer during prelaunch activities. Aside from these irregularities, activation of both guidance systems was normal.The lunar module computer timing was synchronized to the command module timing and the lunar module platform was aligned to the command module platform.Table 8-III is a summary of primary guidance system platform alignments and table 8-IV is a history of the inertial component stability.Table 8-V is a history of the abort guidance system calibrations. The powered descent maneuver was initiated on time. Table 8-VI is a summary of significant events during descent.

Performance during ascent was normal. Velocity residuals at insertion were minus 0.9, minus 1.2, and plus 1.3 ft/sec in the X, Y, and Z axes, respectively, and the orbit indicated by the guidance computer was 50 by 9.1 miles. Insertion errors from various sources are shown in table 8-VII. A vernier adjustment maneuver of minus 3.6, minus 9.0, and plus 1.2 ft/sec was performed to adjust the orbital conditions. Table 8-VIII is a summary of rendezvous solutions from several sources.

TABLE 8-III.- LUNAR MODULE PLATFORM ALIGNMENT SUMMARY

Time hr:min	Type alignment	Alignment mode						Star angle difference, deg	Gyro torquing angle, deg			Gyro drift rates, meru		
		Optiona	Technique b	Detent c	Star	Detent c	Star		X	Y	Z	X	Y	Z
106:56	52	3	-	3	52-Dubhe	1	11-Aldebaran	0.07	0.620	0.695	0.135	-	-	-
108:46	52	3	-	2	16-Procyon	2	11-Aldebaran	0.03	0.056	0.044	-0.028	-1.99	-1.57	-0.99
111:11	57	3	3	3	52-Dubhe	-		0.04	0.051	-0.026	0.097	-1.41	0.72	2.68
181:56	57	4	3	4	36-Vega	-	-	-0.02	0.028	0.014	0.053	-	-	-
184:50	57	4	3	4	36-Vega	-	-	-0.01	0.011	-0.032	-0.014	-0.25	0.74	-0.32

a1 - Preferred; 2 - Nominal; 3 - REFSMMAT; 4 - Landing site.
b0 - Stored attitude; 1 - REFSMMAT + g; 2 - Two bodies; 3 - 1 body + g.
c 1 - Left front; 2 - Center; 3 - Right front; 4 - Right rear; 5 - rear; 6 - Left rear.

TABLE 8-IV.- INERTIAL COMPONENT HISTORY

Parameter	Number of samples	Sample mean	Standard deviation	Flight load	In-flight performance		
					Power-up to surface	Surface power-up to lift-off	Lift-off through rendezvous
Accelerometers							
X - Axis							
Scale factor error, ppm	5	-867	60	-980	-	-	-
Bias, em/sect	5	1.61	0.05	1.64	1.59	1.20[a]	1.21
Y - Axis							
Scale factor error, ppm	5	-444	62	-560	-	-	-
Bias, em/sect	5	1.74	0.03	1.73	1.69	2.06[a]	2.01
Z - Axis							
Scale factor error, ppm	5	-343	49	-469	-	-	-
Bias, cm/sec	5	1.60	0	1.60	1.66	1.60	1.60
Gyroscopes							
X - Axis							
Null bias drift, meru	5	0.4	1.0	0.1	-1.89	-0.15	-
Acceleration drift about spin reference axis, meru/g	5	3.1.	0.6	4.0	-	-	-
Acceleration drift about input							
axis. meru/g	5	12.0	1.0	12.0			
Y - Axis							
Null bias drift, meru	5	0.8	1.2	0.4	-1.17	1.14	-
Acceleration drift about spin							
reference axis, meru/g.	5	5.9	1.0	6.0			
Acceleration drift about input axis, meru/g	5	-2.2	0.8	-4.0	-	-	-
Z - Axis							
Null bias drift, meru . . . . , . .	5	-0.4	0.88	-1.1	-2.10	-1.42	-
Acceleration drift about spin reference axis, meru/g	5	-8.9	0.6	-8.0	-	-	-
Acceleration drift about input axis, meru/g.	5	5.4	1.9	4.0			

[a] A bias update was performed at 182:20.

TABLE 8-V.- ABORT GUIDANCE SYSTEM CALIBRATION HISTORY

Parameter		Preflight performance			In-flight performance				
		Mean of Calibrations [1]	Standard deviation of calibrations	Flight load	System activation	Post-landing	Pre-lift-off	Post-ascent	Post docking
Accelerometers									
Static bias, mu g									
X-axis		74	13.2	62	93	-	-	35	37
Y-axis		-212	9.0	-217	-217	-	-	-241	-242
Z-axis		87	23.1	62	0	-	-	38	37
Gyroscopes									
Gyroscope drift, deg/hr									
X-axis		0.02	0.06	0	-0.32	-0.33	-0.29	-	-
Y-axis		1.07	0.04	1.01	0.87	0.85	0.80	-	-
Z-axis		0.20	0.04	0.23	0.63	0.07	-0.98	-	-

[1]The mean value of the calibrations is based on 36 calibrations of the units during the preflight period.

TABLE 8-VI.- SEQUENCE OF EVENTS DURING POWERED DESCENT

Elapsed time, hr:min:sec	Time from ignition, min:sec	Event
110:09:45	-00.08	Ullage on
110:09:53	00:00	Ignition
110:10:21	00:28	Throttle to full-throttle position
110.11:25	00:32	Manual target update (N69)
110:13:28	03:35	Landing radar velocity data good
110:14:06	04:13	Landing radar range data good
110:14:32	04:39	Enable landing radar updates (V57)
110:17:19	07:26	Throttle down
110:19:15	09:22	Approach phase program selected (P64)
110:19:16	09:23	Landing radar antenna to position 2
110:19:26	09:33	First landing point redesignation
110:19:54	10:01	Landing radar to low scale
110:20:51	10:58	Landing phase program selected (P66)
110:21:58	12:05	Spacecraft landing

TABLE 8-VII.- LUNAR ASCENT INSERTION SUMMARY

Source	Altitude, ft	Downrange velocity, ft/sec	Crossrange velocity (left), ft/sec	Radial velocity, ft/sec
Primary guidance system	60 711	5542	-1	33
Abort guidance system	60 912	5541	-8	34
Powered flight processor	62 001	5541	-9	41

TABLE 8-VIII.- RENDEZVOUS SOLUTIONS

Maneuver	Local vertical coordinates	Computed velocity change, ft/sec		
		Command module computer	Lunar module guidance computer	Abort guidance system
Terminal	Delta Vx	75.9	75.6	75.5
phase	Delta Vy	4.8	5.1	7.2
initiation	Delta Vz	17.6	19.7	17.4
	Total	78.1	78.3	77.7
First	Delta Vx	0	-1.2	-1.1
midcourse	Delta Vy	0.2	0.4	1.9
correction	Delta Vz	-0.3	0.3	-0.6
	Total	0.4	1.3	2.3
Second	Delta Vx	0.1	-0.4	-0.4
midcourse	Delta Vy	-1.4	-0.7	1.9
correction	Delta Vz	-5.4	-1.6	-1.7
	Total	5.6	1.8	2.6

8.9 ENVIRONMENTAL CONTROL SYSTEM

Performance of the environmental control system was satisfactory.

All system components functioned normally except demand regulator A, which caused a rise in pressure in the suit loop while it was unmanned during preparation for the third extravehicular activity. Demand regulator A was placed in the closed position and the pressure rise stopped. Rather than perturbate the extravehicular activity or mission timeline, the regulator was not rechecked, but was left closed for the remainder of the lunar module operations. Regulator A was leaking at a rate between 0.03 to 0.05 lb/hr and was probably caused by contamination between the ball-poppet and seat. Section 15.2.2 contains a discussion of this anomaly.

8.10 CONSUMABLES

All lunar module consumables remained well within redline limits.

8.10.1 Descent Propulsion System

Propellant.- The descent propulsion system propellant load quantities shown in the following table were calculated from known volumes and weights of off-loaded propellants, temperatures, and densities prior to lift-off.

Condition	Quantity, lb		
	Fuel	Oxidizer	Total
Loaded	7521.7	12 042.5	19 564.2
Consumed	7041.3	1 207.6	18 248.9
Remaining at engine cutoff			
Total	480	835	1315
Usable	455	770	1225

Supercritical helium.- The quantities of supercritical helium were determined by computations using pressure measurements and the known volume of the tank.

Condition	Quantity, lb	
	Actual	Predicted
Loaded	51.2	51.2
Consumed	41.6	43.0
Remaining at landing	9.6	8.2

8.10.2 Ascent Propulsion System

Propellant.- The ascent propulsion system total propellant usage was approximately as predicted. The loadings shown in the following table were determined from measured densities prior to launch and from weights of off-loaded propellants.

Condition	Propellant mass, lb			[a]Predicted quantity, lb
	Fuel	Oxidizer	Total	
Loaded	2026.9	3234.8	5261.7	5257.5
Consumed	1918.0	3059.2	4977.2	4946.8
Remaining at ascent stage jettison	108.9	175.6	284.5	310.7

[a] Propellant required for ascent was reduced by 60 lb to account for reaction control system consumption.

Helium.- The quantities of ascent propulsion system helium were determined by pressure measurements and the known volume of the tank.

Condition	Actual quantity, lb
Loaded	13.2
Consumed	8.7
Remaining at lunar module impact	4.5

8.10.3 Reaction Control System Propellant

The reaction control system propellant consumption was calculated from telemetered helium tank pressure histories using the relationships between pressure, volume, and temperature.

Condition	Actual quantity, lb			Predicted quantity, lb
	Fuel	Oxidizer	Total	
Loaded				
System A	107.4	208.2	315.6	
System B	107.4	208.2	315.6	
Total			631.2	631.2
Consumed to:				
Lunar landing			131	157.8
Docking			282	273.0
Remaining at ascent stage jettison			449	473.4
Remaining at ascent stage impact			182.2	157.8

8.10.4 Oxygen

The actual quantities of oxygen loaded and consumed are shown in the following table.

Condition	Actual quantity, lb	Predicted quantity, lb
Loaded (at lift-off)		
Descent stage		
Tank 1	47.71	
Tank 2	47.39	
Ascent stage		
Tank 1	2.36	
Tank 2	2.36	
Total	99.82	
Consumed		
Descent stage		
Tank 1	23.41	22.5
Tank 2	22.94	21.4
Ascent stage		
Tank 1	0.05	0
Tank 2	0.06	0
Total	46.46	43.9
Remaining in descent stage at lunar lift-off		
Tank 1	24.00	24.91
Tank 2	24.45	25.90
Total	48.45	50.81
Remaining at docking (ascent stage)		
Tank 1	2.31	2.36
Tank 2	2.30	2.36
Total	4.61	4:72

8.10.5 Water

The actual water quantities loaded and consumed, shown in the following table are based on telemetered data.

Condition	Actual quantity, lb	Predicted quantity, lb
Loaded (at lift-off)		
Descent stage		
Tank 1	202.6	
Tank 2	204.6	
Ascent stage		
Tank 1	41.3	
Tank 2	42.6	
Total	491.1	
Consumed		
Descent stage (lunar lift off)		
Tank 1	185.7	177.5
Tank 2	189.1	181.5
Ascent stage (docking)		
Tank 1	7.5	6.5
Tank 2	8.1	7.5
Total	390.4	373.0
Remaining in descent stage at lunar lift-off		
Tank 1	16.9	25.1
Tank 2	15.5	23.1
Total	32.4	48.1
Remaining in ascent stage at docking		
Tank 1	33.8	34.8
Tank 2	34.5	35.1
Total	68.3	69.9

9.0 LUNAR SURFACE OPERATIONAL EQUIPMENT

9.1 LUNAR ROVING VEHICLE

The lunar roving vehicle satisfactorily supported the lunar exploration objectives. Table 9-I shows the performance parameters from the lunar roving vehicle. Controllability was good, and no problems were experienced with steering, braking, or obstacle negotiation. The navigation system gyro drift and closure error at the lunar module were negligible. All interfaces between the crew and the lunar roving vehicle and between the lunar roving vehicle and the stowed payload were satisfactory. A detailed discussion of the lunar roving vehicle performance is contained in reference 2. Deployment of the lunar roving vehicle from the lunar module was smooth and no significant problems were encountered. The chassis lock pins did not seat fully, but the crew used the deployment assist tool to seat the pins.

At initial power, up, the lunar roving vehicle battery temperatures were higher than predicted; 95° F for battery 1 and 110° F for battery 2 compared to the predicted temperatures of 80° F for each. This was partially due to the translunar attitude profile flown, and partially to a bias in the battery temperature meter. Following adequate battery cool-down after the first extravehicular activity, temperatures for the remainder of the lunar surface operations were about as predicted.

The following lunar roving vehicle systems

problems were noted:

a. The battery 2 temperature indication was off-scale low at the start of the third extravehicular activity.

b. The right rear fender extension was knocked off prior to leaving the lunar module on the first extravehicular traverse.

The battery 2 temperature indication was off-scale low at the beginning of the third extravehicular activity. This condition continued for the remainder of the lunar surface operation. The most probable cause was a shorted thermistor in the battery. The same condition was noted on ground testing of two other batteries. Electrolyte leakage through the sensor bond, as a result of elevated temperatures, may have caused the short. Temperature monitoring was continued using battery I as an indicator with temperature trends established from data on the first and second extravehicular activities for battery 2.

During the first extravehicular activity at the lunar module site, the Commander inadvertently knocked off the right rear fender extension.

While still at the lunar module site, the Commander taped the extension to the fender.

TABLE 9-I.- LUNAR ROVING VEHICLE PERFORMANCE PARAMETERS

	First extravehicular activity	Second extravehicular activity	Third extravehicular activity	Total	Mission planning value
Drive time, minutes	27	138.5	88	253.5	280
Map distance, kilometers	2.3	19.0	11.0	32.3	32.5
[a] Odometer distance, kilometers				36.0	37.35
Traverse	2.5	20.2	12.0		
Additional	1.0	0.2	0.1		
[b] Traverse mobility rate, kilometers/hr	5.1	8.25	8.20	7.6	
Traverse average speed, kilometer./hr	5.55	8.75	7.50	8.2	8
Energy rate, ampere-hours/ kilometers (lunar roving vehicle only)	1.77	1.54	1.61	1.59	1.8
Energy consumed, ampere-hours				72	85
Lunar roving vehicle	6.2	31.5	19.5		
Lunar communications relay unit	14.8	-			
Navigation closure error, kilometers	0	0	0	-	-
Number of navigation updates	0	0	0	-	-
Gyro drift rate, deg/hr	Small	small	small	-	
[c] Wander factor plus slip, percent .	0	6	10	8.4	-
Maximum speed reported, kilometers/hr .	11	12	18 downhill	-	
Maximum slope reported, deg	-	18° up 20° down		-	

a Odometer distance (traverse) - distance actually drives from traverse starting point (surface electrical properties antenna) to end point (usually surface electrical properties antenna).

b Mobility rate = Map distance/Drive Time

c Wander factor = Traverse odometer - map distance/Map distance

Because of the dusty surfaces, the tape did not adhere and the extension was lost . Lunar surface maps were clamped to the fender (fig . 9-1 I. This fix was adequate.

Figure 9-1.- Crew-manufactured fender extension.

9.2 LUNAR COMMUNICATIONS RELAY UNIT AND GROUND COMMANDED TELEVISION ASSEMBLY

The lunar communications relay unit provided satisfactory support from the lunar surface, and the ground-commanded television assembly produced good quality pictures at all times. Activation was initiated about I hour and I I minutes after crew egress for the first extravehicular activity. Television coverage of crew egress was not available because the capability to televise from the lunar module was eliminated for Apollo I7 to save weight.

The system allowed ground personnel to coordinate lunar surface activities with the crew. The rover fender repair and the deep-core drilling were especially significant in this area. Television coverage, augmented by crew comments, was a valuable asset used in making an early determination of the actual experiment locations, sampling sites, traverse stops, geological features, and the landing point area. Panoramic reproductions of the television pictures were a significant contribution to a preliminary interagency geology report which was issued on December 17, 1972, two days prior to termination of the mission.

The lunar communications system failed to respond to uplink turn-on commands about 36 hours after lunar lift-off. The condition was expected because the lunar environment eventually exceeded the operational temperature limits of the equipment.

Total television operating time was 15 hours and 22 minutes.

9.3 EXTRAVEHICULAR MOBILITY UNIT

The performance of the extravehicular mobility units was good for all three extravehicular periods which totaled 22 hours 4 minutes. The crew had no difficulty in donning or doffing the suits and the portable life support systems operated satisfactorily.

System operation during integrity checks of the extravehicular mobility units prior to each extravehicular activity was normal. Prior to the second extravehicular activity, the Lunar Module Pilot's portable life support system water tanks were reconnected to the lunar module water system to assure their being filled to capacity. The tanks, however, did not require any additional water. Crew comfort was maintained satisfactorily throughout the extravehicular activities with the crewmen adjusting the water diverter valves as required to control cooling. Both crewmen received the necessary feedwater warning tones for the expected depletion of primary feedwater and routinely switched to the auxiliary feedwater supply.

The Lunar Module Pilot encountered some difficulty in operating the sun shade of the lunar extravehicular visor assembly because of lunar dust in the slide mechanism. Dust and scratches on the outer gold visor prompted the crew to operate with the outer visor in the partially raised position during part of the third extravehicular activity.

Both crewmen used the special cover gloves (provided for drilling operations) throughout the first two extravehicular activities, and noted that the gloves were extremely worn. The cover gloves were removed and discarded at the beginning of the third extravehicular activity.

During the pressure regulation check of the Lunar Module Pilot's oxygen purge system after the third extravehicular activity, the regulation pressure was slightly above the specification value. This was attributed to regulator leakage. History of the oxygen purge system regulators has shown that leakage can be caused by either a small amount of contamination (2 to 3 microns) on the seat or by minor seat erosion resulting from the high gas flow. The amount of leakage present was such that the unit was still suitable for a contingency transfer, if needed. However, since only one oxygen purge system was to be returned to support the Command Module Pilot's extravehicular activity, the Commander's oxygen purge system was retained and the Lunar Module Pilot's oxygen purge system was jettisoned with the ascent stage.

The suits and related equipment were returned for postflight inspection. The suits were only slightly worn and leakage was within the specification values.

Oxygen, power, and feedwater consumables are shown in table 9-11.

TABLE 9-II.- EXTRAVEHICULAR MOBILITY UNIT CONSUMABLES

Condition	Commander		Lunar Module Pilot	
	Actual	Predicted	Actual	Predicted
First extravehicular activity				
Time, min	432	420	432	420
Oxygen, lb				
Loaded	1.93	1.86	1.94	1.86
Consumed	1.55	1.26	1.57	1.26
Remaining	0.38	0.60	0.37	0.60
Redline limit	0.37	-	0.37	-
Feedwater, lb				
Loaded	12.19	11.90	12.12	11.90
Consumed	11.23	9.77	10.86	9.77
Remaining	0.96	2.13	1.26	2.13
Redline limit	0.91	-	0.91	-
Battery, amp-hr Initial charge	25.40	25.40	25.40	25.40
Consumed	18.40	18.90	20.20	18.90
Remaining	7.00	6.50	5.20	6.50
Redline limit	3.28	-	3.28	-
Second extravehicular activity				
Time, min	457	420	457	420
Oxygen, lb				
Loaded	1.78	1.81	1.81	1.81
Consumed	1.33	1.16	1.36	1.16
Remaining	0.45	0.65	0.45	0.65
Redline limit	0.37	-	0.37	-
Feedwater, lb				
Loaded	12.79	12.20	12.72	12.20
Consumed	10.20	8.94	10.10	8.94
Remaining	2.59	3.26	2.62	3.26
Redline limit	0.91	-	0.91	-
Battery, amp-hr Initial-charge	25.40	25.40	25.40	25.40
Consumed	19.00	18.90	21.30	18.90
Remaining .	6.40	6.50	4.10	6.50
Redline limit	3.28	-	3.28	--
Third extravehicular activity				
Time, min	435	420	435	420
Oxygen, lb				
Loaded	1.77	1.81	1.81	1.81
Consumed	1.33	1.25	1.43	1.25
Remaining	0.44	0.56	0.38	0.56
Redline limit	0.37	--	0.37	-
Feedwater, lb				
Loaded	12.79	12.20	12.72	12.20
Consumed	11.36	9.66	11.52	9.66
Remaining	1.43	2.54	1.20	2.54
Redline limits	0.91	-	0.91	-
Battery, amp-hr Initial charge	25.40	25.40	25.40	25.40
Consumed	18.00	18.90	20.30	18.90
Remaining	7.40	6.50	5.10	6.50
Redline limit	3.28	--	3.28	--

Apollo 17 crew
Commander Eugene A. Cernan, Command Module Pilot Ronald E. Evans, and Lunar Module Pilot Harrison H. Schmitt

10.0 PILOT'S REPORT

This section discusses the Apollo 17 mission as performed by the crew. Emphasis has been placed upon the operational and hardware differences from previous flights with descriptions and opinions presented on phases of the flight that were deemed operationally interesting. The as-flown flight plan is summarized in figure 10-1 at the end of this section.

10.1 TRAINING

The Apollo 17 crew was thoroughly prepared for this flight in all respects. The experience gained as members of previous flight and backup crews was a major contribution to this preparation. Because of the sophistication and complexity of the lunar-surface and orbital-science equipment, and the variety of lunar terrain to be encountered near the landing site, more time was spent in scientific training than by any previous crew. An even balance between scientific return and operational proficiency was considered a necessity, therefore, no phase of the mission training was compromised.

The successful accomplishment of the mission objectives can be directly attributed to this training, but the preflight and inflight support provided by personnel throughout the NASA organization was a contributing factor. The Apollo 17 support crew provided a most significant contribution.

10.2 LAUNCH

The crew ingress and the prelaunch checkout, lift-off, and powered flight of all three stages was similar in flight characteristics and physiological phenomenon to preceding Saturn V missions.

Following crew ingress, the prelaunch count proceeded normally until T-30 seconds at which time the automatic sequencer initiated a hold.The count resumed several times during the ultimate 2-hour and 40-minute delay, and as a result, the launch azimuth had to be updated and the platform was realigned several times prior to the lift-off.The quickest and easiest way to accomplish the update and realignment is to copy the new launch azimuth as read from the control center and then, with confirmation from the spacecraft test conductor, load and enter the azimuth update in one step.

Except for the hold, the major differences to consider when comparing this launch with previous ones are those associated with the night launch. The crew could see first-stage ignition through the rendezvous window and the hatch window cutout in the boost protective cover from a very few seconds prior to lift-off until a few seconds following liftoff.At staging of the S-IC stage, a bright flash was evident. It was as if the. spacecraft was being overtaken by a fireball at the first stage cutoff. he effect could have been produced by the S-IC stage retrograde motors or the S-IC engines. Escape tower jettison was more spectacular at night than it was in daylight. Again at S-II staging, the crew could see a glow through the windows as the S-II shut down and the S-IVB ignited.This glow could possibly have been the S-II fireball tending to overtake the spacecraft. During the S-IVB firing, the only visual clues came from the auxiliary propulsion system engines as they fired periodically throughout the dynamic firing phase and during attitude hold in earth orbit.

For mission planning purposes, the night-time launch indicated the possible loss of the backup abort capabilities, in the event of an S-IVB yaw hard-over maneuver combined with the loss of both the stabilization control system and the inertial measurement unit. Some effort was expended, during the S-II firing, to examine the possibilities of seeing stars through the window and detecting the presence of the horizon by watching the stars rise from the earth-occulted sky.The lack of night adaptation was a hindrance, even though the lights in the cabin were turned down for short periods of time. No stars were ever visible, nor was the horizon detectable. Had visible stars been required for a mode II or a mode IV abort, it is the Commanders' judgement that the abort would only have been marginally successful because of the small amount of night adaptation that the crew were able to acquire inflight.Although an abort under these conditions considers a second to third order failure, serious consideration must be given to the cockpit lighting configuration for powered flight, if the requirement to see stars during a night launch is valid.The lights were near full bright on Apollo 17 to preclude the possibility of being blinded at engine ignition and during the early seconds after lift-off. The requirement to observe the control panel displays during this period of flight far outweigh the potential requirements of maintaining spacecraft attitude with visual cues during a highly unlikely abort sequence.

There were no systems anomalies during the launch phase. A low battery-current load was noted when the bus ties were turned on. Because of this, the Lunar Module Pilot changed from his normal procedure of monitoring the batteries to monitoring the fuel cells for current fluctuations. When the service propulsion system gimbal motors were activated 6 minutes prior to launch, the fuel-cell currents fluctuated much more sharply than had been noted in the simulator.The apparent reason for the low battery loads was the high efficiency of the fuel cells. During the two staging events of the launch phase, there were no observed systems parameter fluctuations and a particular effort had been made to scan the gages after each major dynamic event to determine if any such fluctuations occurred.

The crew's opinion was that during certain phases of the launch, particularly during the high vibration and high g loads on the S-IC, troubleshooting of systems malfunctions would be very difficult. Launch phase training certainly prepares a person to troubleshoot malfunctions should it become absolutely necessary.

10.3 EARTH ORBITAL FLIGHT

Orbital insertion occurred in total darkness

and the first glimpse of the earth was gained in a spectacular sunrise over the Atlantic Ocean as the spacecraft approached the west coast of Africa. The orbital insertion checklist progressed without incident while being primarily involved with checking the systems, verifying the optics cover jettison and operation, aligning the platform, and preparing for the possibility of a first orbit translunar-injection opportunity. This preparation was complete when contact was made with the Hawaii range station on the first revolution. The next revolution afforded a good opportunity to leisurely make observations from the low earth orbit.

Earth observations pertaining to weather or land mass phenomena requires constant attention. The knowledge of discreet positions on an earth-based map, and the ability to predict the time of upcoming targets is essential to make the observations meaningful. Numerous stereo photographs were taken of the weather phenomena. An analysis of these photographs will be reported in a supplement to this report.

The optics dust covers were jettisoned in daylight and could be seen under reflected sunlight. There is an associated noise that is apparent with the jettisoning. No stars were ever seen through the telescope in the daylight; however, no attempt was made to become fully night-adapted. The pick-a-pair routine worked well with the stars visible in the sextant; however, in the daylight, there was no detectable illumination of the sextant reticle. The sextant cross-hairs showed up as black lines against a light blue background.

A systems problem was noted after earth orbit insertion when numerous spurious master alarm signals occurred, and these for the most part, seemed to be associated with the operation of switches on control panel 2 (sec. 15.1.1). An occasional alarm would occur by simply bumping the panel. These master alarms came in groups throughout the first day and continued into the second day of flight. Another condition noted was a 15-second lag in the mission timer in the lower equipment bay (sec. 15.1.2). The timer was reset and no further time losses occurred during the mission. The only major changes to the insertion checklist were the rearrangement of station acquisition times and the deletion of the television configuration for transposition and docking.

10.4 TRANSLUNAR INJECTION

The translunar injection procedures and maneuver were normal. The most significant change was the update to the backup translunar injection attitude numbers on the cue cards. Final translunar injection preparations began prior to the stateside pass and continuous communications contact was maintained through maneuver completion, using the Apollo range instrumentation aircraft in the final phases. Reception through Apollo range instrumentation aircraft was fair to good onboard the spacecraft and was apparently excellent on the ground.

The maneuver was initiated on time with the instrument unit and command module computer guidance agreeing well throughout the firing. The ignition was in darkness, with shutdown in the daylight. A spectacular sunrise was sandwiched in between the start and end of the translunar injection maneuver. All onboard displays and trajectory monitoring was normal with both the entry monitor system and digital status keyboard confirming a good cutoff. The S-IVB stage sounded and responded very much like it did while going into earth orbit with a noticeable low-frequency buzz that was similar to that reported by crews of previous missions. The acceleration during ignition and its subsequent firing might be likened to that produced by an aircraft afterburner.

Following translunar injection cutoff, numerous large fragments seemed to be coming off the S-IVB stage. They were very bright and tumbled slowly as they moved, generally away from the launch vehicle/ spacecraft combination.

10.5 TRANSLUNAR FLIGHT

10.5.1 Transposition, Docking, and Lunar Module Ejection

Spacecraft /launch vehicle adapter separation

was characterized by an audible pyrotechnic activation plus a noticeable shock or jarring in the command and service module. The shock did not close any isolation valves. The shock broke loose many small particles that appeared to be ice crystals, and these moved radially away from the vehicle. Command and service module translation and attitude maneuvers were audible and the motion of the spacecraft was sensed at the beginning and completion of the transposition maneuver. The estimated closure rate of about 0.1 ft/sec was maintained until docking probe contact. The S-IVB was very stable with no detectable oscillations or motions within the deadbands. The soft docking was normal. with no residual rate damping or attitude corrections being required prior to hard docking. Docking latch actuation was heard as a loud bang followed by an immediate ripple fire of the remaining latches; however, the docking-system-A talk-back indicator remained in the barberpole position. The crew optical alignment sight showed a right yaw of about 1 degree on the target during hard docking. Removal of the tunnel hatch went well and 7 of the 10 latches appeared to be locked during the inspection. On each of the 3 latches which appeared to be unlocked, the power bungee was vertical, but the handle was not locked down, and the red button was showing. The handle was locked on latch 10 by pushing it in the outboard direction. Latches 7 and 9 were each recocked twice and then manually triggered. With the reactivation of latch 9, docking-system-A talk-back indicated correctly. The visible surfaces of the lunar module were extremely clean and remained so throughout the flight.

10.5.2 Translunar Coast Operations

The chlorine ampules were a continuing problem throughout the flight. Between 50 and 70 percent of these ampules either leaked directly or leaked during the injection, into the potable water system. Additionally, attaching the injector to the bayonet fitting did not always cause a puncture of the chlorine ampule and provide a path for the chlorine to flow into the water system (sec. 15.1.6).

Following the S-IVB evasive maneuver, the suits were doffed. The 9-inch addition to the length of the L-shaped bag for stowing the suits proved to be beneficial. Even though it still took some effort, the three suits were stowed with the comfortable feeling that they were not damaged during the stowing process. The 1-hour eating period, which was planned at 7 hours after launch to shorten the long launch day, was of particular importance because of the 2-hour and 40-minute lift-off delay. Also, the planning to include a short sleep period in the first day allowed the crew to return to a normal work-rest cycle.

A slightly higher helium absorption into the service propulsion system oxidizer tank, coupled with a slightly lower-than-normal caution and warning system pressure limit, actuated the service propulsion system pressure light. The caution and warning system was placed in the acknowledge mode until prior to lunar orbit insertion when a manual repressurization of the service propulsion system oxidizer system allowed a return to the normal operational mode.

The ability to continuously observe the earth from a distance offers numerous advantages over the orbital observation of the earth. Of course, as the distance between the earth and the spacecraft increased, the advantage decreased.

10.5.3 Guidance and Navigation

At 17 hours into the flight, the transearth midcourse navigation training was performed to compare the Command Module Pilot's determination of the earth's horizon differential-altitude measurement with the value stored in the computer. The altitude measurement was determined to be 25 kilometers; therefore, because of the small difference, the preloaded erasable memory value of 29 kilometers was retained in the computer. The automatic routine to maneuver to the substellar point was not always precisely on target and the attitude had to be adjusted slightly, using the minimum impulse control mode. This slight error may have been due to spacecraft positioning within the selected deadband. The erasable memory program for cislunar midcourse navigation was utilized throughout the mission and was very worthwhile.

Only one earth horizon was discernible, and it was very easy to see.

The translunar injection maneuver was adjusted to compensate for the delayed lift-off so that !the spacecraft would arrive at the moon at the pre-planned Greenwich mean time for lunar orbit insertion. The flight plan was updated with essentially no changes by eliminating the hour between 46 and 47 hours in the flight plan. Then another hour and 40 minutes of the flight plan was eliminated at 66 hours and rescheduling of the onboard navigation to 67 hours and 40 minutes returned the mission to the original flight plan timeline.

Stars were not visible out of the telescope at any time when the lunar module was attached; however, the pick-a-pair routine worked well and stars were always visible in the sextant. Passive thermal control mode was entered and exited according to the checklist with excellent results each time. In translunar coast, all inertial measurement unit realignment option 1 changes in stable member orientations were accomplished by gyroscope torquing and proved satisfactory. The procedure was changed to use coarse alignment for stable member reorientation when the spacecraft was in darkness during lunar orbit. After coarse aligning, the stars, in all cases were either within or just outside the field of view of the sextant, indicating less than 2 degrees of error.

10.5.4 Lunar Module Checkout and Housekeeping

The probe, drogue, and hatch removal operations were accomplished without incident. A more detailed investigation of the docking latches revealed that latch 4 had not seated properly, although the hook was over the docking ring. The handle was pulled back to the once cocked position, and the hook easily pulled off the docking ring. The latch was left in this configuration until the second transfer to the lunar module when the latch cocking sequence was completed.

The first lunar module checkout proceeded according to the procedures except for a temporary ground communications problem. The second lunar module entry was in accordance with the flight plan for telemetry activation. The lunar module cabin was very clean with the only debris being the usual rivets, screws, etc., totaling not more than a dozen items.

10.5.5 Midcourse Corrections

The only midcourse correction required during the translunar flight was performed at the second option point. The maneuver was a 2-second minimum impulse firing of the service propulsion system, and was performed on single bank A. The maneuver provided a short, quick look at the service propulsion system, which functioned satisfactorily. The gimbal trim checkout, both manual and automatic, was clearly evident by the fuel cell and battery current variations.

10.5.6 Pre-Lunar Orbit Insertion Operations

Preparations and proceedings for scientific instrument module door jettison were normal. There was a definite physiological feeling and sound associated with the jettisoning similar to other spacecraft pyrotechnic functions. A small amount of debris accompanied the door upon jettisoning and the door slowly tumbled in a random manner as it separated from the service module.

Unlike previous flights, Apollo 17 did not go into a penumbra prior to lunar orbit insertion; so daylight prevailed throughout the translunar coast phase. Several hours prior to lunar orbit insertion, a very small crescent of the moon was visible to the crew. This limb grew rapidly and, at about 8000 miles, lunar topographic features were identifiable on the horizon. Observation of the approach towards a 50-mile perilune from several thousand miles away was quite spectacular.

10.6 LUNAR ORBITAL OPERATIONS PRIOR TO DESCENT

10.6.1 Lunar Orbit Insertion and Descent Orbit Insertion

During preparation for lunar orbit insertion, the abort charts and curves were updated to account for the launch delay. This activity required some time to copy and verify, but was handled very well by the ground personnel.

Lunar orbit insertion preparations went well, ignition was on time, and the maneuver was normal. The engine chamber pressure indication was approximately 87 psi on bank A and increased about 5 psi when bank B was armed 5 seconds later. The chamber pressure gradually increased to about 96 to 97 psi, where it remained for the rest of the firing. A slightly low gage bias was known preflight. Pitch and yaw rates were stable; however, roll rates were about 0.2 to 0.3 deg/sec. After the initial transients in the propellant utilization gaging system which lasted about 30 seconds, the system stabilized at 180 pounds decrease and then gradually began diverging. When the unbalance reached 300 pounds, the flow valve was placed in the decrease position. At the time of crossover, the flow had reversed to about 100 pounds decrease where it appeared to stabilize and the switch was then placed in the normal position where it remained for the rest of the firing. Just prior to the completion of the firing, the indicator moved towards decrease again with a resulting unbalance of about 110 pounds decrease.

The first descent orbit insertion maneuver was a 22-second guidance and navigation-controlled service propulsion firing with an automatic shutdown. With no trim, the resulting orbit was 59.1 miles by 14.9 miles.

During the 4-engine 15-second ullage, the spacecraft exhibited low rates about all three axes.

10.6.2 Lunar Landmark Tracking

The first lunar landmark tracked was crater J-3 on the third revolution. This was a training target to the east of the landing site and allowed the Command Module Pilot to become familiar with the telescope shaft and trunnion rates while performing the low-altitude tracking. This provided confidence and enabled the Command Module Pilot to verify that low altitude landmark tracking could be accomplished as planned. Even though the inertial attitude was not exactly correct, a landmark in the area of the landing site was also selected for tracking on the third revolution, and therefore, four or five marks were taken on crater F. Landmark 17-1 was observed in the sunlight and the landmark tracking was shifted to 17-1 for the latter part of this tracking pass. The regular low-altitude landmark track on 17-1 during revolution 14 went well.

The high-altitude landmark tracking was easily accomplished; however, the tracking of landmark RP-3, which was very close to the zero-phase point, disappeared after the point of closest approach.

10.6.3 Lunar Module Activation Checkout

The lunar module activation and checkout was normal, except that the lunar module tunnel venting required 10 to 15 minutes longer than had been anticipated because the command module simulator erroneously indicates an operation of about 1 to 2 minutes.

The crew completed what was probably the smoothest inflight lunar module activation and checkout ever performed. Procedurally, the timeline and activation book, as written, were in excellent condition. However, the following minor conditions were noted.

a. During guidance computer power-up, when the lunar module guidance computer display and keyboard circuit breaker was closed, the restart light did not illuminate and the keyboard displayed a 400 and R2.,

b. During digital autopilot loading and verification, the pitch and yaw pre-set gimbal trim was in the wrong registers; i.e., R1 was loaded with yaw and R2 was loaded with pitch. Since the actual numbers were near equal, the load was not changed and the gimbal was not retrimmed.

c. The secondary glycol loop had small pressure

oscillations and a ragged sound at pump powerup. The sound ceased and the pressure stabilized after about 15 seconds.

Undocking and separation were completed on time. Following separation, the lunar module crew visually tracked the command and service module as it maneuvered for landmark tracking over the landing site (fig. 10-2). The crew obtained a spectacular oblique view of the landing site, while, passing over it on the revolution prior to powered descent initiation. The VHF communications were excellent, resulting in an easy confirmation of the command and service module circularization maneuver prior to the second descent orbit insertion. The second descent orbit insertion maneuver of 7 ft/sec, was smaller than predicted preflight, and was performed easily with the minus X lunar module reaction control system thrusters. The primary navigation guidance system predicted a perilune of 7.0 miles with the abort guidance system being more accurate in this case by predicting 6.7 miles.

10.7 POWERED DESCENT AND LANDING

10.7.1 Preparation for Powered Descent

At acquisition-of-signal on the landing revolution, the primary navigation guidance system state vector was uplinked to the crew much quicker than had been expected or had ever been received in the simulations. After a slight uplink problem due to antenna switching, the uplink was received and the landing maneuver and braking phase program was subsequently

Figure 10-2.- Command and service module over the landing site.

selected for a comfortable and confident countdown to powered descent initiation. Physiologically, both the Commander and the Lunar Module Pilot sensed the ullage ignition; and it was also evident on the computer keyboard. Descent engine start and throttle-up were on time and were verified by the guidance computer. A downrange landing site correction of +3400 feet was inserted into the primary guidance computer 2 minutes into the maneuver. The next major event occurred at 4 minutes when the lunar module was yawed from 290 degrees to 340 degrees with a simultaneous landing radar lock on. Just prior to this yaw maneuver, there was an indication that the radar was attempting to hold a velocity lock. The radar data were accepted and the differential altitude display rapidly converged to zero. The mountain peaks of the lunar terrain, smoothed over by the terrain model, were observed at approximately 30 000 to 32 000 feet as the differential altitude diverged to between 2000 to 2500 feet and then converged back to zero very rapidly. From 30 000 feet down to the lunar surface, the differential altitude remained near zero and the descent closely followed the preflight predicted trajectory. South Massif was easily observed out the Commander's window at an altitude of 13 000 feet. A little extra effort in looking out of the bottom half of the window provided a view of the western part of the valley, the scarp, and the light mantle area. Throttle-down occurred on time, end pitchover followed shortly thereafter at about 7500 feet altitude. When pitchover occurred, it was almost immediately evident that there were better areas in which to land than that selected preflight. As the landing point designator steadied and the trajectory could be verified, the targeted landing site was the pre-mission predicted point. Several pitchup landing point designator commands were made so as to fly the lunar module to the desired spacecraft landing area. When this redesignation was completed to an area several hundred meters south of the targeted landing point, it was evident that boulders and craters were going to be the determining factors in the selection of the final landing point. As the altitude decreased and a more preferable landing spot within the new area could be seen, several additional landing point designator corrections were made to pinpoint the exact landing location. The general slopes of the local terrain were difficult to determine and indications are that they were not a major factor in determining the landing spot itself. Manual takeover occurred just under 300 feet altitude and the landing was completed with near zero lateral velocity and a small forward velocity. The descent rate was approximately 3 ft/sec. The shadow of the spacecraft was used to anticipate touchdown. The somewhat higher-than-normal descent rates after manual takeover, and the freedom in maneuvering the spacecraft provided a comfortable and safe landing approach which was well within the control capability of both man and machine. This is attributed not only to the lunar module simulator, but to the training that was received flying the 'lunar landing training vehicle. This vehicle is considered, as has been stated by previous pilots, an excellent training device for the last 300 or 400 feet of the actual lunar landing phase. During the final phases of the landing, the Commander divided his attention between looking out-the-window and in-the-cockpit. Inside, the velocity crosspointers and the attitude display were monitored. This technique was used in the lunar module simulator and in the lunar landing training vehicle. Dust was first observed at 60 to 70 feet altitude, as indicated on the tape meter.

There were two unexpected incidents noted. During descent, the late ascent battery turn-on followed by slow battery conditioning resulted in one descent battery being turned off early in the powered descent phase and then put back on the line about 5 minutes later; also when the descent engine was armed, the descent quantity light was illuminated. (Note: Neither condition was abnormal). The warning light was reset with the quantity switch and there were no further activations during powered descent. Landing was completed with an indicated 7 to 9 percent of the propellant remaining; however, several minutes after touchdown another descent quantity warning light was triggered, apparently caused by unequal tank quantities due to gravity settling. At touchdown, there were no observed abnormalities in spacecraft systems or associated measurements.

The final attitude of the lunar module after

engine shutdown, was approximately 4 to 5 degrees pitchup, zero roll, and near zero yaw. Subsequent inspection of the area during the extravehicular activities showed that the rear strut of the spacecraft rested near the bottom of a 3- or 4-meter diameter crater and produced the spacecraft pitchup attitude (fig. 10-3). It is assumed that the descent engine shutdown occurred with approximately zero pitch, when the spacecraft hit the surface, it pivoted on the forward (+Z) strut and produced a somewhat harder than anticipated aft impact. The rear strut may have stroked, but the other three did not.

10.8 LUNAR SURFACE OPERATIONS

10.8.1 Preparation and Post Extravehicular Activity

Postlanding activity and preparation for the first extravehicular activity were initiated with the power-down of the spacecraft and completed without incident. The crew was methodical in the portable life support system checkout, verification of suit integrity, and in general housekeeping for the lunar stay activities.

As a result of the methodical operation, the crew was about 30 minutes to 1 hour behind the planned timeline, and they eventually commenced

Figure 10-3.- Landing gear pad in small crater.

the first extravehicular activity approximately I hour late. The subsequent preparations and all post extravehicular activity operations went well.

The surface checklist was followed closely except that the Commander's oxygen hoses were disconnected from the suit during all the preparation activities. The hoses were strapped against the environmental control system bulkhead and were allowed to flow oxygen into the cabin. To provide additional clearance in the portable life support system donning station and in the forward cabin area, the suit hoses were not connected. The liquid cooling garment pump circuit breaker was cycled for cooling purposes and cooling was more than adequate. The Lunar Module Pilot allowed his hoses to remain connected to the suit because it was the most convenient method to handle these short hoses.

The checklist for the preparation and post extravehicular activity operations, as well as the general timeline, had been well considered and the preflight crew training exercises were very worthwhile. Training emphasis was placed on preparations for the first extravehicular activity over all the others, because this was most time-critical, due to the lengthy landing day activities. There was no degradation in performance during the second and third extravehicular activity preparations or the post extravehicular activity operations because of a lesser amount of training in those areas. A partial housekeeping configuration was established before the end of preparations for the first extravehicular activity and the balance of the configuration was completed during preparation for the sleep period. The size of the lunar module made it difficult for two crewmen to freely move around. As a result, all activities in the cabin are accomplished slowly to insure that none of the equipment is damaged. Much of the time required for preparation and post-extravehicular activity operations is a result of the constraining cabin space. A systematic stowage, unstowage, and interim stowage of gear is required during the preparation exercise.

The mission timer was retained in a powered-up configuration for emergency lift-off purposes, although the numeric lighting circuit breaker was disengaged. The computer clock was re-initialized several times during the lunar stay.

Suit donning and doffing was accomplished essentially as planned. Particular care was exercised in the cleaning and lubrication of zippers and other dust sensitive portions of the extravehicular mobility unit.

The dust protectors on the wrist lock-locks were used on each extravehicular activity, but it is difficult to determine how much protection these provided. The conservative approach was to use them; however, by the end of the second extravehicular activity and certainly by the end of the third extravehicular activity, the wrist lock-locks had stiffened considerably. However, the moving parts of the extravehicular mobility unit seemed to be usable for an indefinite number of extravehicular activities providing that proper care was given to them.

The liquid cooling garments were doffed at the end of each extravehicular activity and the constant wear garment was donned for sleep. The crew believed that their sleep was much more comfortable in the constant wear garment partly because it was cleaner than the liquid cooling garment. Also, the constant wear garment removed the general level of pressure that a tight-fitting liquid cooling garment exerts on the body.

Prior to ascent from the lunar surface, the cabin activities included covering all holes in the lunar module floor into which dust had collected or could be swept. Although considerable dust appeared in the cabin upon insertion, taping the holes definitely prevented a major dust problem in zero-g.

10.8.2 Extravehicular Activity

First extravehicular activity.- The intended objectives of the first extravehicular activity were accomplished, including lunar roving vehicle deployment and preparation, experiment deployment, and post-experiment activities. The sole revision to the first extravehicular activity was the shortening of the geology traverse time, necessitated by delays in the Apollo lunar surface experiments package deployment. The shortened traverse time required that station 1, originally planned for the east side of Emory Crater, be moved to a point about one half the intended distance from the lunar module (see figs. 4-1 and 4-2). It was, however, almost on the original planned traverse line. The major sampling objectives of the Emory Crater station were accomplished at a place later referred to as station 1A. The planned investigation of an apparent contact within . the dark mantle unit, and the investigation of Emory Crater itself were not conducted.

The geological objectives met during the first extravehicular activity were connected primarily with the sampling and visual observations of the dark mantle unit in the vicinity of the lunar module and near station 1A. Blocks of subfloor gabbro in this area and along the traverse were also investigated and sampled.

The total duration of the first extravehicular activity was 7 hours and 12 minutes and the lunar roving vehicle traveled approximately 3.5 kilometers. The total number of samples was 18 with a weight of approximately 31 pounds.

Second extravehicular activity.- The plan for the second extravehicular activity was performed essentially as intended. The time extension granted at station 2 plus the addition of a new station (2A), between stations 2 and 3, eventually decreased the time available at station 4. Station 2A was established to provide additional information on a major problem that had been detected with the traverse gravimeter. The extravehicular activity was geologically significant because of the extensive sampling of blocks which were clearly derived from the slopes of the South Massif, at station 2, and because of the sampling and investigation of a mass of orange material on the rim of Shorty Crater at station 4. It appears that the orange material will prove to be one of the most recently exposed geological units sampled on the lunar surface during the Apollo program. The extravehicular activity also provided additional statistical coverage and information on the light mantle unit, the dark mantle unit, and the subfloor gabbro.

At the end of the second extravehicular activity, the Lunar Module Pilot returned to the lunar surface experiment deployment site for additional verification on the deployment of the lunar surface gravimeter.

The second extravehicular activity duration was 7 hours and 37 minutes during which time the lunar roving vehicle traveled approximately 20.4 kilometers. A total of 60 samples were collected with an accumulated weight of approximately 75 pounds. At the end of the extravehicular activity, all hardware systems were operating as expected, except for a noticeable difficulty in the movement of some mechanical parts because of dust permeation.

Third extravehicular activity.- The third extravehicular activity was conducted essentially as planned and met all of the pre-mission traverse objectives. A minor revision was made to accomplish additional closeout activities, connected largely with an effort to solve the problems being experienced by the lunar surface gravimeter. The closeout time for this extravehicular activity was also lengthened to account for additional effort required in dusting the extravehicular mobility units. At the end of the extravehicular activity, several symbolic activities were conducted including the collection of a rock sample for distribution to the foreign nations which were represented by young people visiting the Mission Control Center during the mission, and the unveiling of a plaque commemorating the Apollo 17 lunar landing and all previous landings (fig. 10-4).

The third traverse proceeded normally except for the elimination of station 10, which resulted, primarily, from the increased time required for closeout activities, but also because of additional time taken at other stations prior to station 10. The traverse provided a comparative investigation of blocks of material derived from the slopes of the North Massif, blocks of material at least spatially associated with the Sculptured Hills, and the investigation and sampling of what appeared to be an extremely young impact crater. There was also continued sampling and investigation of dark mantle materials and

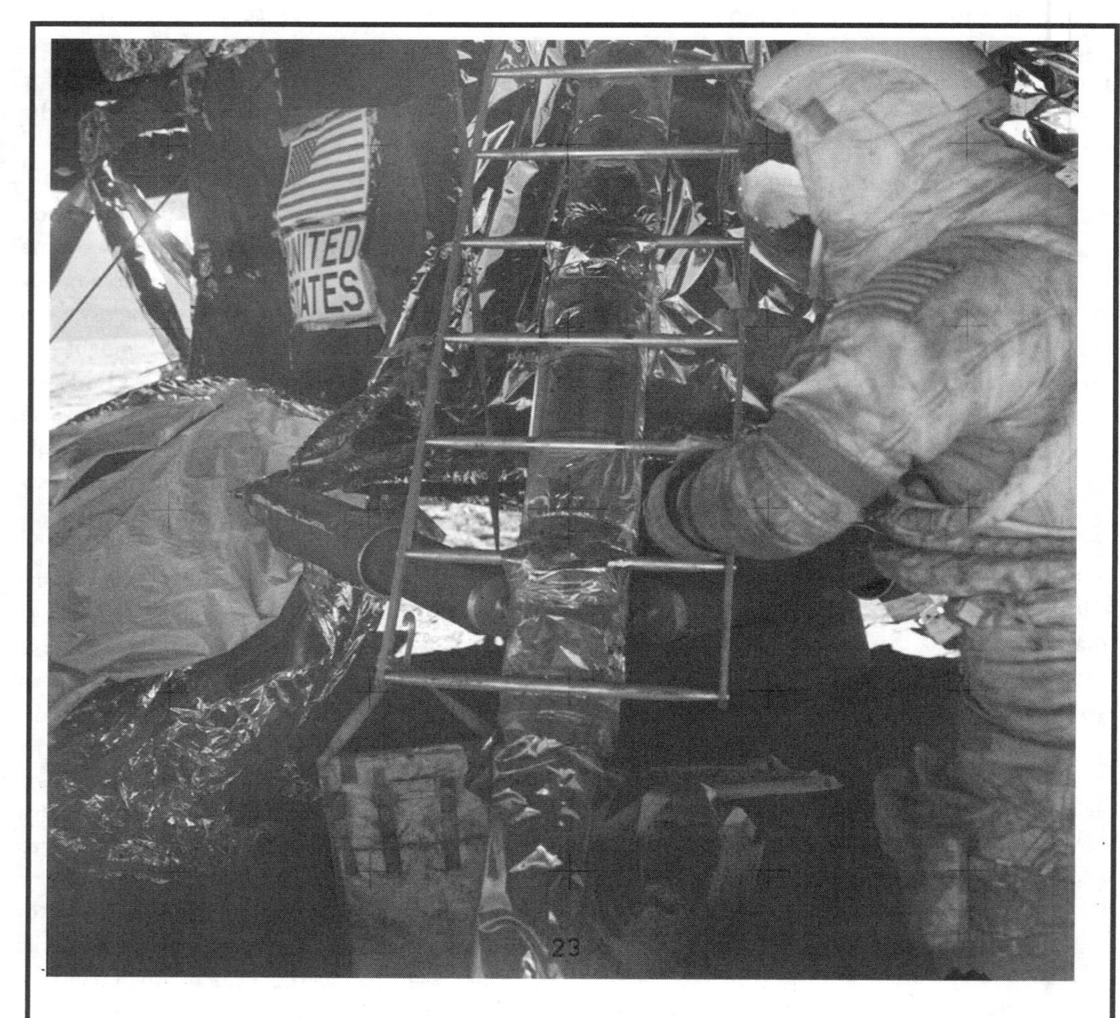

Figure 10-4.- Commander unveiling plaque on lunar module.

subfloor gabbro. The objectives at station 10, which were eliminated, were largely fulfilled by other activities during all three of the extravehicular activities.

The total time for the third extravehicular activity was 7 hours and 15 minutes with a traverse distance covered of approximately 12.1 kilometers. A total of 63 samples were collected and amounted to about 137 pounds.

In summarizing the operations on all three extravehicular activities, one of the most important ingredients for total efficiency was the crew staying near one another and working together, remaining in close proximity when, required to work independently, and complementing each other's activities. Only in rare instances was there separation to the extent that the crew could not correlate each other's geological observations and/or physically come to one another's assistance. The total distance driven with the lunar rover was approximately 36.0 kilometers at an average speed of 8.2 kilometers/hour with an occasional peak speed of 18 kilometers per hour. The total weight of the returned payload was 243 pounds.

10.8.3 Lunar Surface Hand Tools and Auxiliary Equipment

The tools provided for lunar surface operations, particularly those required in the sampling of lunar materials, all performed essentially as expected. The geological hammer was particularly useful to the Commander and usable when required by the Lunar Module Pilot. However, for future designs it is recommended that tools which required extensive gripping be custom fit to the dimensions of the crewmen's hands. The hammer was too large for use by the Lunar Module Pilot; however it was the right size for the Commander.

The lunar surface scoop was the primary sampling tool used by the Lunar Module Pilot and it worked well. However, by the beginning of the third extravehicular activity, dust in the scoop-locking mechanism prevented extensive use in any of the multiple detents. Only the 45-degree position was used during most of the third traverse.

For special sampling 'tasks, the Commander used the tongs effectively and the Lunar Module Pilot used the rake, as required. The Lunar Module Pilot auxiliary staff, used for mounting maps, and the lunar roving vehicle sample bag, both worked as planned and fitted the desired position exactly. The rover sampler met all requirements for utility in sampling from the rover, and in auxiliary sampling while walking in the areas of the Apollo lunar surface experiments package and surface electrical properties transmitter. The dust brush was probably one of the most often used pieces of equipment. It was employed on the rover thermal surfaces and reflectors, for cleaning the television camera lens, and by both crewmen in an attempt to minimize the dust carried into the cabin.

The geopallet was used as planned for the first two extravehicular activities. By the start of the third extravehicular activity, most of the moving parts of that pallet had begun to bind because of dust permeation along interfacing surfaces.

10.8.4 Lunar Roving Vehicle

Lunar roving vehicle deployment and preparation for loading the tools and experiments proceeded normally. The only discrepancy was that several of the yellow hinge pins had to be forcefully set to assure that the fore and aft chassis was locked. The manual deployment went well; however, take-up reels or closed-loop deployment cables would have improved the operation.

The lunar roving vehicle loading proceeded well with the ground controlled television assembly, lunar communications relay unit, all pallets, and the experiments being easily attached. Difficulty was experienced in mating the surface electrical properties receiver electrical plug to the vehicle, but this has been previously experienced with this type of electrical connector.

All lunar roving vehicle systems functioned properly during powerup and the short test drive. The only unexpected condition was a slightly high battery 2 temperature, and this was attributed to the delay on the launch pad and slight differences in the nominal trajectory and sun attitudes during translunar coast. This ultimately caused little or no problem because of the cool-down capability of the radiators during some of the long station stops.

Double Ackermann fore and aft steering was used throughout all extravehicular activities, and it greatly enhanced the maneuverability of the vehicle when negotiating craters and rocks. The acceleration was about as expected with a slightly lower average speed, possibly because of the heavier loaded rover on this mission. Slopes of up to 20 degrees were easily negotiated in a straight-ahead mode. While climbing such slopes at full power, the vehicle decelerated to a constant speed of 4 to 5 kilometers per hour. Coming down these slopes, the vehicle was operated in a braking mode with no indication of brake-fading, or feeling that the rover could not be controlled. Side slopes were negotiable, but not necessarily comfortable. During the second and third extravehicular activity, familiarity with the rover, both in riding and in driving, allowed the crew to go places and negotiate side slopes that engendered a great deal more caution during the first extravehicular activity.

The batteries were not dusted until well into the second extravehicular activity; however, after that time, the battery covers were brushed clean at every stop. The cleanliness of the batteries is attributed to the fact that the covers were continually dusted and kept clean. Dusting was time-consuming, but it was no greater problem than anticipated preflight, and it was part of the overhead in system management that leads to successful vehicle operation.

The right rear fender was accidentally knocked off by catching it with a hammer handle. This resulted in breaking about 2-inches off of the inside rail on the permanent fender. The fender extension was replaced and taped into position; however, tape does not hold well when placed over dusty surfaces. The fender extension was lost after about an hour's driving. Prior to the second extravehicular activity, a temporary fender was made from maps and taped and clamped into position, where it worked satisfactorily. Loss of the fender created concern that the dust problem would severely limit the crew's operation and the capabilities of the rover systems, not only thermally, but mechanically.

In summary, the lunar roving vehicle operation was approached cautiously;' but during the second and throughout the third extravehicular activity the vehicle was pushed to the limits of its capability. Although the crew believed that the rover could have negotiated slopes of 20 to 25 degrees without great difficulty, side slope operation never became comfortable. The rear wheels broke out only on exceptionally sharp turns when the vehicle was moving at high speeds. Slippage seemed to be minimal. During the second, and principally the third extravehicular activity, a large portion of driving time on the rover was spent negotiating boulder and crater fields, with one of the four wheels bouncing off the surface at regular intervals. The rover is an outstanding device which increased the capability of the crew to explore the Taurus-Littrow region and enhanced the lunar surface data return by an order of magnitude and maybe more.

10.8.5 Experiments

General.- The majority of the Apollo 17 lunar surface science requirements were successfully fulfilled during the three extravehicular activity periods. A new lunar surface experiments package was deployed and activated. Of this package at the writing of this report, only the lunar surface gravimeter appears to have failed to perform its primary functions. In addition to the Apollo lunar surface experiments package, a cosmic-ray experiment and a subsurface neutron-flux experiment was activated and returned. Three traverse experiments, which took advantage of the mobility, and navigation potential of the lunar roving vehicle, performed successfully in all cases. However, thermal degradation prevented the completion of all the planned data collection for the surface electrical properties experiment. All major field-geology traverse objectives were achieved or adequate substitutes were found. Only the polarimetric photography requirement of the field-geology experiment was not performed because of unanticipated and higher priority activities.

The 142 lunar samples returned weighed 110.4 kilograms (243 pounds). This number included 115 documented rock and soil samples, 12 large rock samples, 2 special magnetic glass samples, 1 permanently shadowed soil sample, 2 east-west split samples, 2 boulder-shielded soil samples, 3 double-core tube soil samples, 2 single-core tube soil samples, 1 vacuum core sample, 1 vacuum contamination sample and a 3-meter deep drill core sample. The Apollo lunar surface experiments package central station was deployed and activated (fig. 10-5) ; however, three problems prolonged the time required for this activity. These problems are discussed in the following paragraphs.

Fuel cask dome removal

After rotating the fuel cask dome removal tool. about 90 degrees, the dome removal tool came loose from the dome.

Figure 10-5.- Central station after activation.

Upon inspection and further rotation, the corner of the tool interface with the dome was found to be bent outward. The dome was partially separated from the cask by rocking the tool and complete separation was accomplished using the chisel end of the hammer. For further details, see section 15.4.4.

Central station Leveling

Leveling of the central station was difficult because of the presence of about 3 t o 5 centimeters of very loose top soil at the deployment site (fig . 10-6). The leveling was finally accomplished by working the south edge of the station down to a level below this layer of top soil and placing a large, relatively flat rock under the northwest corner. In the process of leveling, about 30 percent of the upper surface of the station sunshield was covered with a thin (approximately 0.5-mm) layer of dust, and no attempt was made to remove this dust. Also, soil became banked against the southeast and southwest corners and against the south edge of the station. Later, upon request from ground personnel, this soil was removed by clearing a 15- to 20-centimeter wide moat around those portions of the station. Some dust and soil still adhered to the sides of the station, but the white thermal coating was visible through most of this dust.

Antenna gimbal leveling.- During antenna gimbal leveling, both the north-south and the east-west level bubbles appeared to be sticky and prevented precise leveling of the antenna gimbal. The north-south bubble eventually became free-floating, but the east-west remained at the east end of the fluid tube. Precise antenna pointing was not verified, but ground personnel reported that the signal strength appeared to be adequate.

With minor and generally predictable difficulties, the deployment and activation of the individual Apollo lunar surface experiments package experiments went, as planned. Specific comments with respect to each experiment are provided in the following paragraphs.

Lunar surface gravimeter.- There were no known anomalies in the deployment of the gravimeter fig. 10-7) that would account for the problems encountered upon the commanded activation of the experiment (see sec. 15.4.1).The gimbal was observed to be free swinging after the initial release and after all subsequent jarrings and shakings. A small amount of dust fell off the universal handling tool into the gimbal housing during the final jarring of the gimbal, but all final alignments and leveling by the crew were normal.

Lunar mass spectrometer.- The deployment of the lunar mass spectrometer fig. 10-8) was as planned. A small amount of dust (approximately 0.1-mm thick) covered about 30 percent of the north-facing surface of the experiment.

Lunar ejecta and meteorites experiment.

The lunar ejecta and meteorites experiment fig. 10-9 was deployed west of a small low rim crater about 3 meters in diameter and about 3 meters northeast of a boulder that projected about 0.3 meter above the surface and was about 1 meter wide in the section it presents to the

Figure 10-7.- Lunar surface gravimeter.

experiment. The discarded packaging material for the Apollo lunar surface experiments package was placed about as planned and is shown in figure 10-9.

Heat flow experiment.

Although the drilling of the probe holes required more time than anticipated, because of presently unknown soil characteristics at the deeper levels , the emplacement of the two heat flow probes was as planned (fig . 10-10).

Lunar seismic profiling experiment

No major difficulties were encountered in the deployment of the antenna or the geophones for the seismic profiling experiment (f i g . 10-11). The geophones were oriented within a few degrees of vertical and were completely buried, except for their top plate. The geophones were located near their planned position, with a considerable amount of slack in all lines. As a consequence of the ineffectiveness of the anchoring flag, the geophone module overturned during geophone deployment. In the subsequent confusion, the geophone 4 line crossed under the line for geophone 1.

Photography

Photography of the Apollo lunar surface experiments package was curtailed because of time constraints. However, in the course of the three extravehicular activities, much of the desired photography was obtained.

Special experiments.

Two special hardware experiments were performed on the lunar surface. Deployment and retrieval of these experiments was as planned or as revised by the ground controllers. The shade plate of the cosmic ray experiment (fig . 10-12) was deployed on the forward lunar roving vehicle walking hinge, and the sun plate (fig. 10-13) was deployed on the strut of the lunar module minus Z landing gear. Neither plate was touched during deployment and there was no observed contamination on either plate. Upon request of the Mission Control Center personnel, the experiment was terminated at the beginning of the third extravehicular activity.

The neutron flux experiment was properly activated and easily inserted to the full depth of the deep drill hole (fig . 10-14). The experiment was deactivated within a minute or two after

Figure 10-8.- Lunar mass spectrometer.

Figure 10-9.- Lunar ejecta and meteorites experiment.

withdrawal from the hole, and was in the shade of the lunar module within about 3 minutes.

In the area of long-term experiments, two selected pieces of discarded Apollo lunar surface experiments package hardware were positioned in the lunar soil , orienting t heir broadside reflective surfaces to deep space. Photographs were taken showing the locations of these two articles relative to the experiments package array. The long-term effects of the lunar environment on the surfaces will be compared with the long-term effects of similar surfaces exposed for similar periods on earth. (Table 13-1). The Lunar Module Pilot's camera lens was not available for positioning on the lunar roving vehicle seat; consequently, the 500-mm lens was positioned horizontally and facing south (approximately 192 degrees) in the Commander's seat pan with the seat up.

Traverse experiments

The lunar seismic profiling experiment and the traverse gravimeter operated as planned. The operational and thermal problems of the surface electrical properties experiment did not prevent the collection of data dong two legs of the second extravehicular activity traverse.

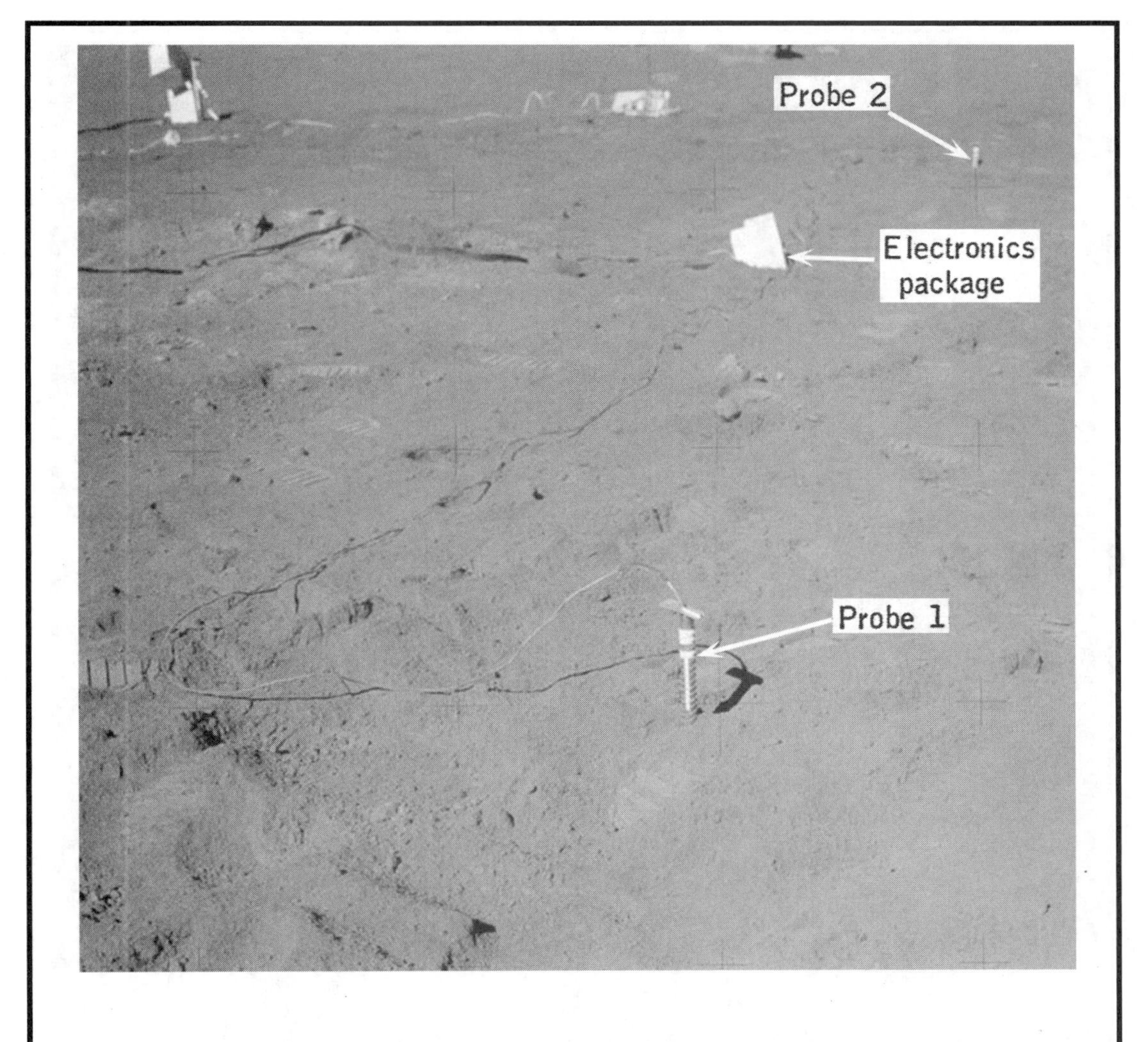

Figure 10-10.- Heat flow probes emplaced in lunar surface.

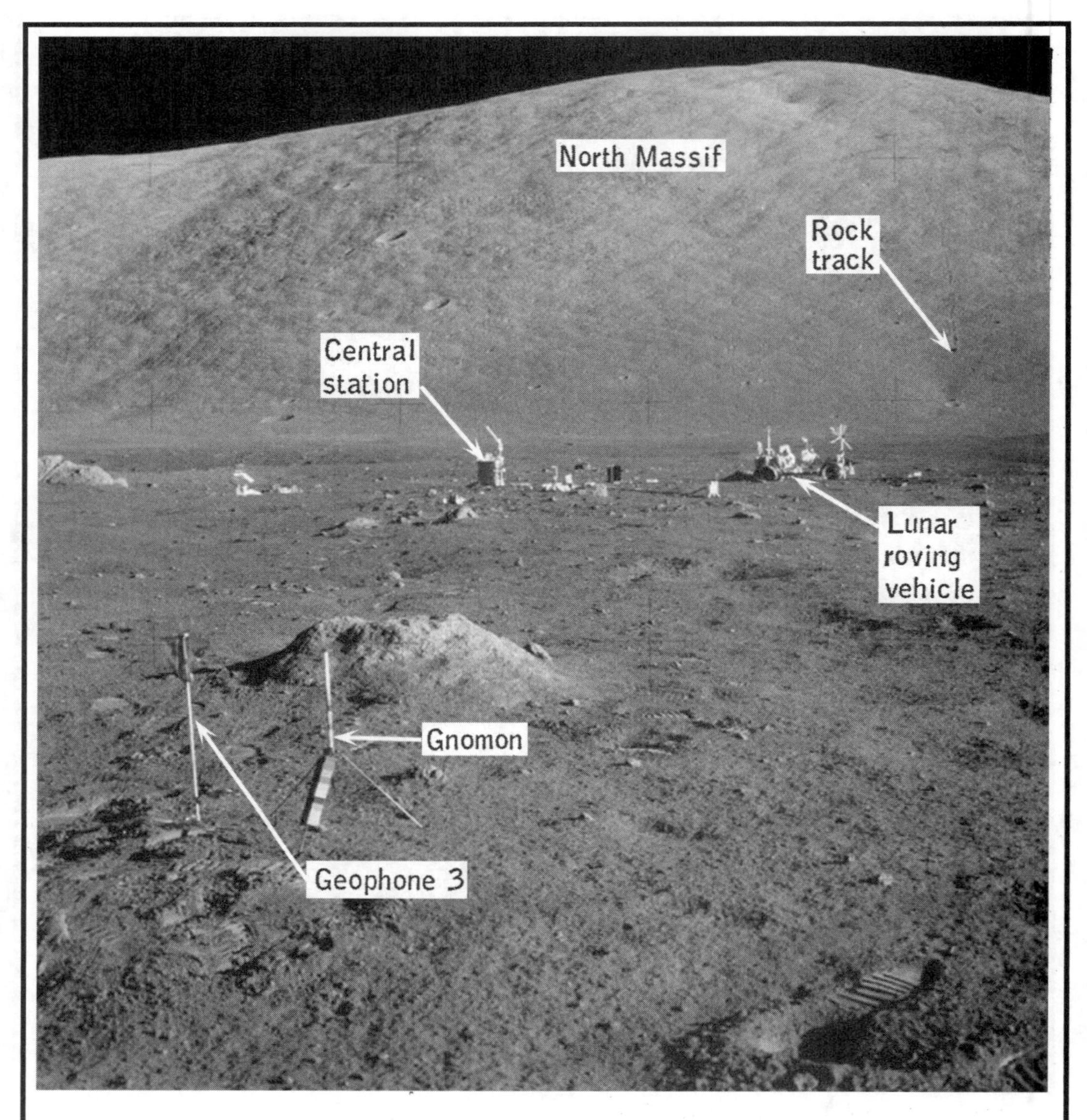

Figure 10-11.- Apollo lunar surface experiments site with geophone flag in foreground.

Figure 10-12.- Cosmic ray experiment showing shade plate.

The deployment points for the seismic profiling charges (fig. 10-15) were as required by pre-mission plans or as redirected in real time . Near complete or complete photographic panoramas at each charge site should establish their exact positions.

At least in the case of charge 1 (Victory Crater) and 5 (Van Serg Crater), there was no line-of-sight view to the antenna at the Apollo lunar surface experiments package site. Charges deployed within line-of-sight of the Apollo lunar surface experiments package were deployed in shallow depressions.

The traverse gravimeter (fig. 10-16) was operated normally, except where slopes were steep and measurements were obtained with the instrument mounted on the lunar roving vehicle. All but two planned gravity measurements were taken, and these were a second measurement at the Apollo lunar surface experiments package site and a measurement at station 7 near the North Massif.

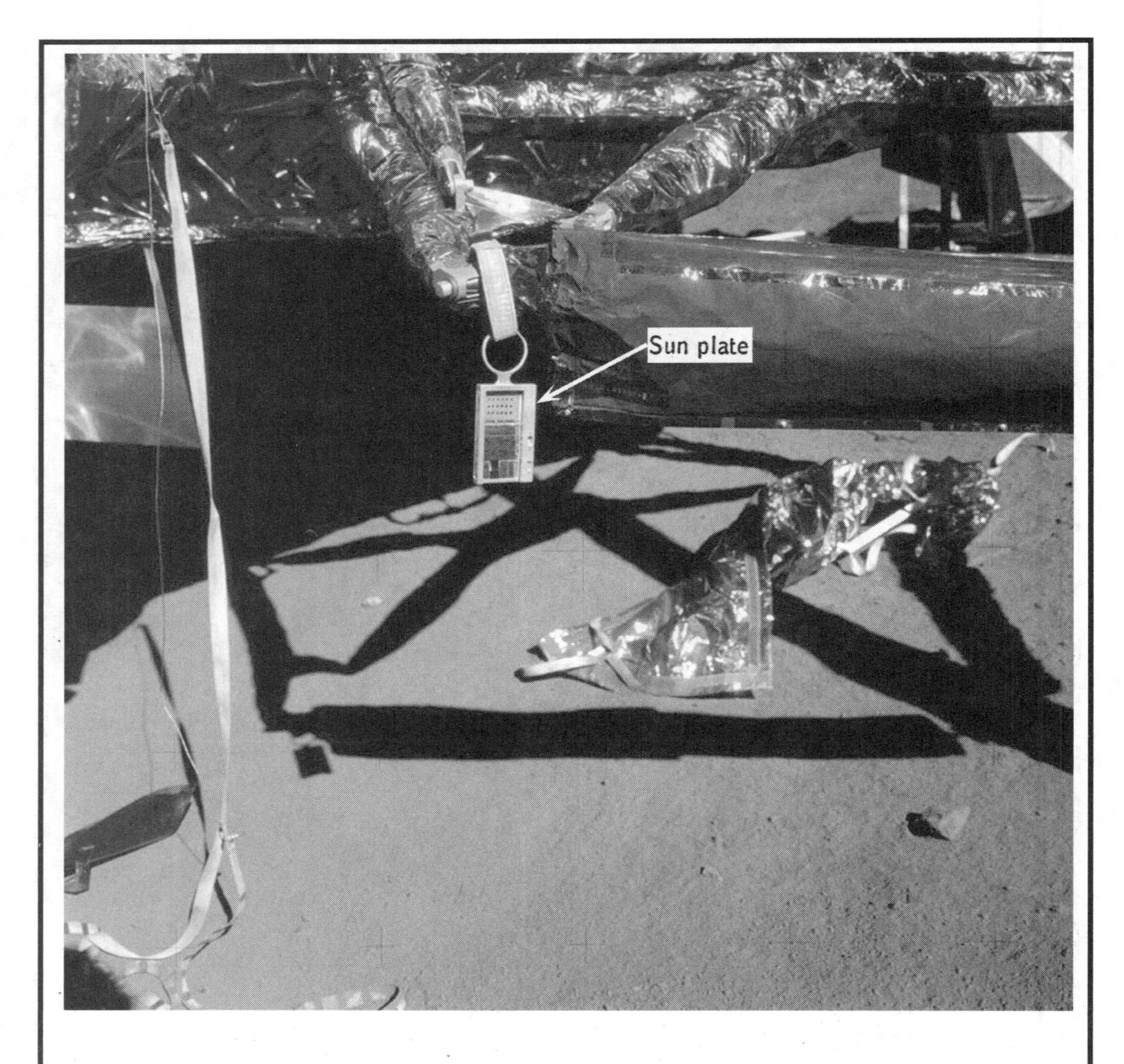

Figure 10-13.- Cosmic ray experiment showing sun plate.

The collection of data for the surface electrical properties experiment was limited by temperatures in excess of the operating limit of the tape recorder in the receiver (fig. 10-17). The operating limit was apparently exceeded at station 5 near Camelot; however, several subsequent attempts were made to collect data. For one of these, the traverse from the lunar module to station 6, the receiver was probably placed in the standby mode rather than in the on mode as was requested by the Mission Control Center. The tape recorder was removed from the receiver at station 9 (Van Serg) and stowed under the lunar roving vehicle seat until it was transferred into the equipment transfer bag and ultimately into the lunar module cabin.

There were minor deployment problem with the surface electrical properties transmitter (fig. 10-18). Three of the antenna reels fell from their stowage locations when the reel retainer was opened. Stiffness and memory in the wires between the fold-out solar panels required the panels to be taped at the hinge to provide a more nearly flat surface. Neither of these problems prevented the normal deployment and operation of the transmitter.

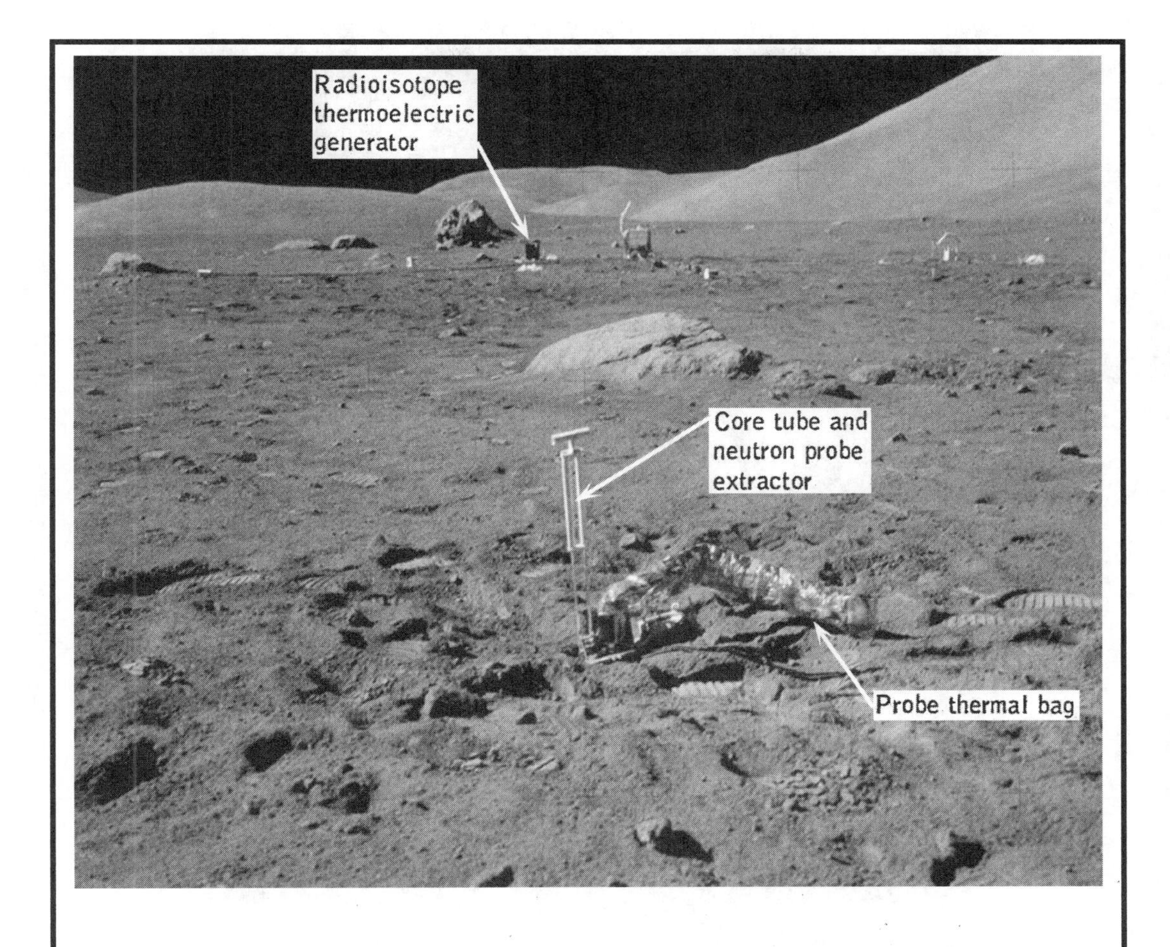

Figure 10-14.- Neutron flux experiment emplaced in lunar soil.

10.9 SOLO OPERATIONS IN LUNAR ORBIT

10.9.1 Command and Service Module Circularization Maneuver

The circularization maneuver was performed on the backside of the moon and the firing was normal.

The procedures employed were the same as for all previous service propulsion system firings.

Mission rules dictated that the lunar module crew be informed of the successful circularization maneuver prior to their initiation of the second descent orbit insertion maneuver. Simplex A VHF communications were maintained and the Command Module Pilot utilized the voice-operated relay mode for transmitting to the lunar module. Atimeline difference between the Apollo 17 mission and previous missions was that the inertial measurement unit realignment /crew optical alignment sight calibration was delayed until after the circularization firing because this task was originally within the time frame for lunar module separation. The Command Module Pilot timeline for 7 to 8 hours following lunar module separation was not overly crowded; however, there was no scheduled eating period during this time, and activity prevented suit removal.

Figure 10-15.- Lunar surface profiling experiment explosive package.

Figure 10-16.- Commander place the traverse gravimeter on the lunar surface.

Figure 10-17.- Lunar surface electrical properties receiver on lunar roving vehicle.

10.9.2 Visual Observations

The human observer's capability to describe and interpret what is being observed greatly enhances all existing, or proposed, imaging devices. The human eye can see and identify geological features which either fail to appear or which appear, but can not be recognized on photographs. This is particularly true when attempting to describe subtle color differences.

The eye can determine the subtle color differences much better than any onboard photographic device that was available. Another advantage of the human eye is the ability to almost simultaneously observe features in a wide range of illumination. In the low sun-angle observations, the eye could see detail in the shadows. This is primarily due to reflected illumination from the other side of the craters; however, detail in shadow almost never shows in the photographs. Good viewing of a particular spot on the lunar surface is limited to about 1 minute. The observer must, therefore, develop the capability to observe and describe at the same time. When utilizing the comand module as a visual observation platform, there are some compromises necessary; however, placement of the windows is such that usually one or two windows are directed toward the lunar surface. The technique of observation and simultaneously using a handheld camera requires that the crewman stabilize his body with his feet because both hands are always busy manipulating the camera. In making visual observations, it is easy to position the body as required to observe a particular target. Surface target recognition is more readily accomplished in vehicle attitudes

Figure 10-18.- Surface electrical properties transmitter showing taped panels.

which allow observation as the spacecraft approaches the target. Forward viewing along the velocity vector provides' rapid target recognition and also permits increased time for observation. Conversely, attitudes which do not allow target recognition until the spacecraft is either over, or has passed beyond the target, provide the least amount of time for useful observation.

The importance of establishing a preflight visual observation plan and then following it throughout the flight cannot be overstressed. This was accomplished on the Apollo 17 mission by assigning 10 specific visual targets. The visual targets were well planned preflight, and all visual targets were discussed during the course of the mission.

The moon is a very intriguing body and the little spare time was spent observing or just looking out the windows and picking out unusual geological characteristics, descriptions of which were recorded on the data storage equipment tape or related over the air-to-ground network.

Preflight geological training for visual observation was thorough. The preflight training must be thorough enough to enable instantaneous recall of the general geography of the lunar surface. The effectiveness of the observational techniques can be improved by the use of optical devices with image stabilization and greater magnification.

The Apollo 17 landing site was unique in that the greatest intensity of earth illumination, or earthshine, was available during this mission Visual observations in earthshine must be restricted to. differences in albedo characteristics; crater rims are very visible, and the slopes and slumping associated with the crater walls can be seen in the larger craters. The ejected ray material is not visible for more than about a quarter of the crater diameter beyond the crater. However, anything that shows up as an albedo difference in sunshine shows up as an albedo difference in earthshine. The illumination of the lunar surface in earthshine is brighter than the brightest moonlit night observed on earth. An attempt was made to use the onboard color wheel to determine colors and albedos of the lunar surface; however, none of the color chips approached the actual colors on the moon. Any further attempts to label the colors of the moon should be accomplished with a color wheel that varies from very light tan to light browns. It should also have the various colors of light gray to blue-gray and tan-gray to a darkish gray. One big problem of utilizing a color wheel is that the color wheel should be exposed to the same light as the object on the lunar surface.

A thin cloud of small particles could always be seen following the spacecraft from spacecraft sunrise and sunset to the lunar surface terminator crossing. These particles had the same general appearance of waste water droplets and would show up as a bright blur in the sextant. Attempts to observe them with the binoculars gave essentially the same appearance as the unaided eye. They were assumed to be ice particles and were always with the spacecraft. During the waste water dumps, they were originally very dense, but would later disseminate to a wider area. This particle concentration was relatively low; however, it remained constant after a steady-state period was reached following the waste water dumps.

10.9.3 Photography

One sunrise solar corona photographic sequence and three sunrise zodiacal light photographic sequences were accomplished. The sunset photography sequence was cancelled when it interfered with the trim adjustment maneuver prior to the lunar orbit plane change. The procedures for the sunrise sequences were workable and could be memorized just prior to operation. Photographs of the lunar surface in earthshine were outstanding. The technique used was to expose two frames with 1-second settings and an f stop of 1.2, then photograph the same target for 0.5, 0.25, 0.125 and 0.0625 second. The pictures of the craters Eratosthenes and Copernicus indicates that the 0.25-and 0.125-second exposures were the best.

Some photographs of the earth-set terminator around the crater Orientale are outstanding. These were taken with the lens fully open and a 1/2-second exposure. Extra black and white film in the dimlight magazines were used for nearside and far-side photography. The outstanding results of these photographs are attributed to the capability of through the-lens viewing and to the internal light-meter adjustment. These are the only good far-side terminator pictures. The camera settings for the 70-mm near-side and far-side terminator were mostly good; however, all could have been good if there had been a capability to view the photographic target through the lens and set the exposure by means of an internal light meter.

All scientific photographic targets were completed, and two extra 70-mm magazines were exposed on targets of opportunity. At least one more magazine of color exterior film could have been used. Several good near-terminator photographs were taken of the lunar surface with the 35-mm camera and color interior film.

10.9.4 Scientific Instrument Module

Operation of the scientific instrument

module experiments was normal. All preflight planned tasks were accomplished and additional ultraviolet passes were obtained. The laser altimeter was turned on for one complete sleep period and as a result, 3769 altitude measurements were made on the flight.

The lunar sounder antennas presented some difficulties. On HF antenna 1, the barberpole indicating full retract did not work correctly; also HF antenna 2 seemed to stall during extension. Not withstanding the extension and retraction problems, the booms were fully extended or fully retracted when required, and all scheduled lunar sounder operations were completed. The extension and retraction of the mapping camera required about 4 minutes instead of about 2 minutes. Although it did fully extend and retract, the number of cycles was changed to prevent excessive operation. After lunar rendezvous, the mapping camera reaction control system plume shield door appeared to be fully open when the mapping camera was fully retracted (sec. 15.1.7). During the transearth extravehicular activity, the mapping camera and the plume shield door were fully retracted. No visible evidence of previous binding could be observed during the extravehicular activity. Also, there was no visible service module reaction control system contamination in the vicinity of the scientific instrument module bay.

10.9.5 Flight Planning

Part of the preflight training was devoted to checking flight plan development in the Cape simulator. Several small errors became evident in the preliminary flight plans during these training sessions, and they . were readily corrected. This method of flight plan development helped considerably in having a final flight plan that was in excellent shape. Very few changes had to be made during the mission, and the small number that were incorporated were minor in nature. Utilization of time during the Command Module Pilot's solo activities was very efficient.

10.9.6 Orbital Trim and Plane Change Maneuvers

The plane change maneuver was performed the day of lunar lift-off and rendezvous. The command and service module orbit did not decay at the planned rate; therefore, a 9 ft/sec reaction control system height adjust maneuver was performed 1 hour prior to the lunar orbit plane change to attain an optimum command and service module orbit for the rendezvous. The reaction control system firing was accomplished with no problems. After maneuvering to the planned firing attitude, a 30-degree roll was initiated to keep the sunlight away from the panoramic camera. The plane change maneuver was successfully accomplished with about a 20-second firing of the service propulsion engine.

10.10 ASCENT, RENDEZVOUS, AND DOCKING

10.10.1 Ascent

The lunar module activation procedures for ascent proceeded smoothly and were on time through lift-off. Lift-off provided a dynamic physiological effect with no feeling of spacecraft settling. The primary navigation guidance system appeared to be flying the lunar module along the prescribed trajectory through insertion and all displays were converging at shutdown. Residuals in the X axis were less than 2 ft/sec; therefore, no trim maneuver was required. The vernier adjustment maneuver was performed shortly after insertion with -4.0 ft/sec in the X axis, -9.0 ft/sec in the Y axis, and +1.0 ft/sec in the Z axis, residuals of which were all nulled to less than 0.2 ft/sec. This out-of-plane buildup was seen throughout the ascent on the abort guidance system, and it now appears that the abort guidance system would have inserted the ascent stage into a slightly better in plane orbit. Following the trajectory adjustment, the approach path was nominal for tracking and the subsequent terminal phase initiation. Uplink communications with the ground were lost following pitchover; however, it was subsequently learned that the, ground monitored all the down-link transmissions. The problem was determined to be ground station operations.

10.10.2 Rendezvous

Tracking from the lunar module was initiated with a rendezvous radar automatic lock-on. Raw range, range rate, shaft, and trunnion data were verified on the display keyboard prior to accepting marks. The first few marks produced acceptable results. Automatic abort guidance system updates were accepted with no subsequent variation from normal in either system. The computer recycle at 17 marks produced the expected solutions. Continuous manual plotting of position throughout the tracking sequence until just before terminal phase initiation followed the planned direct rendezvous trajectory. The sunlit command module was visible from the lunar module at approximately 110 miles. After entering darkness, the command module was lost and was not seen again until the flashing light became faintly visible at a distance of 30 to 40 miles. Approximately 8 minutes prior to terminal phase initiation, with about 25 marks into the computer and an almost equal number of marks into the abort guidance system, final computation were initiated and the lunar module was manually maneuvered in yaw and roll into the firing attitude for terminal phase initiation. The solutions, given by the ground computer, the abort guidance system, and the command module computer are shown in table 8-VIII. After the terminal phase initiation maneuver, the residuals were somewhat higher than expected; approximately +7 ft/sec in the X axis and 4 ft/sec in Y and Z axes, respectively. These were quickly trimmed to less than 0.2 ft/sec. Radar lock with the command and service module was maintained throughout the terminal phase initiation maneuver.

The command module attained VHF tracking of the lunar module immediately after insertion at a range of approximately 155 miles. The VHF ranging broke lock during the lunar module vernier maneuver, was reset, then broke lock again. After the second reset, it remained locked on for the remainder of the rendezvous. The lunar module could not be seen through the sextant or telescope in daylight. Once the lunar module was in darkness, the tracking light was visible in the sextant as a star about 1/4 the magnitude of Sirius. The light was not visible in the telescope until the range was reduced to about 60 or 70 miles.

Following the terminal phase initiation maneuver, the lunar module was allowed to automatic track throughout the midcourse data gathering period. The midcourse solutions were all less than 1.6 ft/sec, with very good agreement between the primary navigation guidance system and the abort guidance system. The command module computer midcourse solutions also compared well with the exception of an S ft/sec difference in differential velocity in the Z axis for the second midcourse correction. The lunar module relative motion plot agreed with the center of the nominal trajectory and had the expected range rates. Braking gates were performed as planned and this provided a comfortable rendezvous through station-keeping. A slight ;innovation trained for, but not included procedurally, was the incorporation of three manual marks into the abort guidance system after the second midcourse correction. The objective was to maintain the credibility of the abort guidance system state vector throughout the braking phase. The success of this technique is evident in the 2 ft/sec range rate and less than 1/2 mile in range at the termination of braking.

10.10.3 Docking

After a fly-around inspection proved the scientific instrument module bay (fig. 10-19) to be in order, the lunar module was maneuvered to the docking attitude and station-keeping responsibility was given to the command and service module. The closure rate was very slow, probably less than 0.1 ft/sec, with some small translation corrections required because of the lunar module dead-banding. Capture was not accomplished on the first contact, probably due to the slow closure rate. By increasing the closing velocity, capture was positive on the second attempt. Because the lunar module was drifting in pitch and yaw, attitude hold was reinitiated momentarily in the lunar module to allow the command and service module to null all rates prior to probe retraction. All 12 docking latches fired and seated properly.

Figure 10-19. Scientific Instrument module bay view from the Lunar Module.

10.11 LUNAR ORBITAL OPERATIONS - POST-DOCKING TO TRANSEARTH INJECTION

10.11.1 Post-Docking Activities

The post-docking activities in the lunar module and the command and service module were accomplished as planned with the checklist as an inventory list and as a backup to common sense. The vacuum cleaner was operated continuously in the lunar module to remove dust floating in the cabin. As a result of this operation and the special attention paid to the bagging and sealing of the samples prior to transfer, the command module remained remarkably dust free. During vacuum cleaner checkout, a main bus B under-voltage light was illuminated; however, there were no caution and warning lights when the vacuum cleaner was used for the lunar module post-docking activities. Preparation for lunar module jettisoning was normal through hatch closure.

10.11.2 Lunar Module Jettison

Tunnel venting and pressure suit integrity checks were slow and tedious, but were accomplished in an orderly fashion resulting in the lunar module being jettisoned on' time. The lunar module drifted away from the command and service module and it was evident that the attitude control system was functioning properly.

10.11.3 Orbital Observation/Scientific Instrument Module Bay Operations

The planned extra day in lunar orbit was the first time the potential of three full-time lunar orbit observers was exercised. It became readily apparent that the combined operation of the scientific instrument module bay with effective visual observation of the lunar surface would require an orderly system. Therefore, the Commander stayed primarily in the left seat handling spacecraft operations and systems and keeping the mission revolving around the flight plan timeline, while the Command Module Pilot and Lunar Module Pilot used their geological background for orbital observations and operating the scientific instrument module bay as dictated by flight plan and procedures.

10.11.4 Transearth Injection

The only noteworthy item, during an otherwise perfect transearth injection maneuver, was the unexpected observance of 30.4-degree roll rate across the deadband. Pitch and yaw rates were stable. The service propulsion system again operated normally with the propellant utilization gaging system in decrease for the entire maneuver and gradually stabilizing the oxidizer/fuel ratio to near zero at the end of the firing.

10.12 TRANSEARTH FLIGHT

Operation of the scientific instrument module bay ultraviolet spectrometers and infrared scanning radiometer was the most time consuming activity during the transearth coast phase of the mission. To add to the science information gained in lunar orbit, several targets were selected for observation during the transearth coast phase. This activity required the spacecraft to be maneuvered repeatedly to several different attitudes and required close monitoring because of the proximity of the attitude to gimbal lock. However, there was never a problem other than the requirement for the crew's continuous attention to the maneuvers. The spacecraft was in attitudes for long periods of time that were not ideal for passive thermal control, and as a result, the cabin temperature dropped to an uncomfortably cold level and the side hatch interior became blanketed with moisture. At the conclusion of the experiments, a normal attitude for passive thermal control was used and the thermal problems disappeared.

During checkout for the final midcourse correction, the entry monitor system did not pass the null bias check (see sec. 15.1.5). A reaction control system firing of 2 ft/sec was performed successfully, although the entry monitor system provided an erroneous reading.

10.12.1 Extravehicular Activity

The transearth extravehicular activity preparation and the checkout was normal. The counterbalance was removed from the side hatch by backing out the pin which locks the two bell crank assemblies together. Removal and reinstallation of this pin was simple. Using the pin allows the hatch to be closed by pulling on the D-ring that hooks onto the counterbalance assembly. Cabin depressurization was nominal with no debris noticed going towards the hatch depressurization valve. The hatch had a tendency to open as soon as the latches were disengaged, because of the stabilized residual pressure in the cabin produced by the flow through the Command Module Pilot's suit. While the hatch was held in a partially open position, numerous droplets of ice streamed through the opening.

The visibility during the extravehicular activity was good with the hatch open. The sun was shining from off of the Command Module Pilot's right lower quarter as he stood in the extravehicular activity slippers. Raising the gold visor on the lunar extravehicular visor allowed adequate vision into the shadows of the scientific instrument module bay. Moving about during the extravehicular activity was easy, but it should be said "take it nice and slow--don't be in any hurry, and don't try to make any jerky fast motions." The umbilical did not torque or maneuver the Command Module Pilot, provided adequate restraint was available. Retrieval of the lunar sounder cassette, and the panoramic and mapping camera film cassettes was easy and was accomplished as it was in training.

The Lunar Module Pilot found that there was little or no problem in receiving the cassettes from the Command Module Pilot in the hatch area. However, the Lunar Module Pilot noted a restriction to rotational and lateral movements because of the stiffness of the spacecraft hoses.

The workload requirements on the Command Module Pilot were never too great when moving about slowly with everything working well. Unless workloads demands are beyond the Apollo 17 experience, gas cooling with the oxygen umbilical is completely adequate.

The caliber, amount, and type of training for the transearth extravehicular activity was adequate to produce a comfortable and complete job.

10.13 ENTRY, LANDING, AND RECOVERY

Entry stowage preparation had been underway for the day and a half following the transearth extravehicular activity, and on the night before entry almost everything was stowed and tied down with the exception of the large bulky bags which covered the still necessary stowage containers. Final stowage was completed in accordance with the checklist on the morning of entry with some effort taken to organize and inventory the stowage locations. This proved to be very beneficial to the recovery crew. The only problem associated with entry stowage evolved around the suit bag. The lunar extravehicular visors and helmets would not fit in the bag along with the suits, rocks, and other miscellaneous gear. Two visors, with helmets, were strapped to stowage containers on the right-hand side of the cockpit and they were adequately restrained during entry.

The command and service module separation was normal. While using the stabilization and control system minimum impulse to control the command module after separation, there was an apparent trim away from the plus yaw and minus pitch axes; i.e., a constant correction had to be applied in the yaw right and pitch down directions to maintain the desired entry attitude. The command module thrusters were always audible no matter which one was fired.

During the entry monitor system entry test, the 0.05-g light was illuminated in test position 1. Because of the previous accelerometer anomaly along with this discrepancy, the entry monitor system was left in the ENTRY and STANDBY position until 0.05-9 time when it was switched to NORMAL. The 0.05-9 light illuminated on time, and the entry monitor system operated normally. Entry simulations should be accomplished without ink in the scroll pen, because the only scribe visible on the entry monitor system pattern is a slight tear on the pattern itself.

The command module oscillations, once the drogue parachutes were deployed, were moderate to violent with higher than expected amplitudes. The command module reaction control system isolation valves were closed while still on the drogue parachutes. There was no subsequent venting of the lines through hand controller actuation. Main parachute deployment was on time with one parachute disreefing several seconds after the other two had fully inflated.

The landing (at an altimeter reading of 100 feet) was sharp with no discomfort or disorientation. The main parachute release circuit breakers were closed within 1 or 2 seconds after touchdown, the parachutes were jettisoned, and the spacecraft remained in the stable I attitude. Once the hatch was opened, the recovery crew passed in a bag containing the Mae West's, a temperature measuring unit, and a container in which to transfer the cosmic ray crystal: In the future, if an instrument such as a temperature measuring unit is used after landing, a simple bungee strap should be provided to secure the unit. Two of the three Mae West's had to be inflated manually.

The Command Module Pilot wore a orthostatic countermeasure garment which was donned 2 hours prior to entry. It was pumped up before the Command Module Pilot left the couch after landing. The pressure was maintained in the garment throughout the recovery operation and through about 3 hours of the post-recovery medical debriefing. The garment which was slightly tight on the legs became somewhat uncomfortable at the end of the 3 hours.

The entire recovery activity, including the medical protocol and ship-board obligations was handled well.

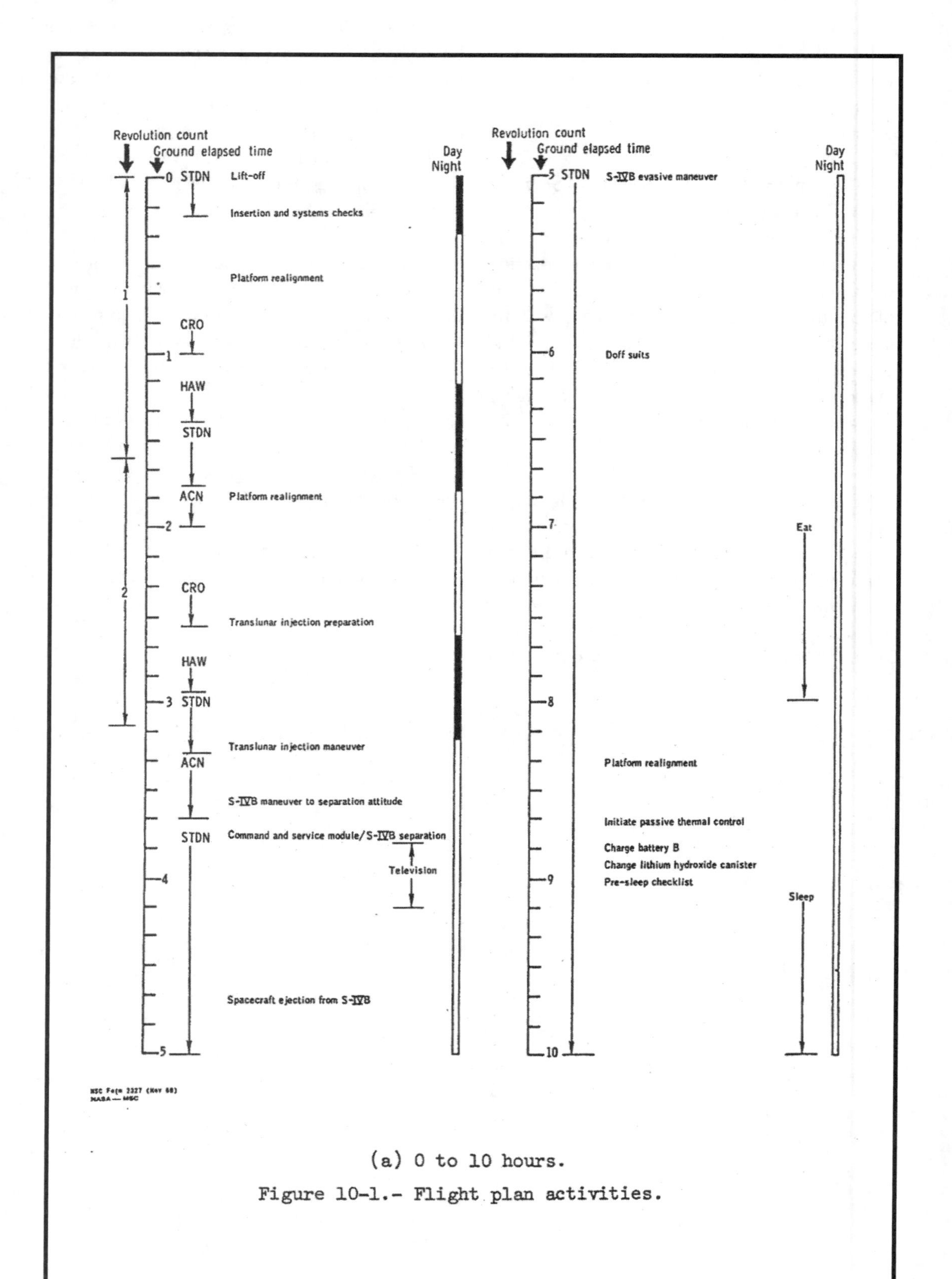

(a) 0 to 10 hours.

Figure 10-1.- Flight plan activities.

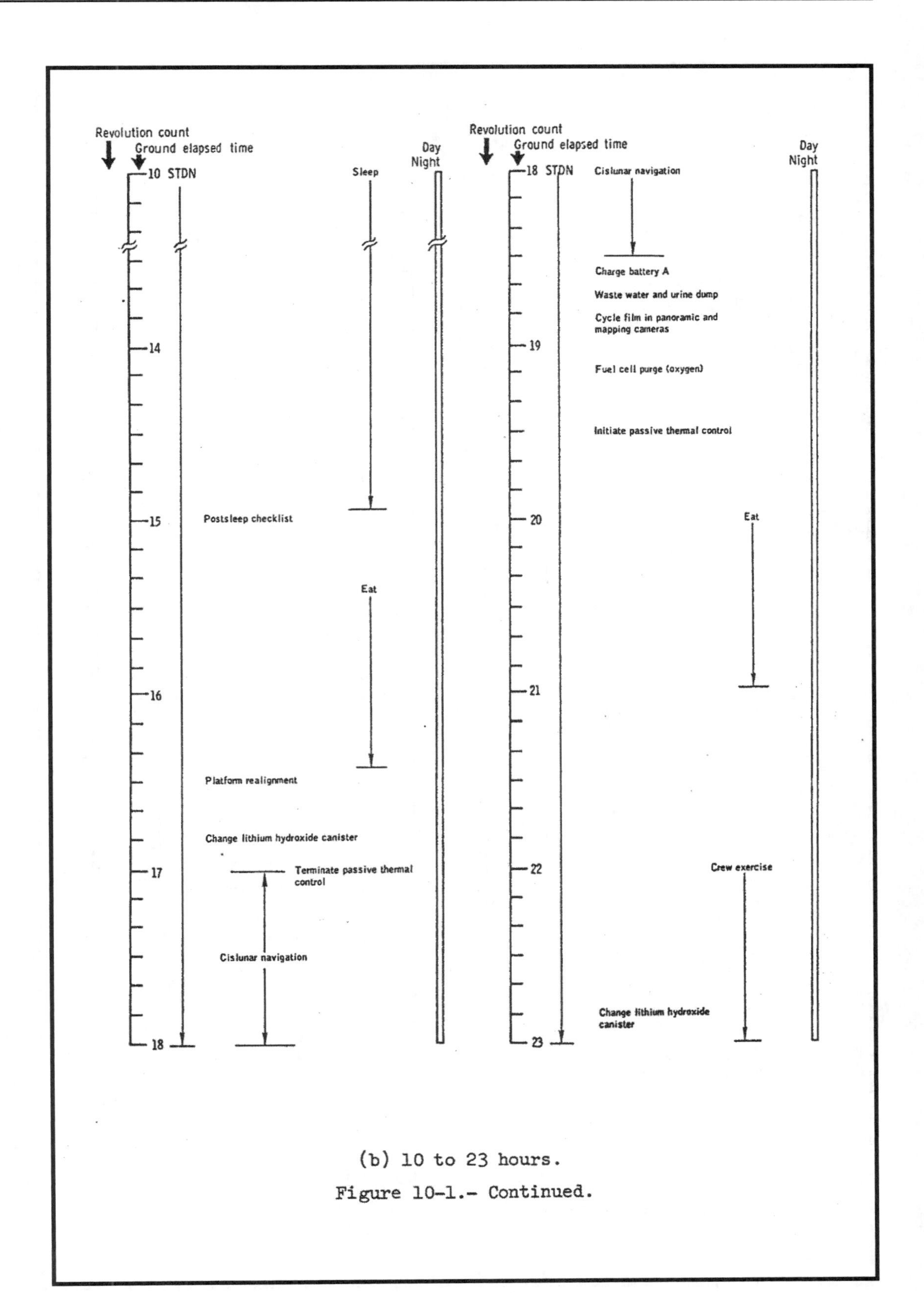

(b) 10 to 23 hours.

Figure 10-1.- Continued.

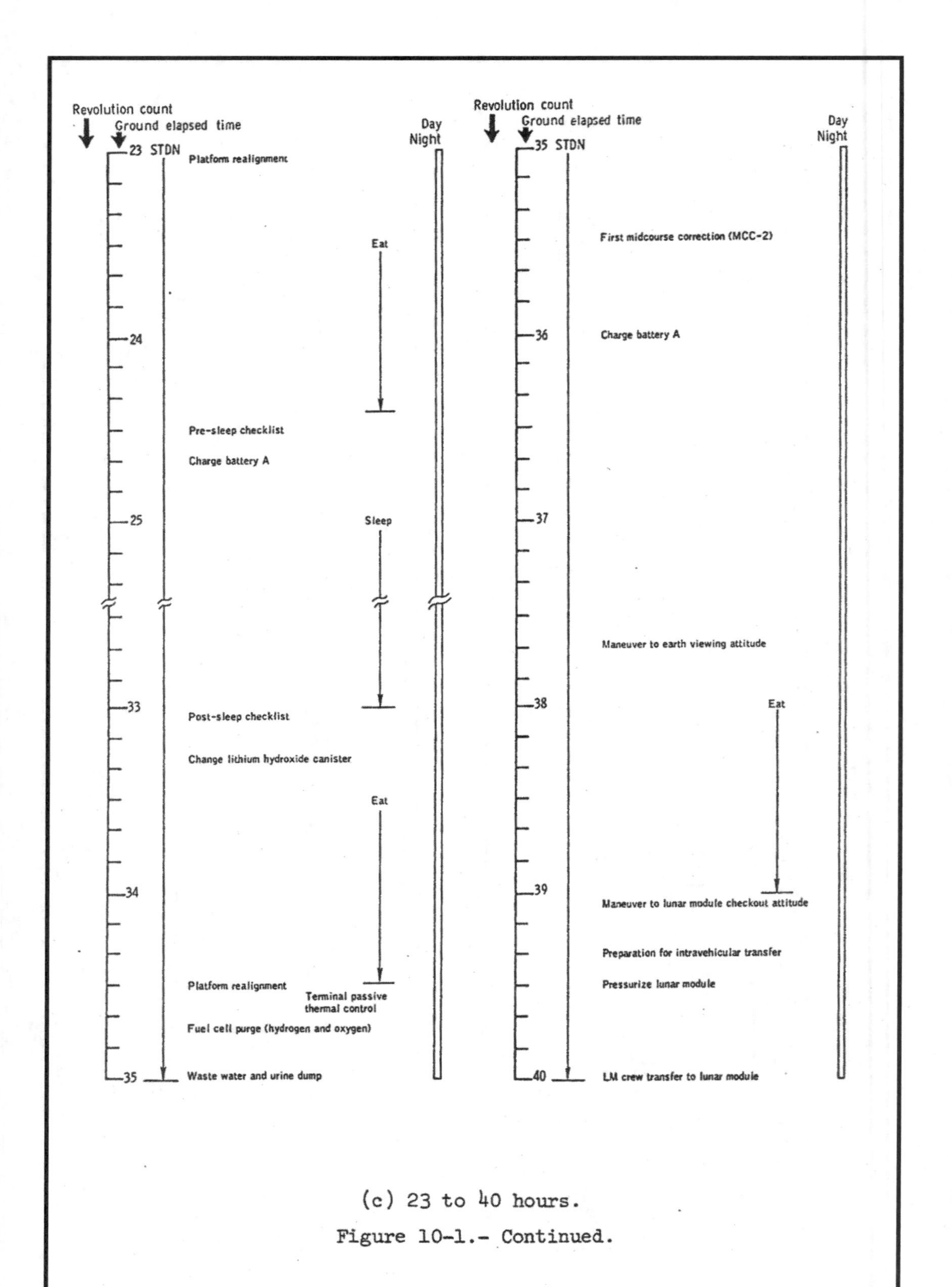

(c) 23 to 40 hours.

Figure 10-1.- Continued.

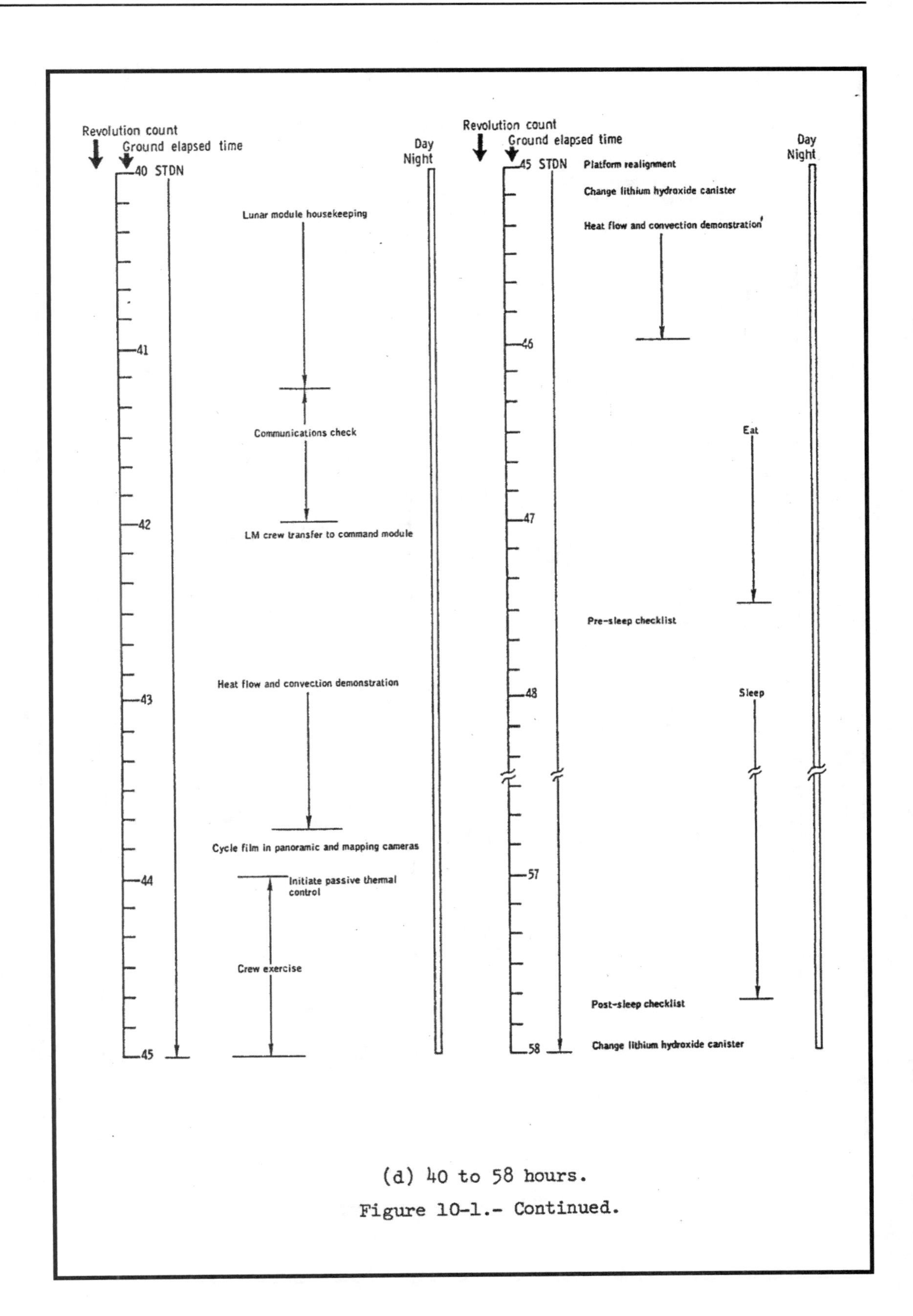

(d) 40 to 58 hours.

Figure 10-1.- Continued.

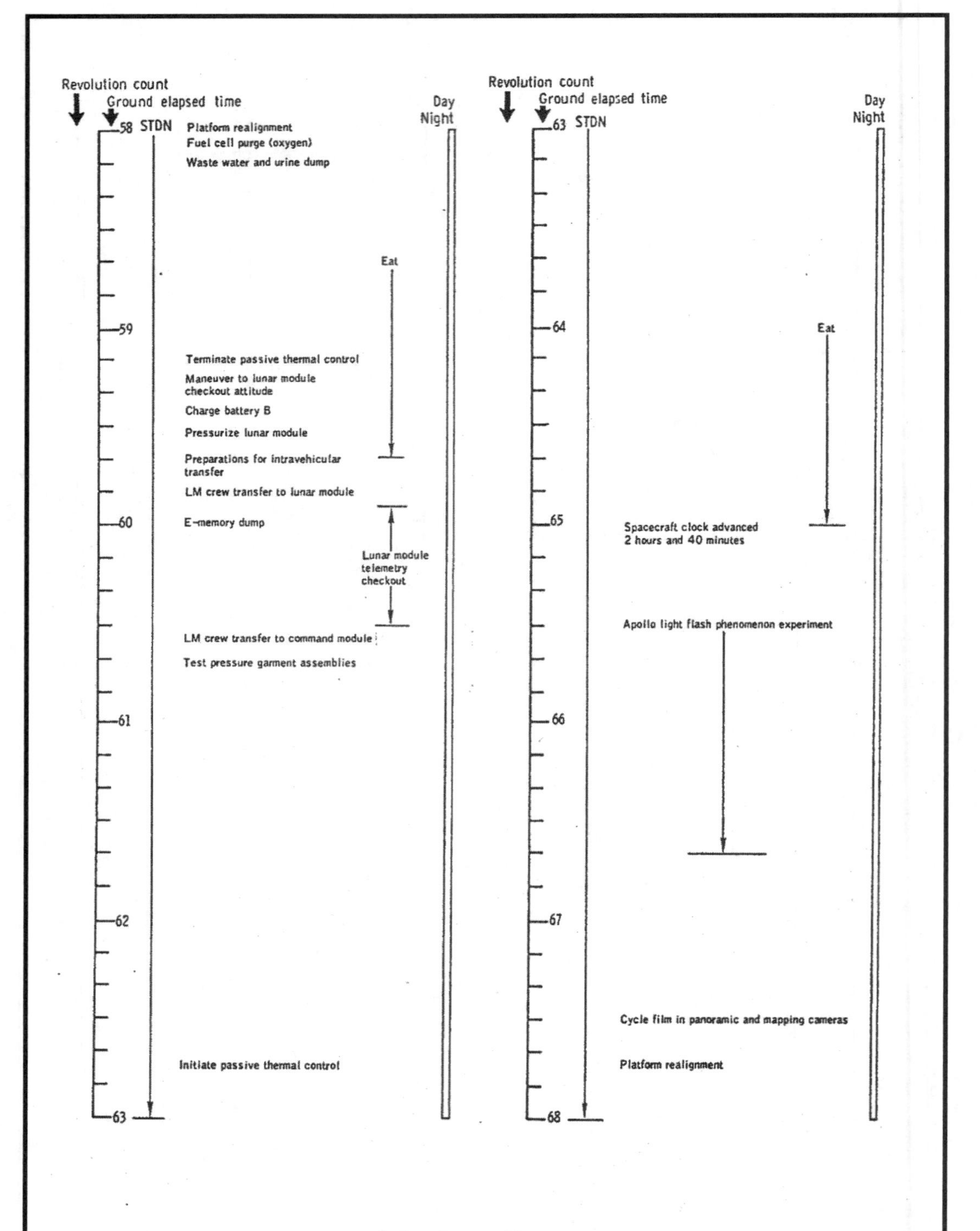

(e) 58 to 68 hours.

Figure 10-1.- Continued.

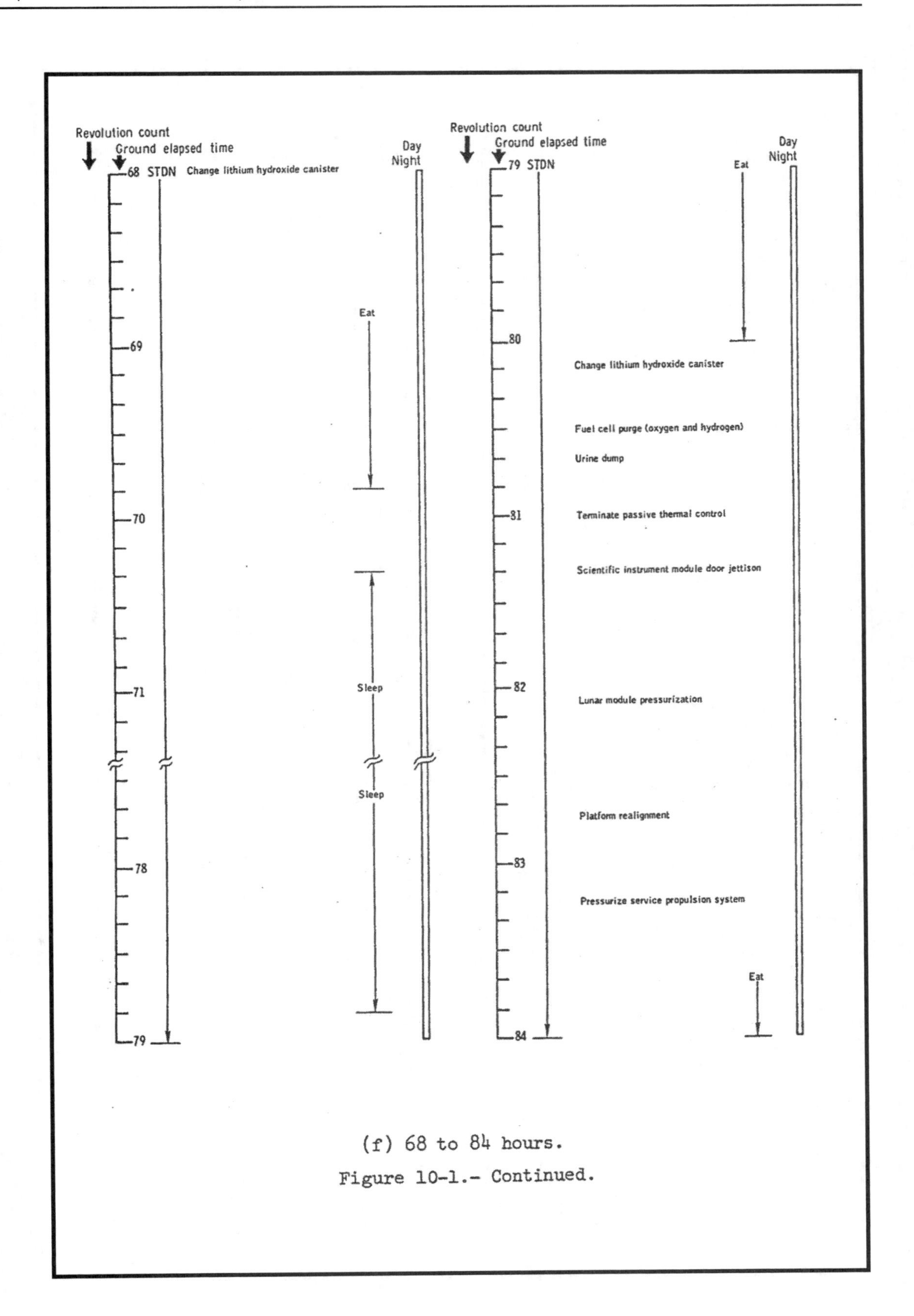

(f) 68 to 84 hours.

Figure 10-1.- Continued.

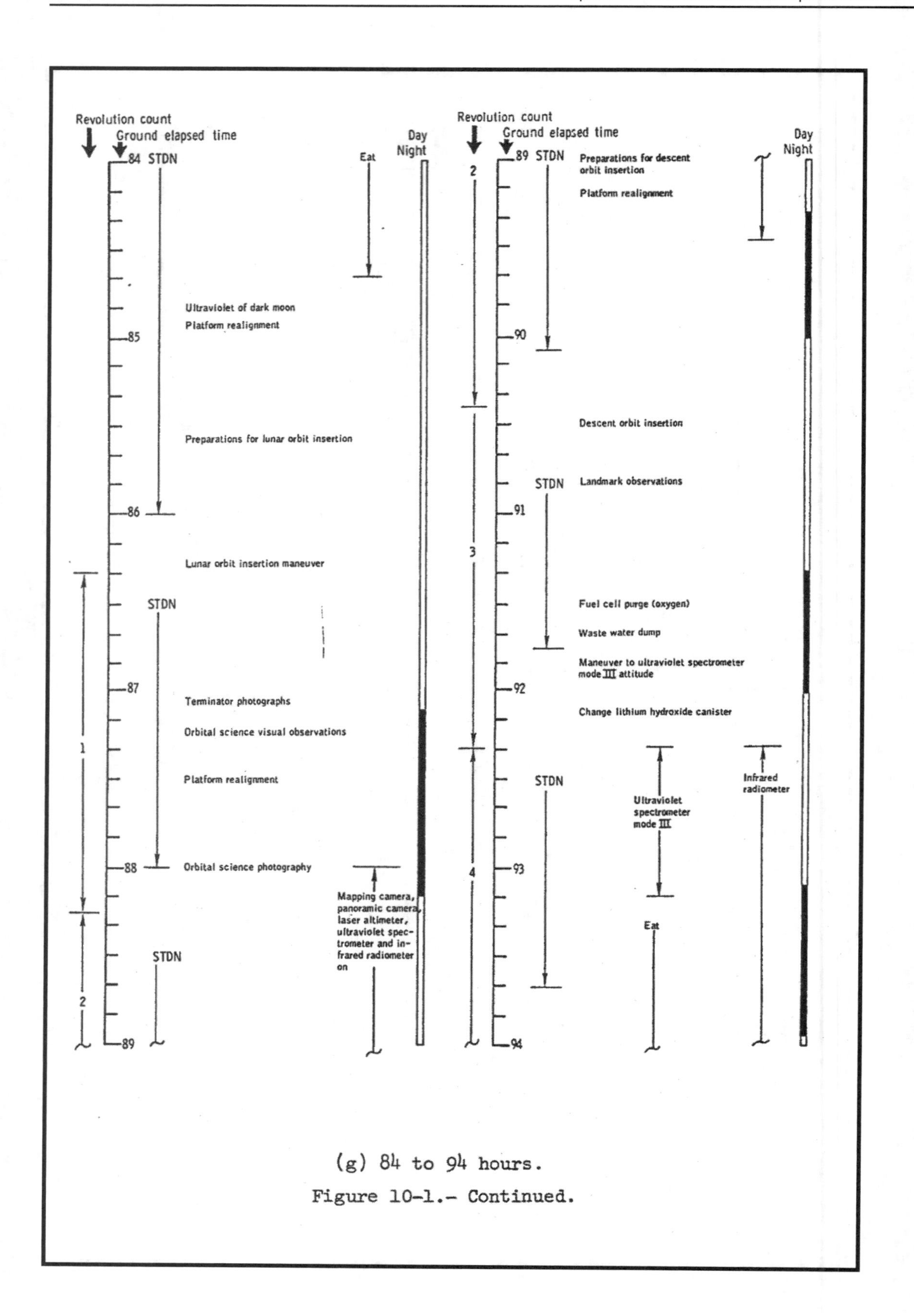

(g) 84 to 94 hours.

Figure 10-1.- Continued.

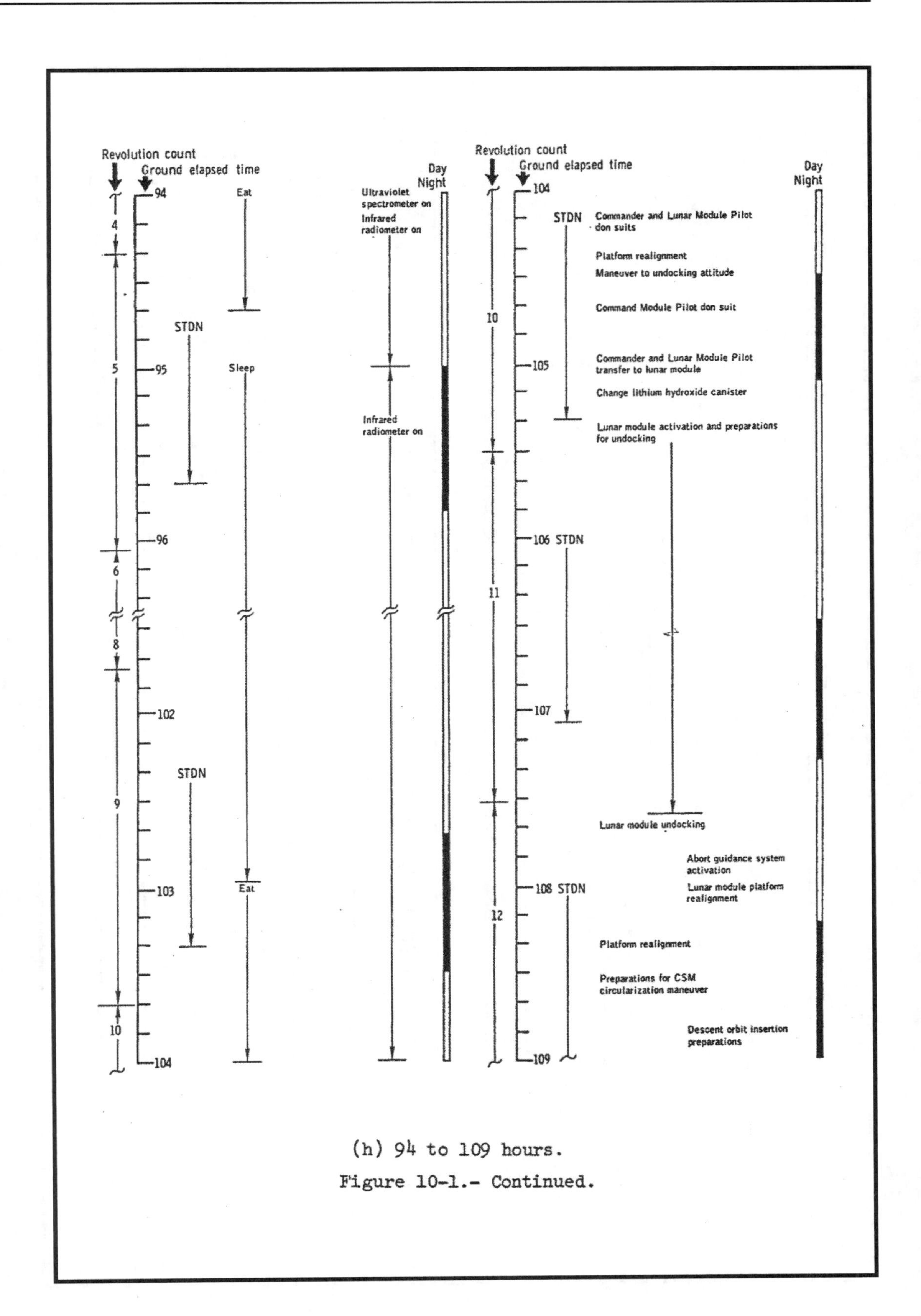

(h) 94 to 109 hours.

Figure 10-1.- Continued.

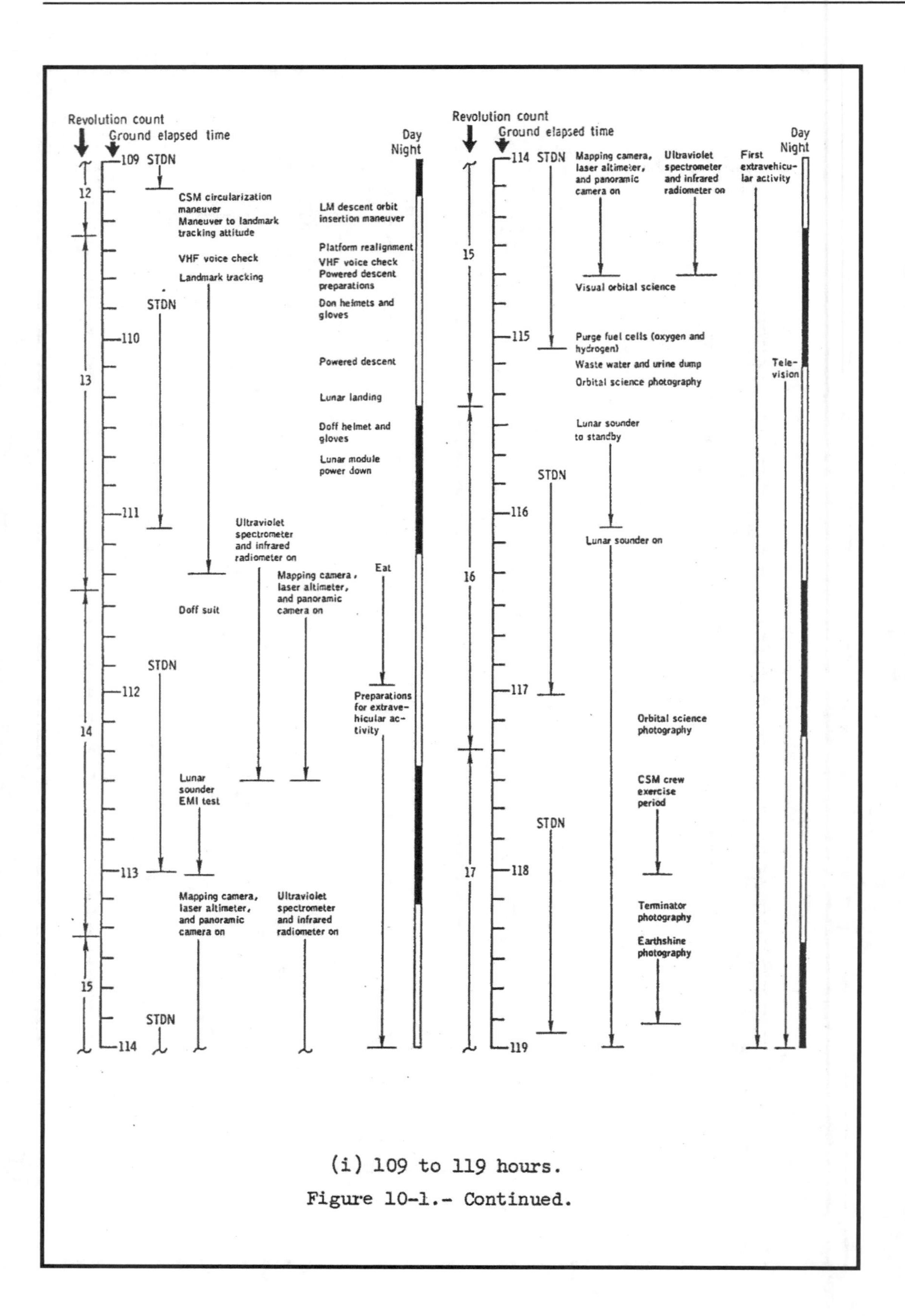

(i) 109 to 119 hours.

Figure 10-1.- Continued.

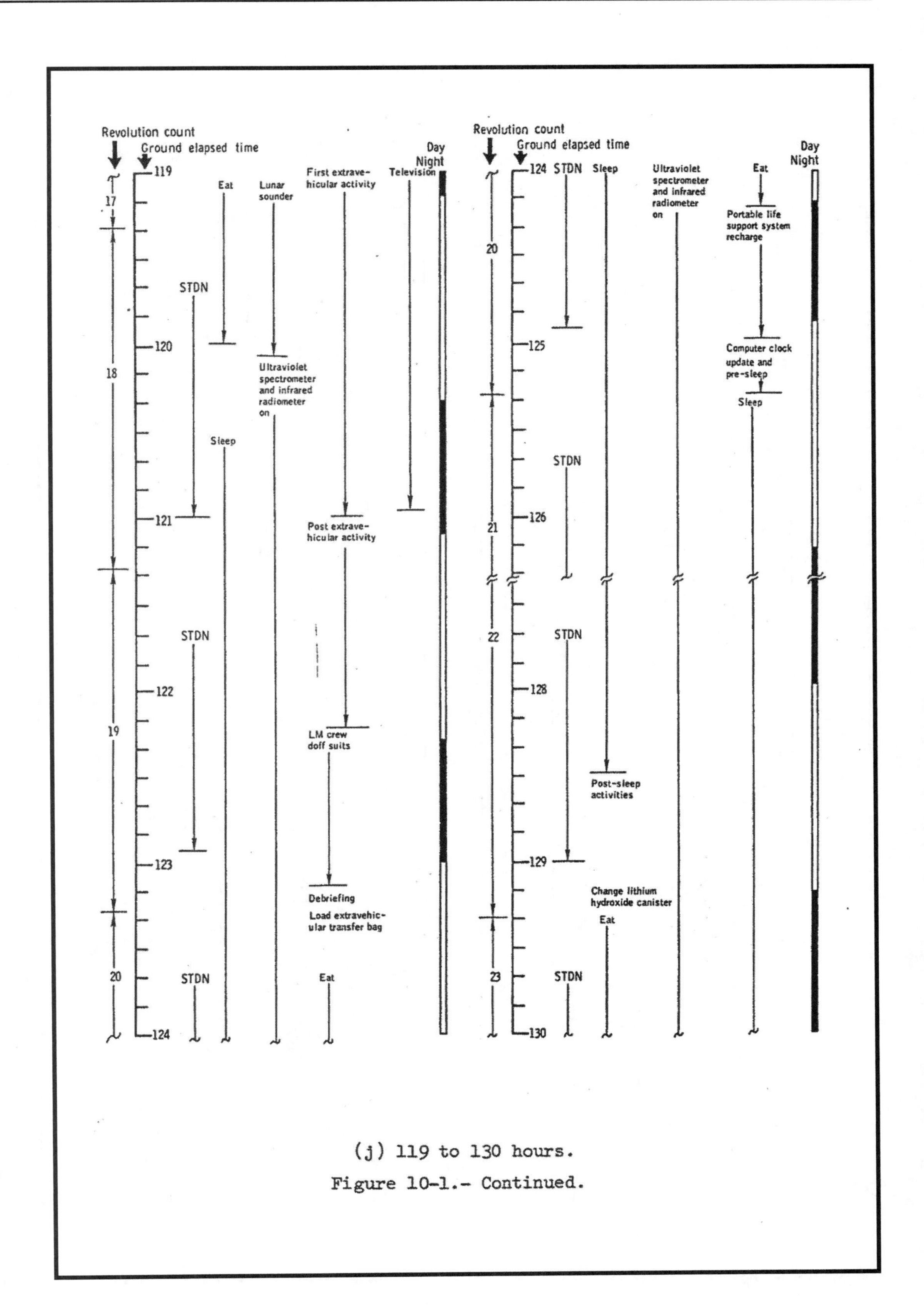

(j) 119 to 130 hours.

Figure 10-1.- Continued.

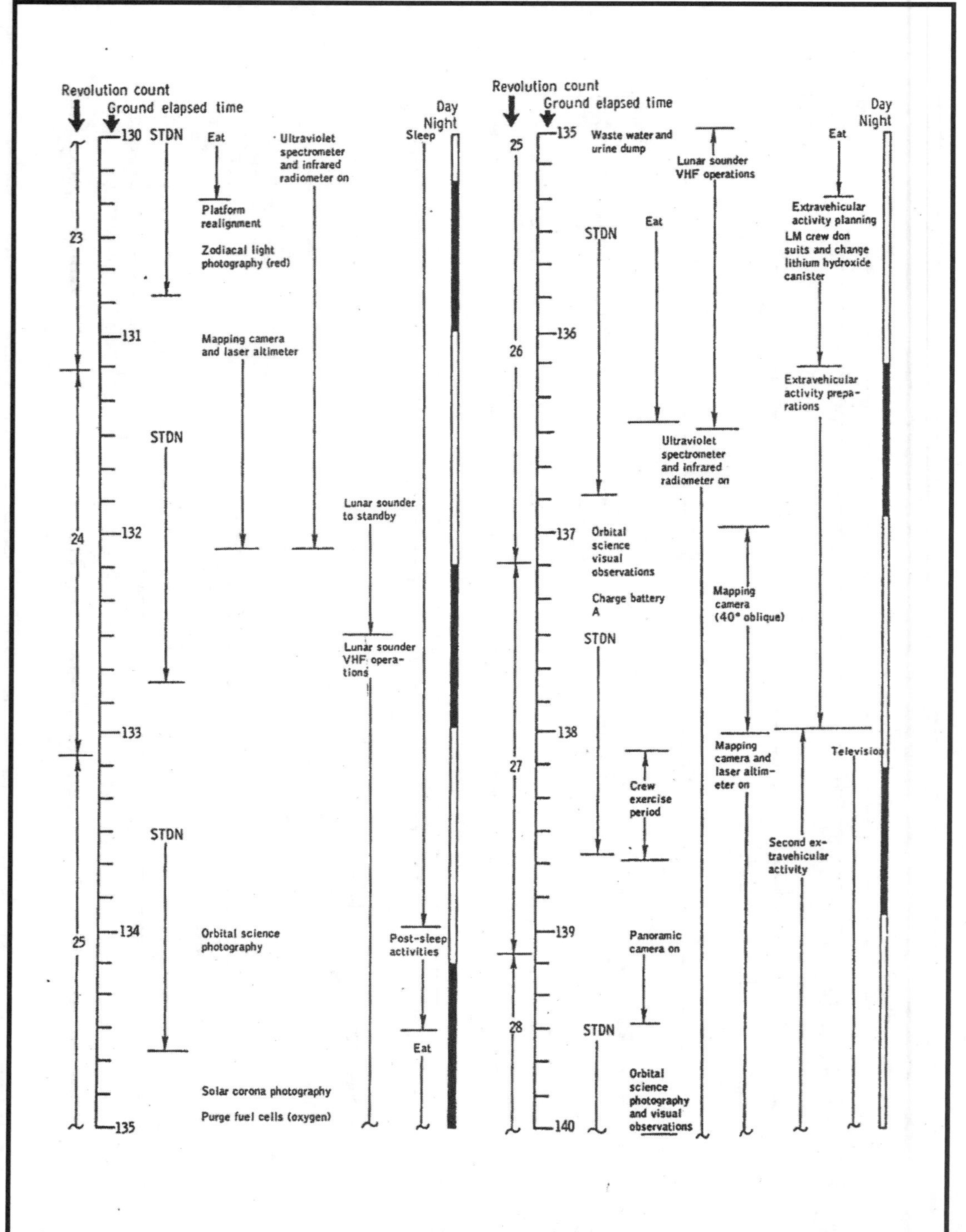

(k) 130 to 140 hours.

Figure 10-1.- Continued.

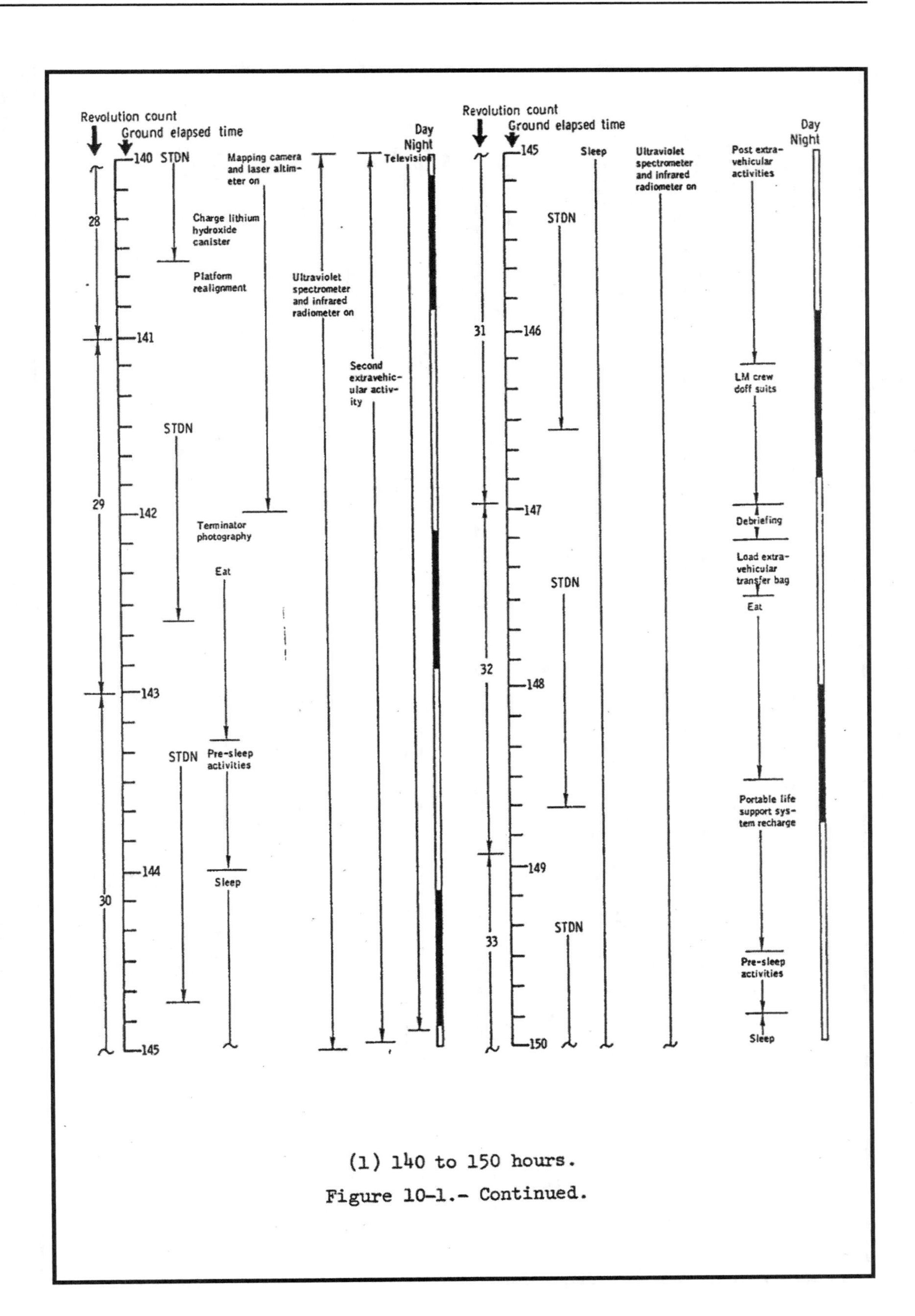

(l) 140 to 150 hours.

Figure 10-1.- Continued.

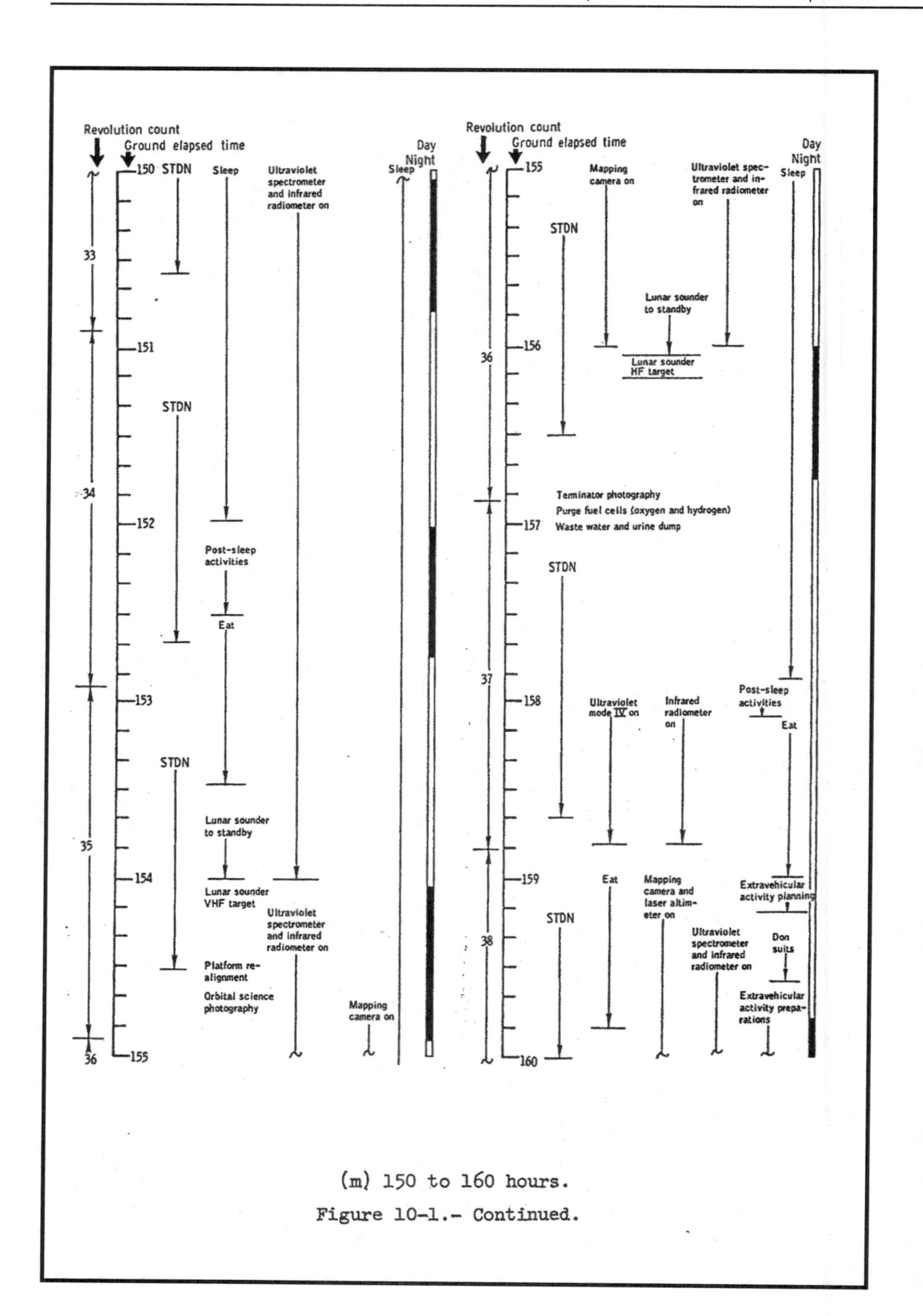

(m) 150 to 160 hours.

Figure 10-1.- Continued.

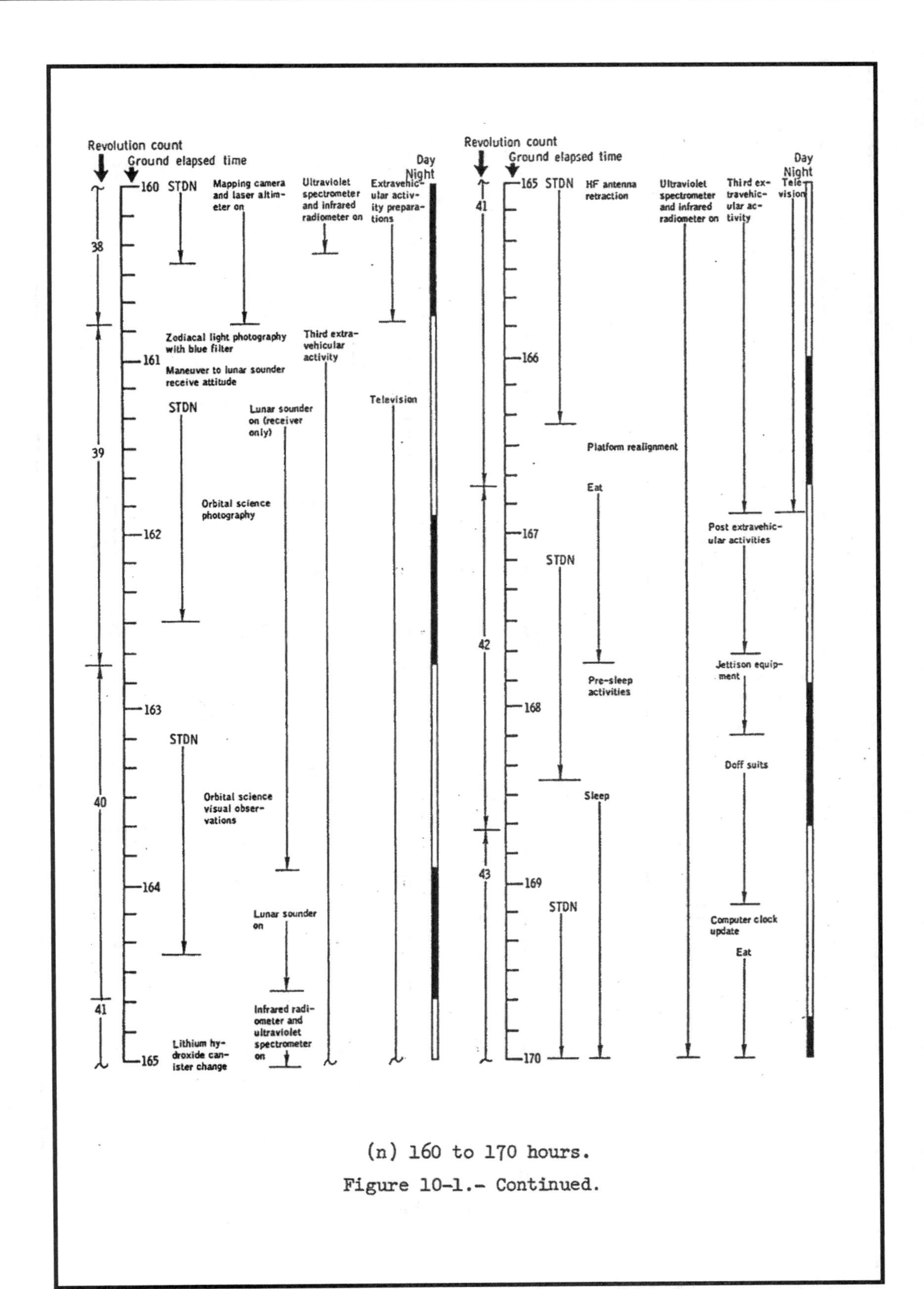

(n) 160 to 170 hours.

Figure 10-1.- Continued.

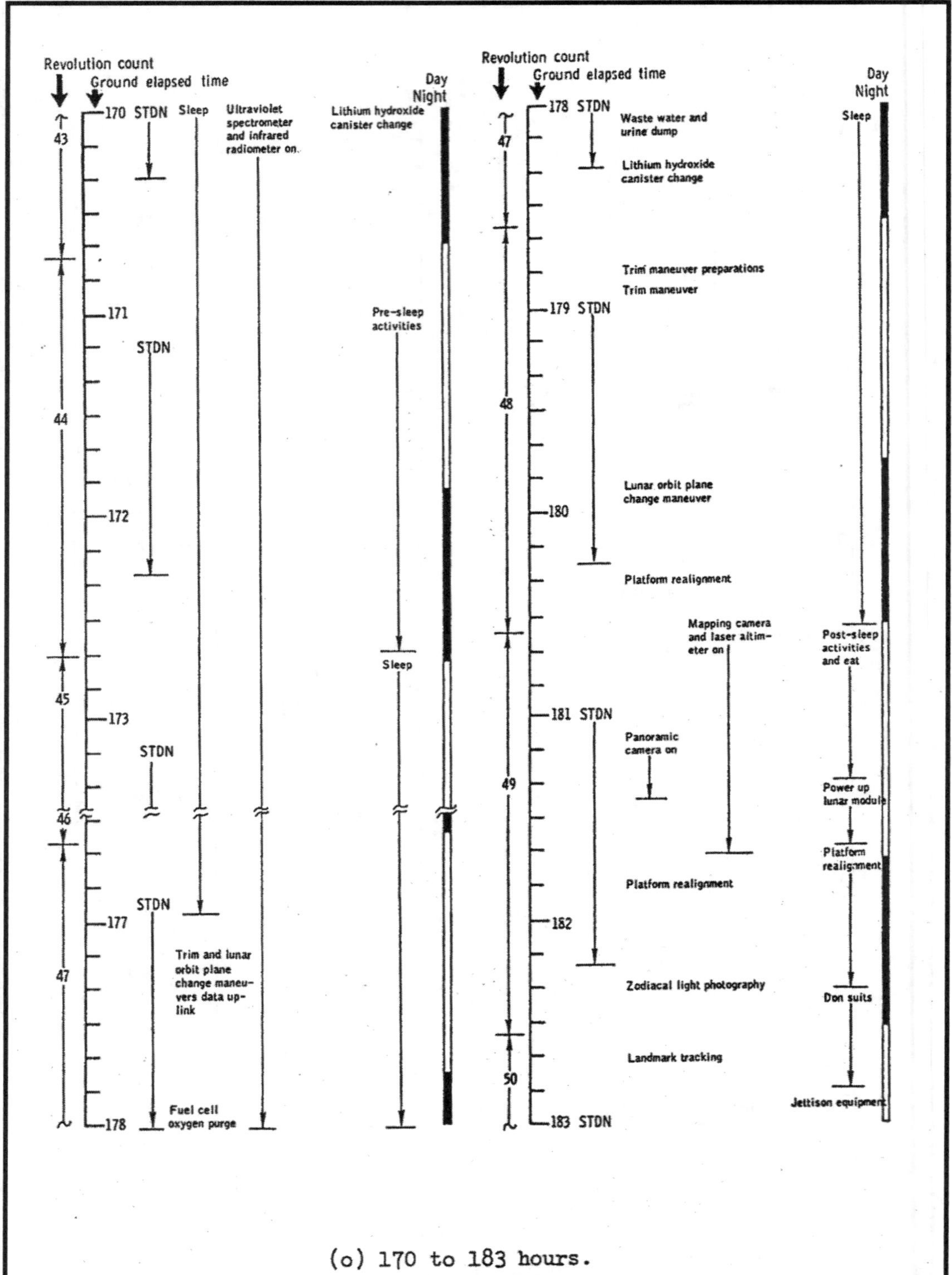

(o) 170 to 183 hours.

Figure 10-1.- Continued.

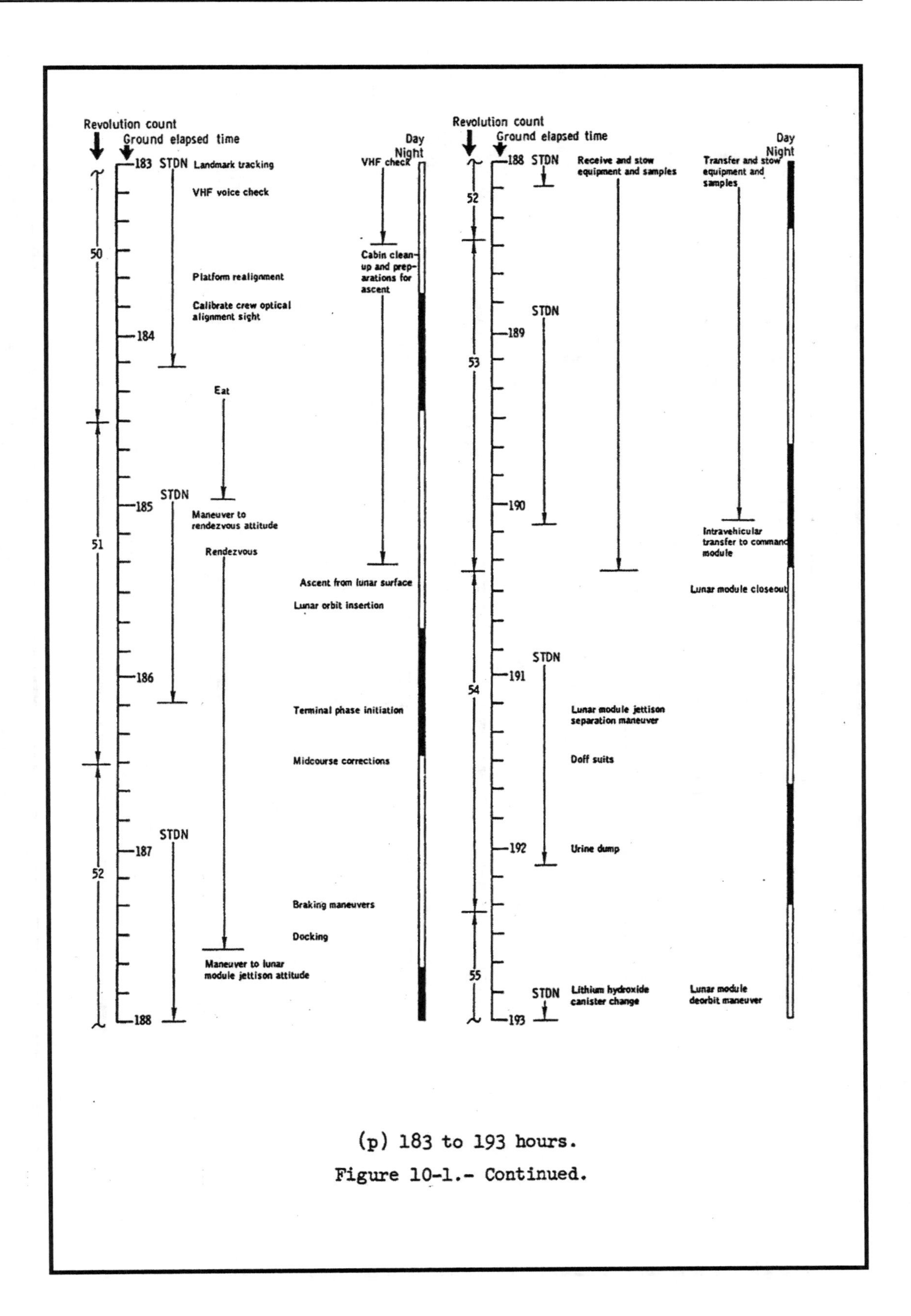

(p) 183 to 193 hours.

Figure 10-1.- Continued.

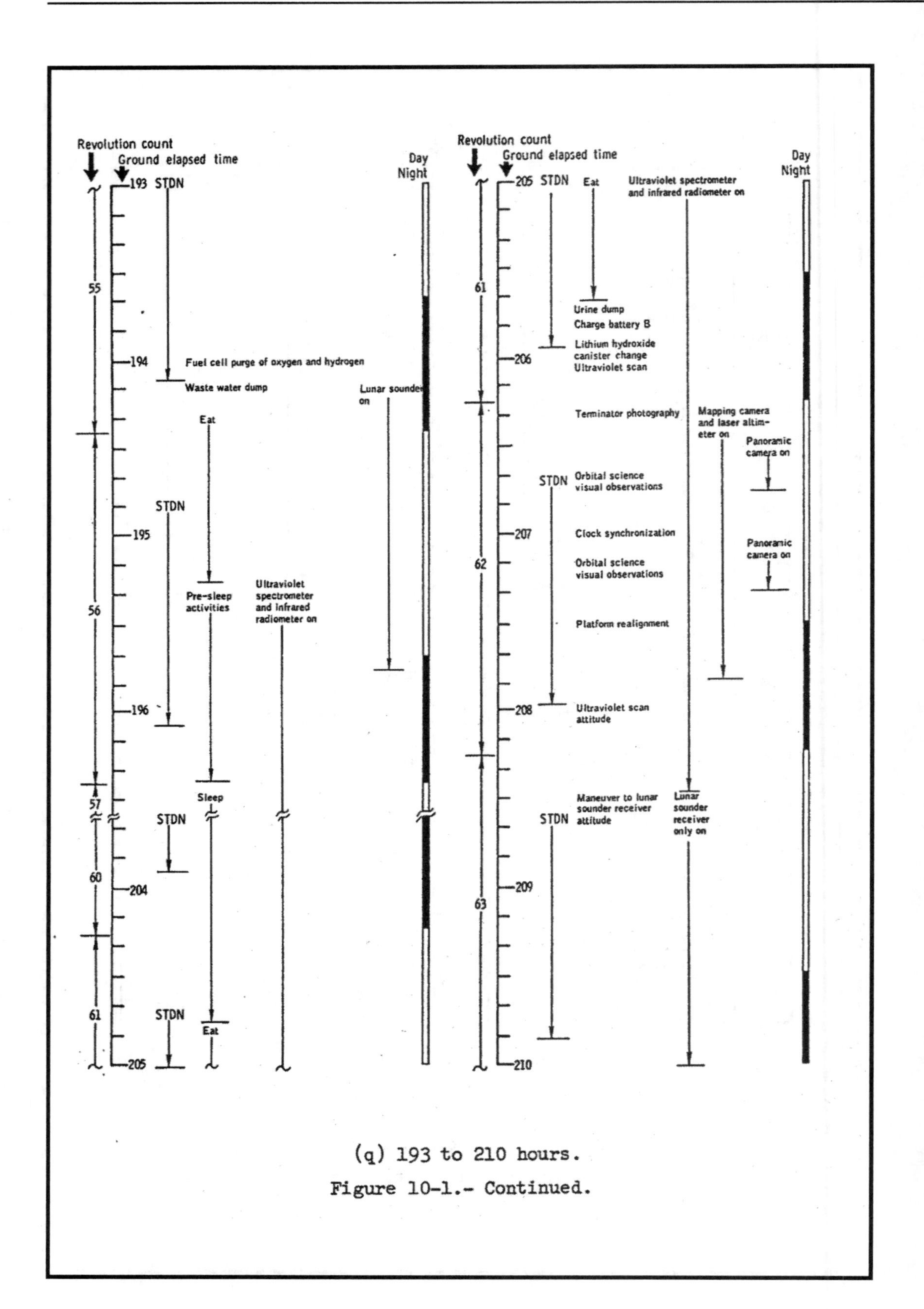

(q) 193 to 210 hours.

Figure 10-1.- Continued.

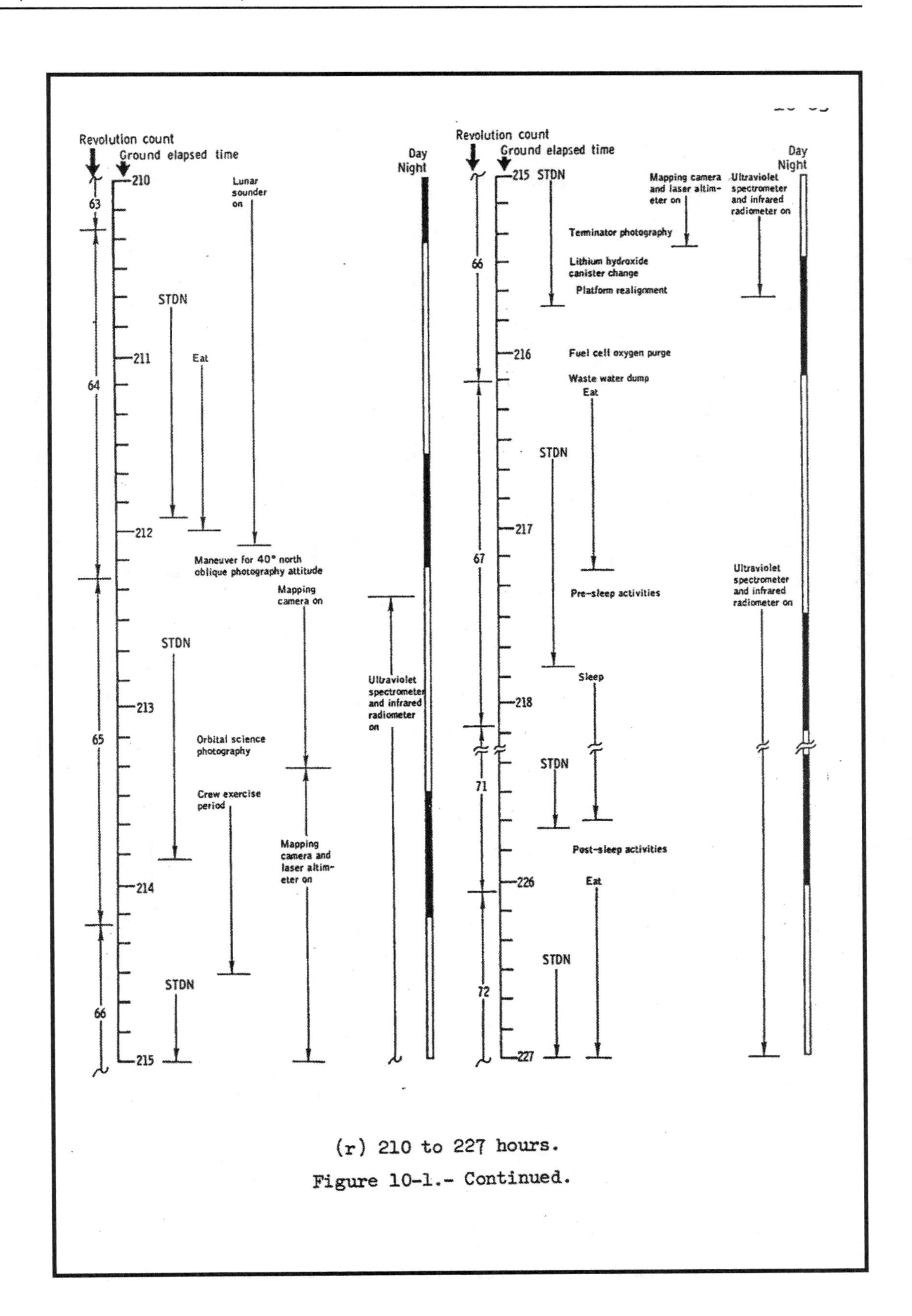

(r) 210 to 227 hours.

Figure 10-1.- Continued.

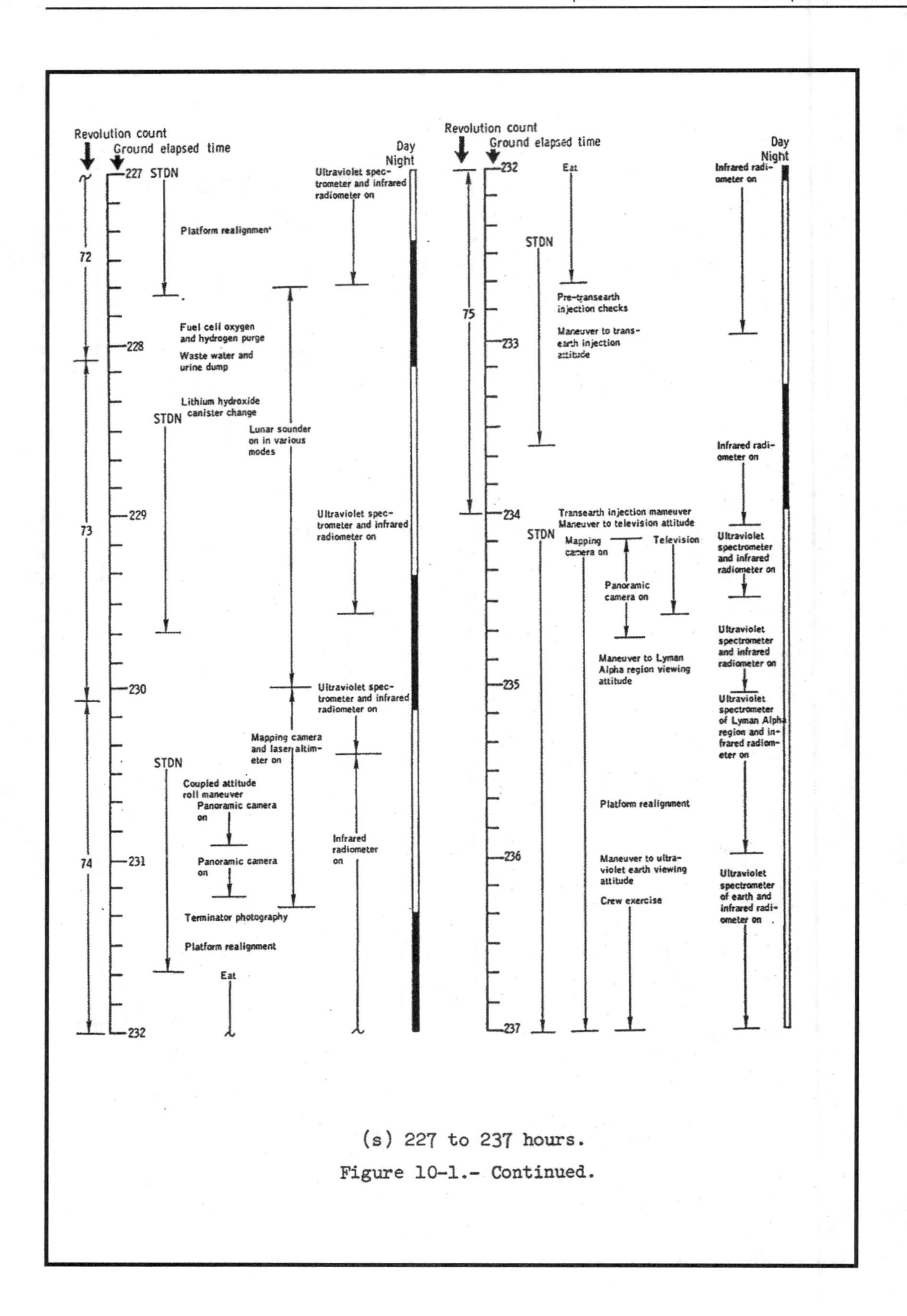

(s) 227 to 237 hours.

Figure 10-1.- Continued.

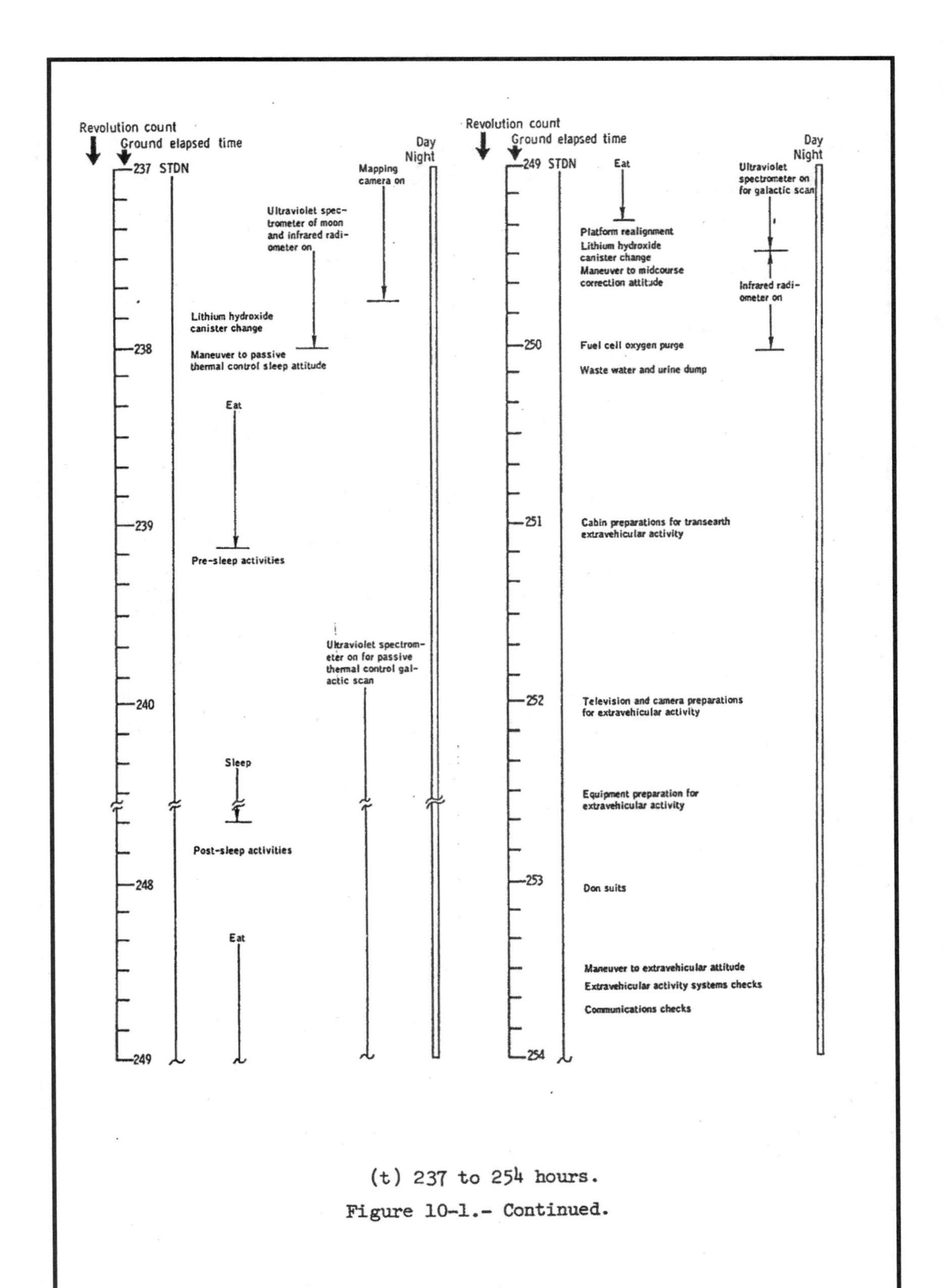

(t) 237 to 254 hours.

Figure 10-1.- Continued.

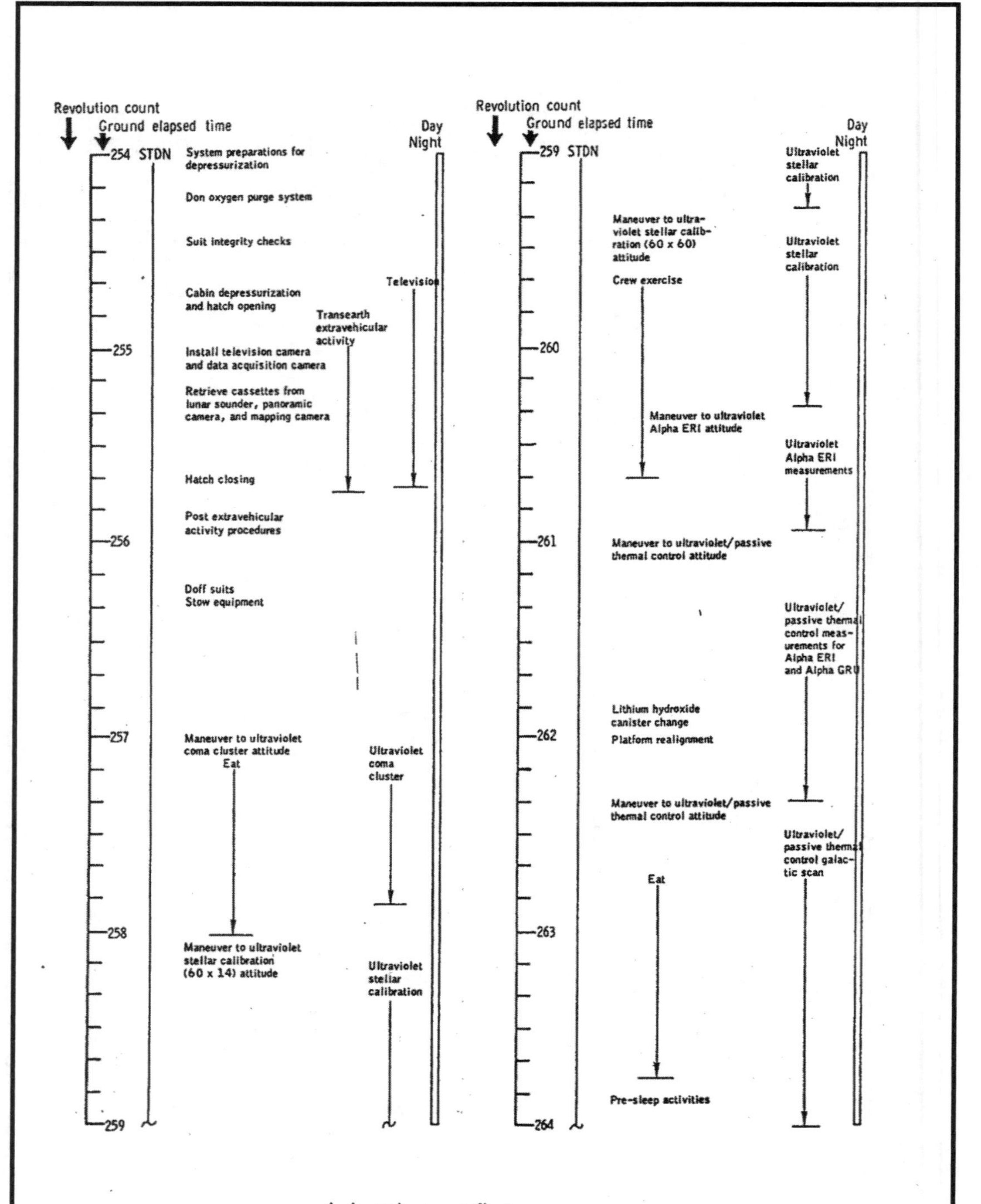

(u) 254 to 264 hours.

Figure 10-1.- Continued.

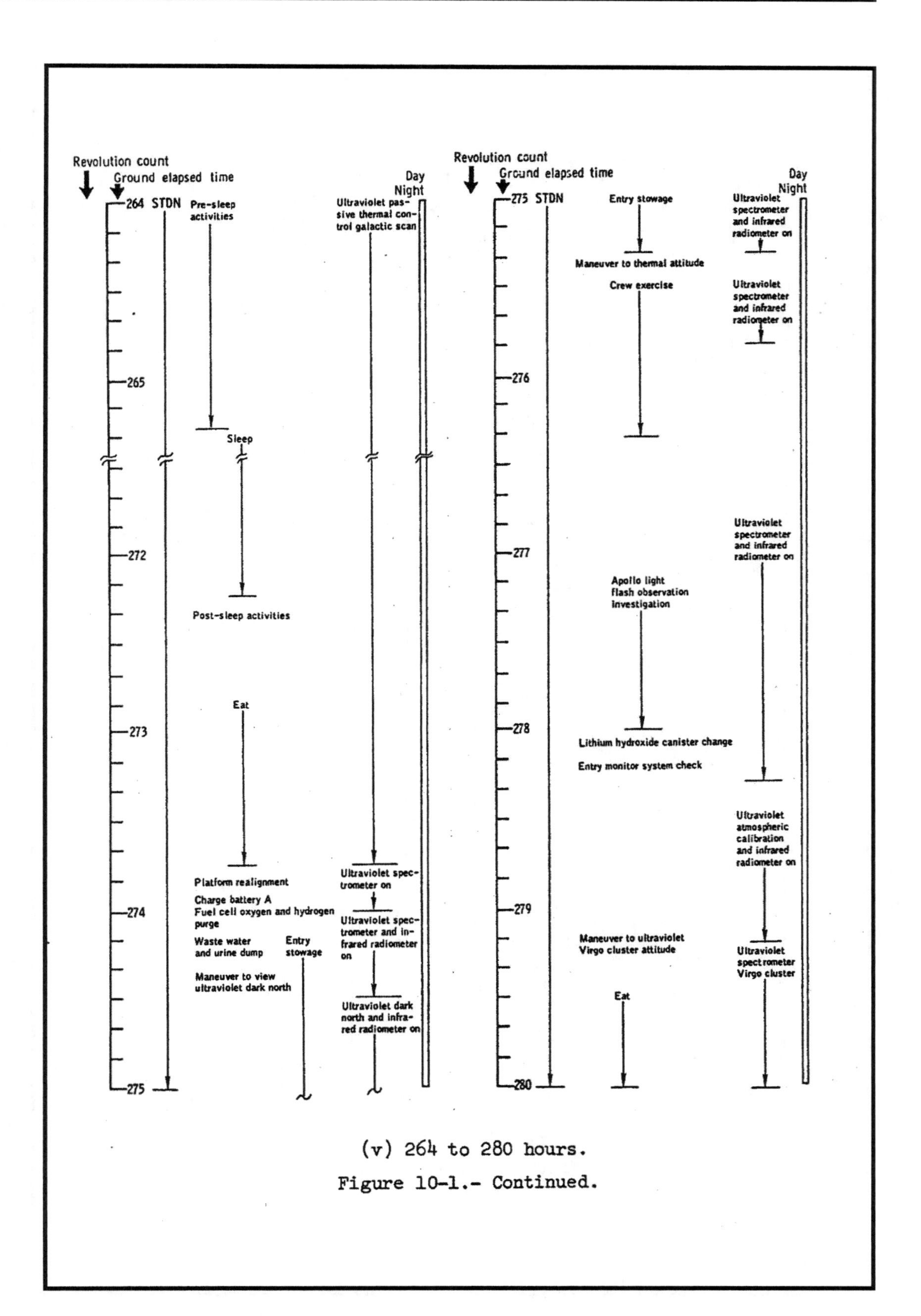

(v) 264 to 280 hours.

Figure 10-1.- Continued.

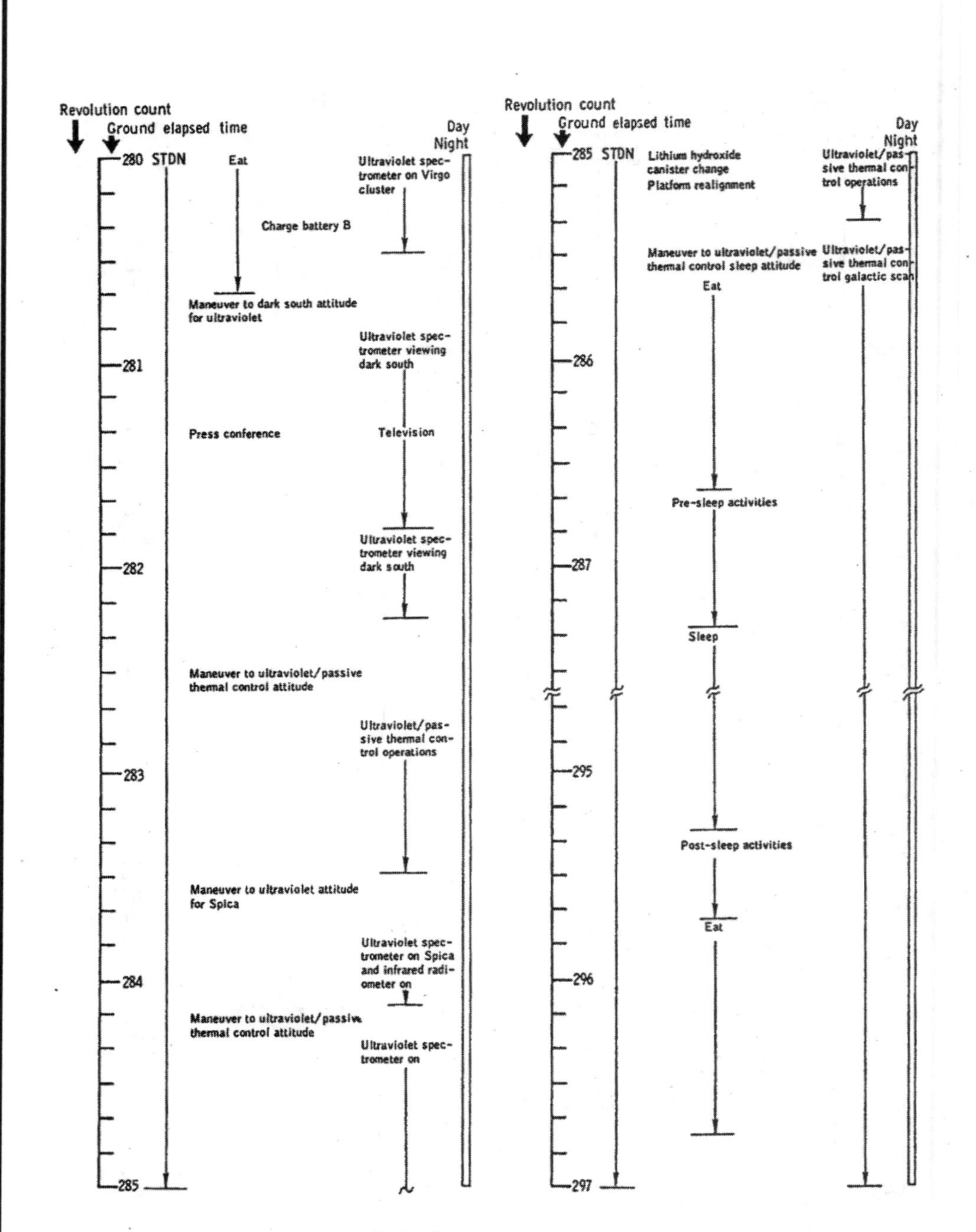

(w) 280 to 297 hours.

Figure 10-1.- Continued.

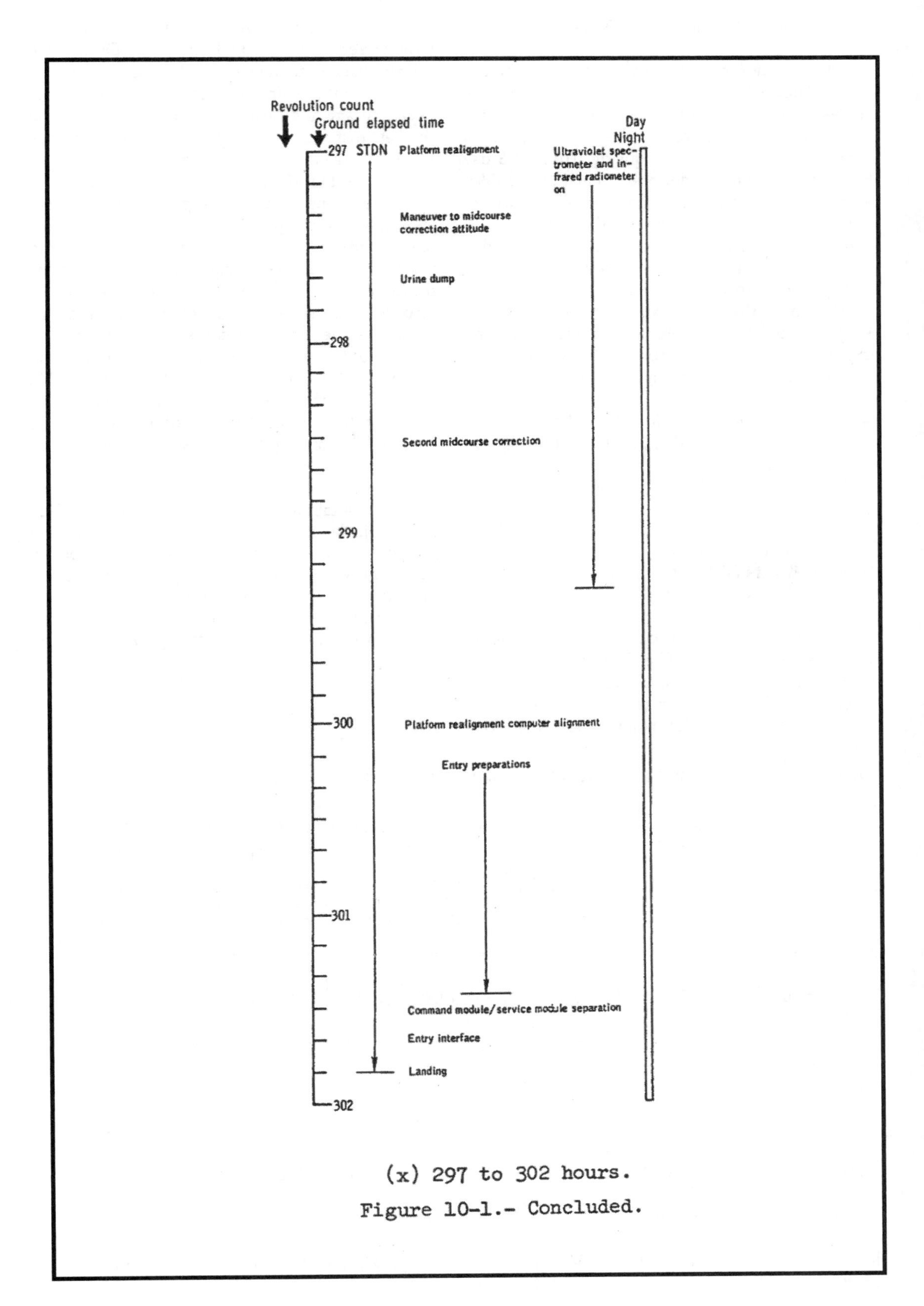

(x) 297 to 302 hours.

Figure 10-1.- Concluded.

11.0 BIOMEDICAL EVALUATION

This section summarizes the medical findings of the Apollo 17 mission based on a preliminary analysis of the biomedical data. The three astronauts accumulated 913.5 man-hours of space flight experience during the 12.6-day mission. The Commander and Lunar Module Pilot each accumulated approximately 22 hours of extravehicular activity time on the lunar surface and the Command Module Pilot accumulated 1 hour and 6 minutes of extravehicular activity during the transearth coast phase. All inflight medical objectives were successfully completed. All physiological parameters remained within the expected ranges, and no medically significant arhythmias occurred during the mission.

Two biological experiments and a light flash investigation were conducted during the mission to gather data on solar radiation effects on biological systems. These are reported in section 6.

11.1 BIOMEDICAL INSTRUMENTATION AND PHYSIOLOGICAL DATA

All physiological measurements remained within expected limits. The Commander's heart rate during lunar descent and ascent is shown in figure 11-1. The metabolic energy expenditure of the crewmen was determined for the four periods of extravehicular activity and is summarized in tables 11-I and 11-II. During the lunar surface extravehicular activities, the average metabolic rates were obtained by measuring the oxygen usage and heat loss of the crewmen. The best estimate of the metabolic rate for discrete activities during the lunar surface extravehicular activities is based on heart rate. The change in heart rate along with the change in metabolic rate measured both preflight and postflight are referenced to the average metabolic rate and heart rate measured during each extravehicular activity (see figs. 11-2 through 11-4). Heart rate is responsive to several parameters in addition to work rate; therefore, the correlation of heart rate to metabolic rate for a single point during an extravehicular activity can involve considerable error. Only the heart rate was available for assessing the metabolic rate of the Command Module Pilot during the transearth extravehicular activity.

The metabolic rates during the lunar surface extravehicular activities were a little higher than predicted, but well within the range experienced during previous missions. The predicted average rate was 892 Btu/hr, and the measured rate was 951 Btu/hr. The metabolic rate during the transearth extravehicular activity was near the predicted value.

TABLE 11-I- AVERAGE METABOLIC RATE AND HEART RATE DURING ALL EXTRAVEHICULAR ACTIVITIES

Activity	time, Elapsed hr: min		Duration, min	Commander		Lunar Module Pilot		Command Module Pilot	
	Start	End		Heart rate, beats/ min	Metabolic rate, Btu/hr	Heart rate, beats/min	Metabolic rate, Btu/hr	Heart rate, beats/min	Metabolic rate, Btu/min
First extravehicular activity	114:21	121:33	432	102.3	1090	102.4	1074	- -	
Second extravehicular activity	137:55	145:32	457	103.0	821	88.5	835	- -	
Third extravehicular activity	160:53	168:08	435	100.6	929	90.0	942	- -	
Transearth extravehicular activity	254:54	256:00	67	-	-	83.3	569	115	<1200

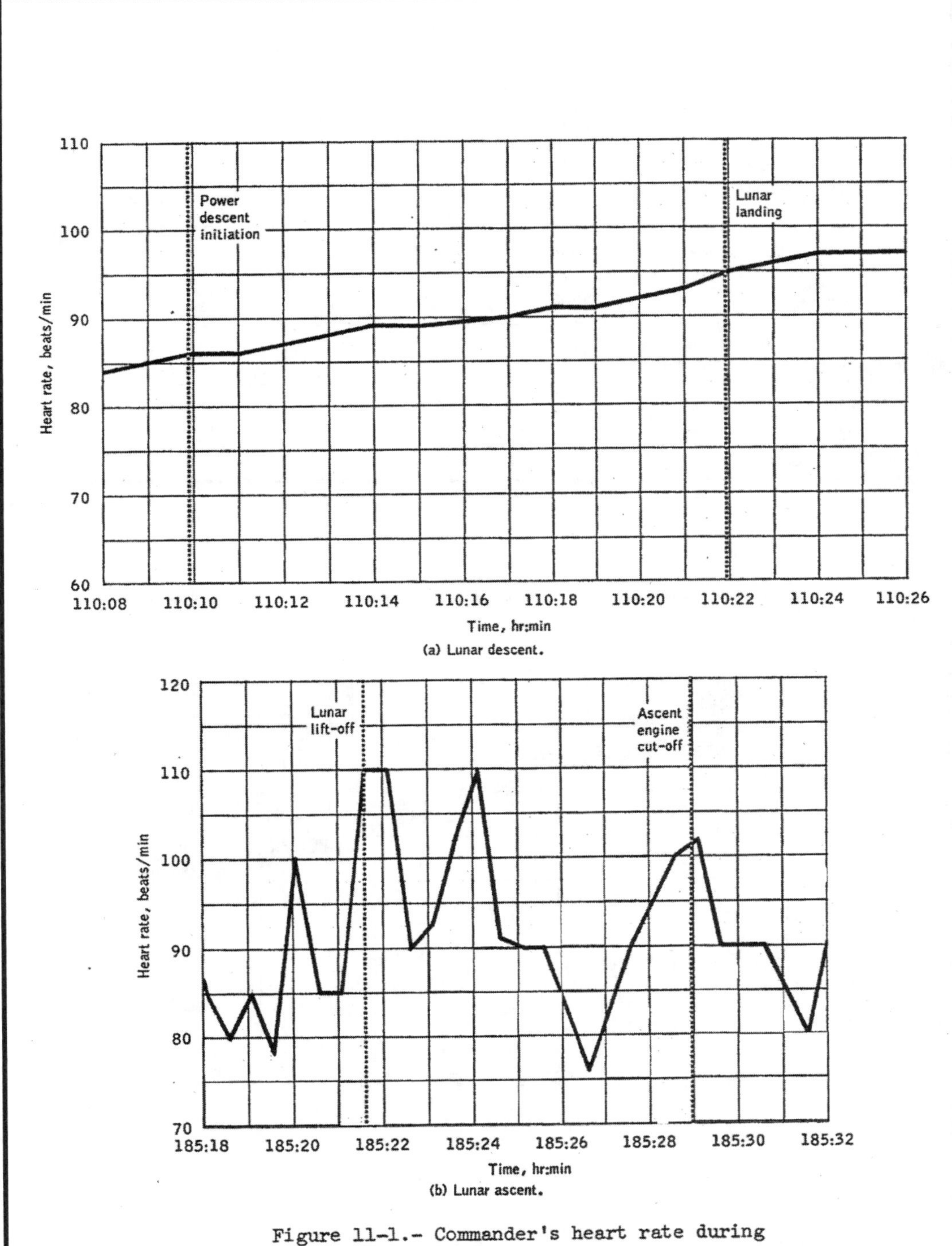

Figure 11-1.- Commander's heart rate during lunar descent and ascent.

TABLE 11-II- METABOLIC ASSESSMENT SUMMARY DURING ALL SURFACE EXTRAVEHICULAR ACTIVITY

Activity[a]	Commander		Lunar Module Pilot	
	Actual, Btu/hr	Prelaunch prediction, Btu/hr	Actual, Btu/hr	Prelaunch prediction, Btu/hr
Lunar roving vehicle traverse	479	550	447	550
Geological station activities	1036	950	1189	950
Overhead	1200	1050	1130	1050
Apollo lunar surface experiments package activities	1129	1050	1104	1050
All activities	946	892	950	892

[a] Averaged values.

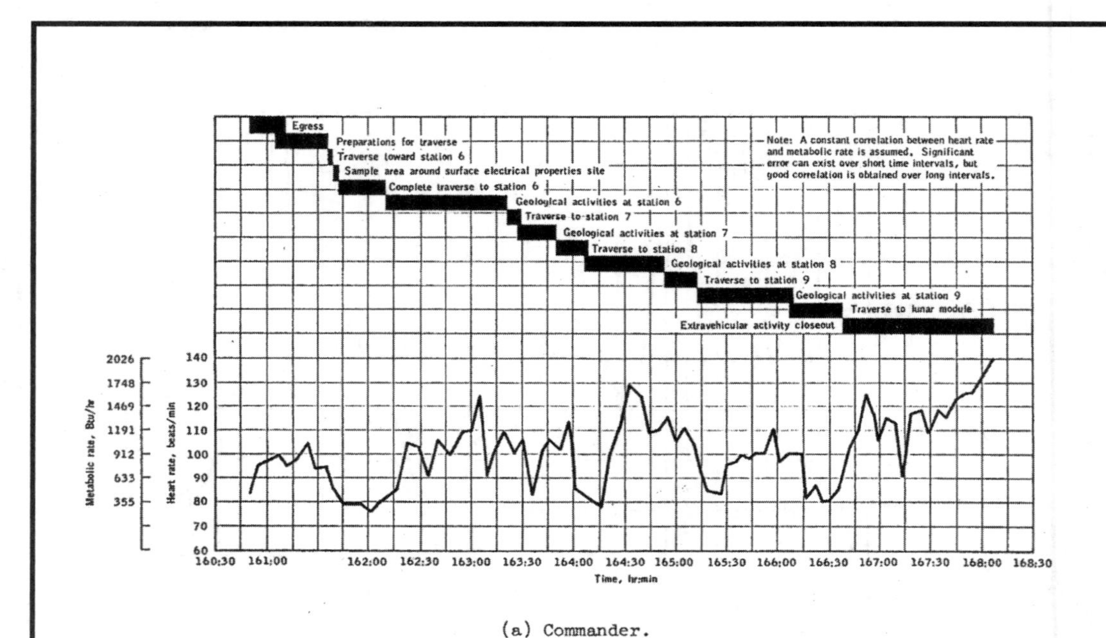

(a) Commander.

Figure 11-4.- Heart rates and calculated metabolic rates during third extravehicular activity.

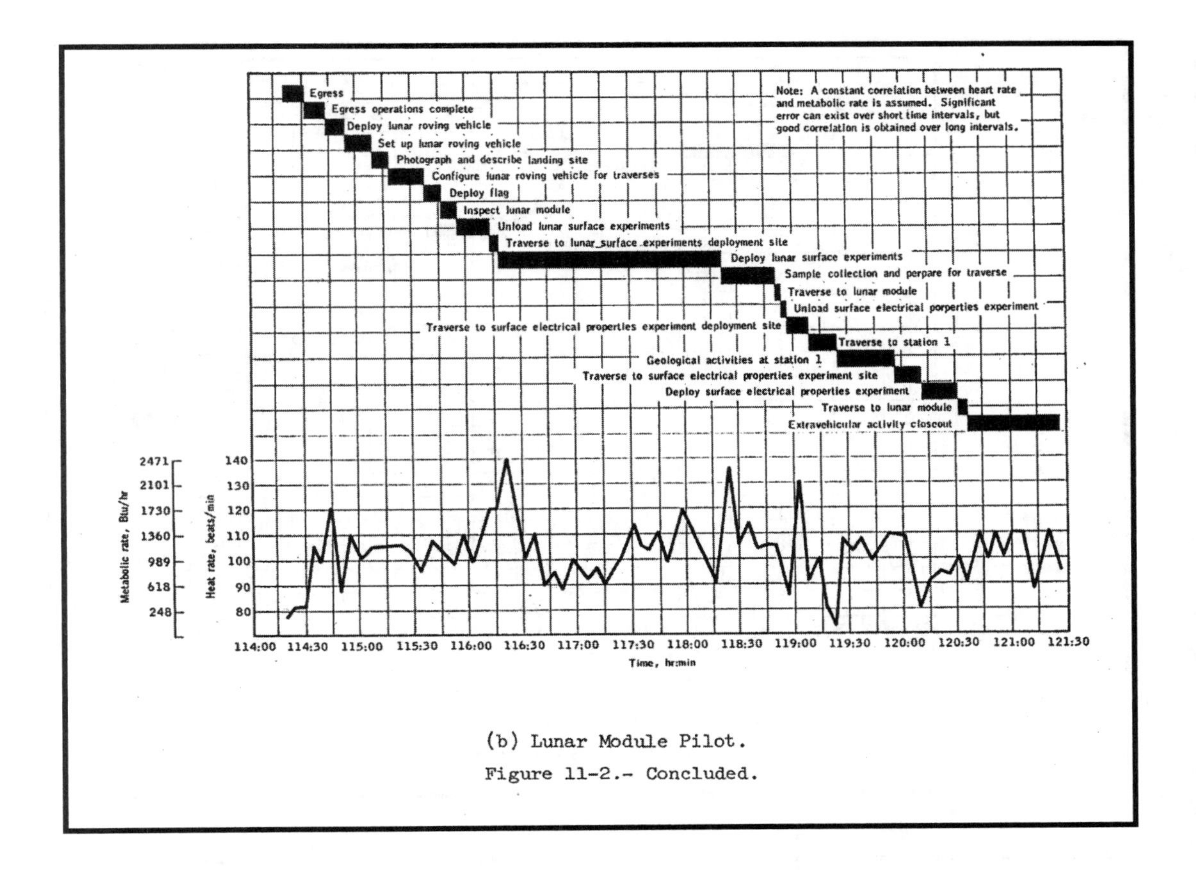

(b) Lunar Module Pilot.

Figure 11-2.- Concluded.

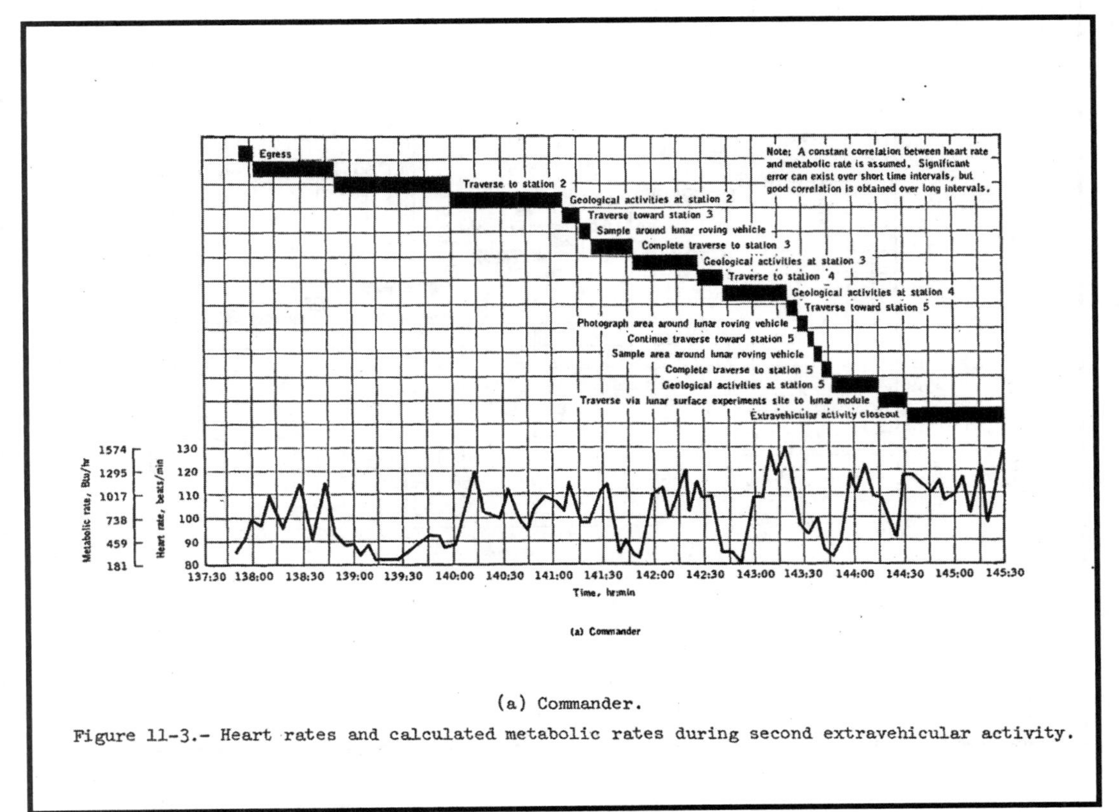

(a) Commander.

Figure 11-3.- Heart rates and calculated metabolic rates during second extravehicular activity.

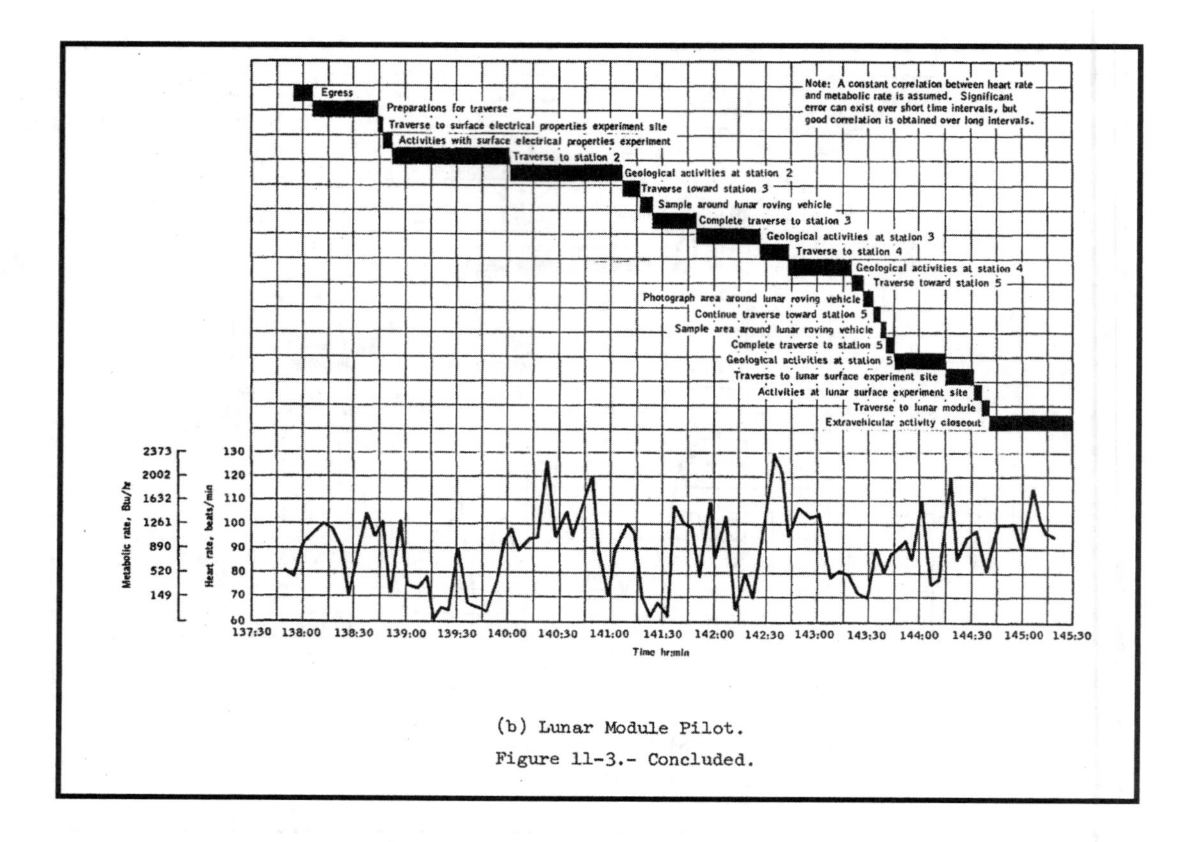

(b) Lunar Module Pilot.

Figure 11-3.- Concluded.

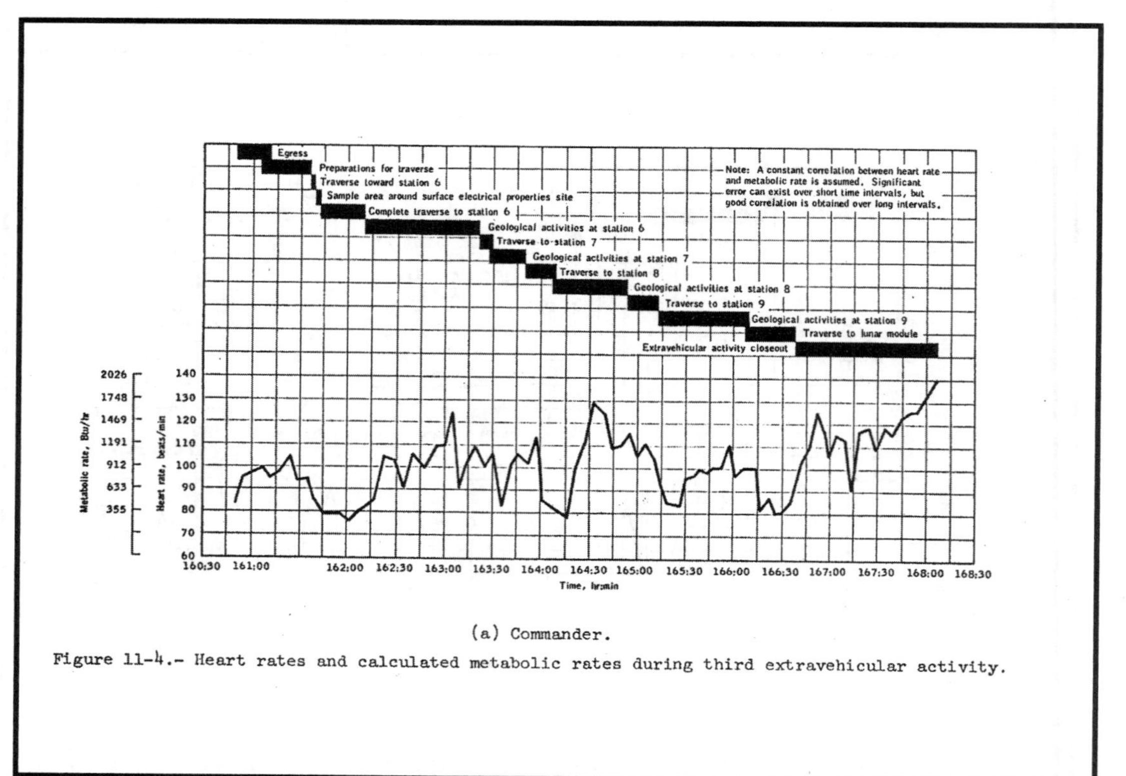

(a) Commander.

Figure 11-4.- Heart rates and calculated metabolic rates during third extravehicular activity.

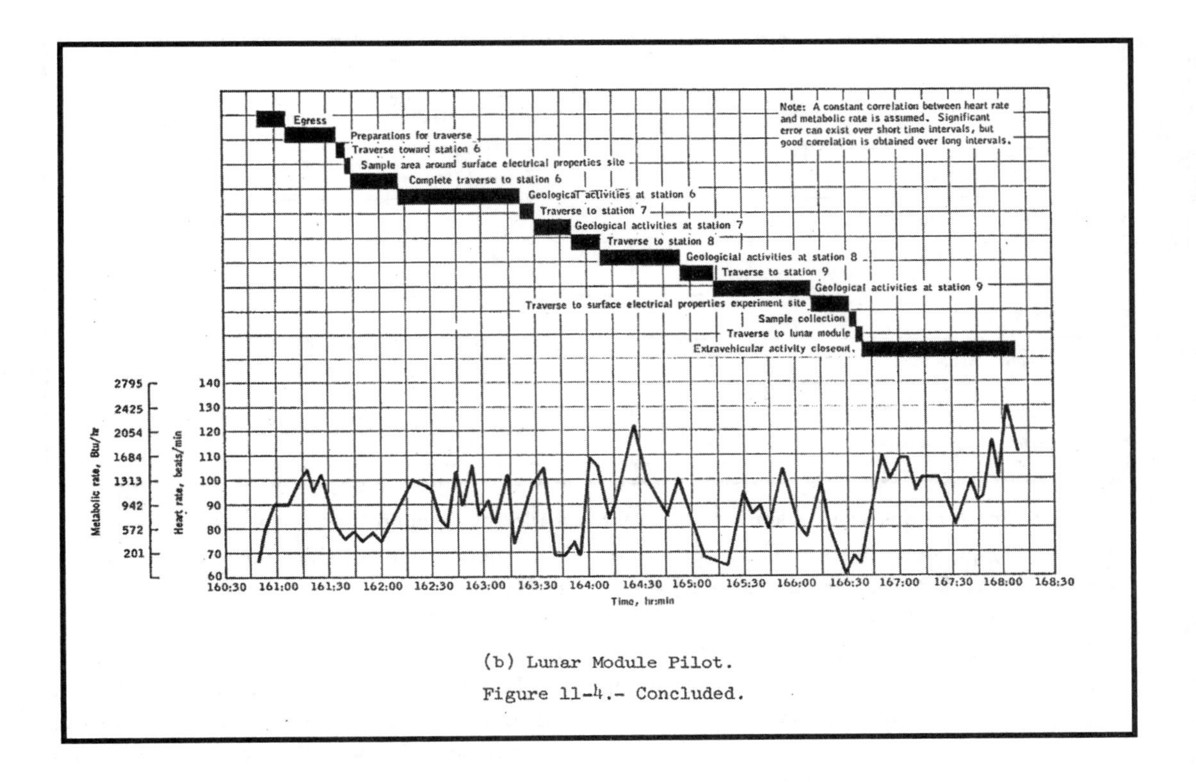

(b) Lunar Module Pilot.

Figure 11-4.- Concluded.

11.2 MEDICAL OBSERVATIONS

11.2.1 Adaptation to Weightlessness

Two of the three crewmen experienced the typical fullness-of-the head sensation, one immediately after earth-orbital insertion and the other after a 6-hour exposure to weightlessness.

Based on the crew's postflight comments, the adaptation to weightlessness required approximately 1 1/2 days. There were no instances of nausea, vomiting, or disorientation; however, all three crewmen did experience the need to limit their movements and perform the necessary movements slowly during the initial period of adaptation. In addition, all three crewmen had varying degrees of "stomach awareness" and a decreased appetite for the first 1 1/2 days of flight. Once adapted, the crewmen were able to perform all types of movements without restrictions. No readaptation to weightlessness was required after residing for 75 hours on the moon at 1/6-g.

11.2.2 Medications

The medications taken by each crewman are presented in the following table:

Medication	Commander	Units taken Command Module Pilot	Lunar Module Pilot
Seconal (Sleep)	3	7	
Simethicone (Antiflatulence)	29	1	0
Dexedrine scopolamine (Antimotion sickness)	1	0	0
Aspirin (Headache)	0	0	4
Lomotil ; (Antiperistalsis)	0	4	1
Afrin nose drops (Decongestant)	0	3	0
Actifed (Decongestant)	0	1	0

More medications were taken on this flight than on any previous mission. The intermittent use of Seconal for sleep by all three crewmen and the daily use of Simethicone for symptomatic relief of gastrointestinal flatulence by one of the crewmen were the principal factors contributing to the high intake of medications.

The Command Module Pilot and the Lunar Module Pilot experienced one loose bowel movement each, on the eleventh and twelfth days of flight, respectively. In each case, Lomotil was taken as a preventative measure and was effective.

The Commander used one Dexedrine-Scopolamine tablet on the second day of flight as a substitute for the Simethicone tablets which he could not locate initially.

In addition to the medications cited previously, skin cream was used by all three crewmen to reduce irritation at the biosensor sites.

11.2.3 Sleep

As on previous missions, displacement of the terrestrial sleep cycle contributed to some loss of sleep. In addition, changes to the flight plan occasionally impacted previously planned crew sleep periods. In general, however, an adequate amount of good sleep was obtained by all crewmembers. The estimates of sleep duration made by ground personnel were in general agreement with the crew's subjective evaluations.

All three crewmen averaged approximately six hours of sleep per day throughout the mission. Only during the first sleep period was the amount of sleep obtained (approximately three hours) inadequate from a medical point of view.

The crew reported that the Seconal effectively induced sound and undisturbed sleep for a period of four to five hours. Sleep restraints were used for every sleep period by all three crewmen. The Commander also emphasized the importance of, programming an eight-hour sleep period each day.

11.2.4 Radiation

The personal radiation dosimeters showed the total absorbed dose to the crew was slightly less than 0.6 rad at skin depth. This is well below the threshold of detectable medical effects.

The Lunar Module Pilot's personal radiation dosimeter read higher than the other two crewmens' units, due to atypical performance of the dosimeter at low dose rates.

11.2.5 Cardiac Arrhythmias

No medically significant arrhythmias

occurred during the mission, but isolated premature heart beats were observed in two of the three crewmen. The fact that the frequency (less than one per day) and character of these prematurities remained consistent with electrocardiographic data obtained on these same crewmen during ground-based tests, clearly indicates that they were not related to or resultant from space flight. In addition., one crewman demonstrated ten isolated intermittent contractions of the auricles that were not transmitted to the ventricles (non-conducted P-waves). This benign conduction defect was associated with an increased discharge of the vagus nerve (vagotonia). (Note: A non-conducted P-wave causes the heart to skip a beat). This condition commonly occurs in athletes and is not medically significant. The same arrhythmia was observed on this crewman during extravehicular training and the countdown demonstration test.

11.2.6 Water

Crew comments indicated that the potable water from both the command module and lunar module systems tasted good. However, there was a large amount of dissolved gas in the command module water.

A modified Skylab beverage dispenser was used for measuring the crew's potable water intake during the mission.

All postflight chemical and microbiological data indicated that the potable water for the command module remained within acceptable limits, and a measurable free chlorine residual of 0.01 mg/liter was obtained.

11.2.7 Food

The menus for this mission were designed to meet physiological requirements of each crewmember as well as requirements of the food compatibility assessment study. This study was implemented to determine the following:

a. Metabolic requirements of space flight.

b. Compatibility of menus with respect to gastrointestinal function. c. Endocrine control of metabolism.

Interpretation of the data required that nutrient intake levels be constant and that all food and fluid intake be measured for the flight as well as for a short period before and after flight. This controlled study included the daily amounts specified in the following table:

Nutrient	Daily range
Protein, g	100 - 120
Calcium, mg	750 - 850
Phosphorus, mg . . .	1500 - 1700
Sodium, mg	3000 - 6000
Potassium, mg . . .	Not less than 3945
Magnesium, mg . .	300 - 400

Some of the beverages were fortified with potassium gluconate to meet the minimum daily requirement of 3945 mg of potassium. The daily energy intake specified was approximately 300 kilocalories less than the calculated one-g energy requirements for each crewman. Individual menu design was based on age, body weight, and adjusted according to dietary performance on the six-day control study. A daily caloric intake of 2660 kilocalories for the Commander, 2520 kilocalories for the Command Module Pilot, and 2550 kilocalories for the Lunar Module Pilot was provided. Preliminary estimates of food consumption indicate that an average of 1902, 2402, and 2148 kilocalories per day were consumed by the Commander, Command Module Pilot, and Lunar Module Pilot, respectively.

A limited number of pantry items (snacks and beverages) were supplied and could be used to substitute or supplement the normal meal items as long as the nutrient intake for a 24-hour period was not significantly altered. In-suit food bars and water for the in-suit drinking device were supplied and used by the suited crewmen while on the lunar surface. Also, each crewman had an in-suit drinking device filled with water for consumption during the prelaunch countdown period.

New foods for this mission included rehydratable tea and lemonade, thermostabilized/irradiated ham, nutrient-complete fruitcake, and asceptically packaged butterscotch pudding in aluminum cans. For the first time, the bread was kept frozen until it was stowed on the spacecraft approximately 24 hours before launch. This change in procedure was initiated to allow packaging several weeks prior to flight without loss in break quality; however, the procedure was not as effective as expected.

After menu selection was completed, each crewman participated in a six-day food control study to establish baseline data on the excretory levels of electrolytes, nitrogen, and various hormones. During this six day period, one crewman experienced gastrointestinal discomfort following each meal. This discomfort persisted and was attributed to gastrointestinal gas and his inability to eliminate the gas. Modifications were made to the menus to eliminate usual gas-forming food items. In addition, Simethicone tablets were provided to alleviate symptoms, if they occurred in flight.

The principal medical problem experienced during this mission was the presence of a greater amount of gastrointestinal gas than anticipated, particularly in the one crewman who had these symptoms preflight. On the second day of flight, similar symptoms of gastrointestinal distress occurred. The crewman consumed Simethicone tablets following each meal and this medication relieved, but never completely eliminated his symptoms.

In addition, belching provided further relief. Mild gastrointestinal distress was experienced by the other two crewmen; however, they were able to effectively eliminate the gas, and at no time were the symptoms severe enough to interfere with the operational duties of the crew. The exact cause of the gastrointestinal flatulence during the mission is unknown, but the problem was probably aggravated by the oxygen gas noted in the potable water supply of the command module.

As on previous flights, negative nitrogen and potassium balances occurred and confirmed a loss in total body protein. In addition, a loss of body calcium and phosphorus was noted. Although some of the observed weight loss can be attributed to changes in total body water, the hypocaloric regimen, in conjunction with the well-known tendency to lose body tissue under hypogravic conditions, indicates that a considerable portion of weight loss is from fatty and muscle tissues.

Water intake and output data were generally consistent throughout the flight. However, when insensible water loss is considered, the crew on this mission were in a state of mild negative water balance. These data are consistent with water-balance data from Apollo 16. During the immediate postflight period, only the Lunar Module Pilot's urine volume was significantly decreased; the other two remained normal. This postflight finding, along with the slight decreases in total body water, confirms the normal to decreased level of antidiuretic hormone. This observation differs from the Apollo 15 mission where high urine volumes and increased levels of antidiuretic hormone were observed. The more complete data from this mission', suggests that the major component of weight loss was tissue mass rather than total body water. The lack of weight gains during the first 24 hours after the flight provides additional evidence for the fact that fatty and muscle tissues mass were lost.

The endocrine data shows a normal to increased excretion of glucocorticoids and mineral conserving hormones. Previous Apollo data showing an increased free urinary hydrocortisone level in the presence of a low to normal total 17-hydroxycorticoids was confirmed. The metabolism of steroid hormones appears to be altered by space flight, but the exact mechanisms are unknown.

A sodium conserving steroid hormone, aldosterone, was decreased during the early portion of flight, but increases were noticed later in the flight. These data support the hypothesis that aldosterone levels were the major cause of the slightly negative potassium balance. The significantly decreased levels of serum potassium during postflight tests are consistent with these data. As on the Apollo 15 mission, a decrease in total body exchangeable potassium was observed. Nonetheless, the elevated dietary potassium, coupled with an adequate fluid intake, were apparently sufficient to preserve normal cellular electrolyte concentrations. This contrasts to previous Apollo observations in which alterations in cellular homeostatis were manifested.

11.3 PHYSICAL EXAMINATION

Comprehensive physical examinations were performed on the Commander and Command Module Pilot 30 days prior to launch and on the Lunar Module Pilot 26 days prior to launch. These examinations were repeated 15 days and 5 days before launch and daily thereafter. The comprehensive physical examinations conducted shortly after recovery showed the crew to be in good general health and physical condition. Body weight losses during the mission were 10-1/4 pounds for the Commander, 4-1/2 pounds for the Command Module Pilot, and 5-1/2 pounds for the Lunar Module Pilot. The Lunar Module Pilot suffered moderately severe skin irritation caused by prolonged contact with the biosensors, and the Commander evidenced a much milder reaction. Both of these crewmen exhibited subunguial hematomas from the pressure suit gloves, but these were more extensive and apparent on the Lunar Module Pilot. The Commander had a herpetic lesion (fever blister on the right side of the upper lip, and it was approximately 72 hours old at the time of recovery.

Only one of the crewmen was within his preflight baseline during postflight bicycle-ergometry tests. The other two crewmen returned to their baseline by the second day after recovery.

Orthostatic tolerance was decreased and returned to preflight baseline values by the second day after recovery.

11.4 VESTIBULAR FUNCTION

Two different types of vestibular function tests were performed preflight; the first, to test postural stability with and without the aid of vision, and the second graphic recordings of nystagmus (eye movement) induced by caloric irrigation of the right and left ear canals.

The results of these tests indicated that postural equilibrium, with the eyes open or closed, was within normal limits for all crewmen. One crewman exhibited poor eyes-closed postural equilibrium, especially at the examination 5 days prior to flight. Nystagmus responses to caloric irrigation were normal for all crewmen. No vestibular function tests were conducted postflight. On the basis of crew comments, it is known that all three crewmen experienced very mild vestibular hypersensitivity during the first two days of flight. However, the symptoms dissipated with adaptation to weightlessness and did not reappear.

11.5 SKYLAB MOBILE LABORATORIES OPERATIONAL TEST

An operational test of the Skylab Mobile Laboratories was conducted in conjunction with the Apollo 17 recovery. The objective of this test was to perform all operational procedures associated with Skylab recovery operations, both for a normal end-of-mission recovery (primary recovery ship) and contingency recovery (C5A aircraft). All test objectives were satisfied and many equipment and procedural changes resulted from the test. During the Apollo 17 recovery, all six Skylab Mobile Laboratories were used to conduct the initial postflight medical examinations and debriefings.

12.0 MISSION SUPPORT PERFORMANCE

12.1 FLIGHT CONTROL

Flight control provided satisfactory operational support during the mission. Only anomalous operations are described; discussion of routine support activities are not included. A number of problems that were encountered are discussed elsewhere in this report. Only those problems that are unique to flight control or have operational considerations not previously mentioned are presented in this section. The launch delay due to the automatic sequencer failure caused no significant problems. The 2-hour 40-minute late lift-off was subsequently compensated for (during the translunar coast) by a resetting of the mission clocks on the spacecraft and in the Control Center.

During the first lunar module activation at about 40 hours, the lunar module cabin repressurization function was initiated unexpectedly. The command module environmental control system normally raises the cabin pressure of both vehicles to 5.0 psi during the lunar module checkout period. However, the activation procedure was being completed ahead of schedule and the lunar module cabin pressure had not reached the minimum limit of 4.0 psi when the lunar module emergency repressurization system was activated during the procedure. Consequently, the system automatically initiated the repressurization and resulted in the premature use of about 1.7 pounds of descent stage oxygen.

The crew did not respond to the morning wakeup call on the third day. The Command Module Pilot had accidentally turned off the audio power switch, and the ground calls and crew alert tone were not getting through. The crew woke up about 1 hour late and re-established communications. The planned activities through translunar coast, lunar orbit insertion, and the lunar landing were accomplished without any significant difficulties.

The first lunar surface extravehicular activity was about 1 hour behind the timeline because of the crew leaving the lunar module late and the longer time required to complete the initial traverse preparation activities and Apollo lunar surface experiments package deployment. As a result, it was necessary to eliminate some of the planned activities and to shorten the geology traverse to make up for the lost time.

During the second extravehicular activity, the surface electrical properties receiver and the lunar roving vehicle batteries operated at higher temperatures than expected. The traverse during this second extravehicular activity was particularly interesting because it utilized the equipment to its operational limit. The distance to station 2 (7370 meters) was almost at the maximum limit; the times of departure from stations 4 and 5 also were at the limit; and all planned activities were accomplished.

The third extravehicular activity was conducted in a routine manner, although some replanning was necessary due to getting behind the planned timeline and the extremely high temperature of the surface electrical properties receiver.

Preparations for lunar module ascent and rendezvous were accomplished normally. However, an additional unplanned command and service module orbit shaping maneuver was made prior to the planned plane change. The command and service module orbit had been perturbed in such a way that it was more eccentric than expected, and it was decided to make the orbit more clearly circular for the rendezvous.

During lunar module ascent, the ascent feed was terminated early by request of the lunar module flight controller because of a potential mixture ratio problem. The rendezvous and the remainder of the mission was routine and according to plan, with the exception of some inadvertent switch activations by the crew.

12.2 NETWORK

The Mission Control Center and the Spaceflight Tracking and Data Network supported the Apollo 17 mission satisfactorily. Although no network problems caused significant mission impact, the following problems were experienced.

During the terminal countdown, two

television channels were shown at once on one console in the Mission Control Center. At the same time, the channels on all other consoles could not be switched. The simultaneous failures were caused by a malfunctioning clock driver module and a switch module. The modules were replaced and normal operation was restored prior to launch.

An antenna on the Vanguard tracking ship began a high-frequency oscillation during the launch phase, causing the loss of some spacecraft data at this site. Since this station was not prime, no data were lost at the Mission Control Center. The problem disappeared and tracking was reacquired. The problem was caused by a noise burst in the antenna feed system.

At acquisition-of-signal on the first lunar orbit, 4 minutes were required to establish two-way communications with the spacecraft. The problem was caused by improper pointing of the prime antenna at Goldstone. A handover was made to the Jet Propulsion Laboratory wing antenna and normal operations were resumed.

On lunar module ascent, two-way lock with the lunar module transponder was lost. This resulted in a 4-minute loss of uplink voice, and tracking data during ascent. It was necessary to have the Command Module Pilot pass comments from the ground to the lunar module crew during this period.

12.3 RECOVERY OPERATIONS

The Department of Defense provided recovery support in accordance with the Apollo 17 mission planning. Recovery ship support for the primary landing area in the Pacific Ocean was provided by the aircraft carrier USS Ticonderoga. Active air support consisted of four SH-3G helicopters and one E-1B aircraft from the primary recovery ship and one HC-130 rescue aircraft staged from American Samoa. Three of the helicopters carried underwater demolition team personnel. The first, designated "Recovery", also carried the flight surgeon and was used for both command module and flight crew retrieval operations. The helicopter, designated "Swim" served as a backup to "Recovery" and aided in the recovery of the forward heat shield. The third helicopter, designated "ELS" (earth landing system), aided in the retrieval of the main parachutes. The fourth helicopter, designated "Photo", served as a photographic platform for both motion picture photography and live television coverage. The E-1B aircraft, designated "Relay" served as a communications relay. The HC-130 aircraft, designated "Samoa Rescue 1", was positioned to track the command module after it had exited from S-band blackout, as well as to provide pararescue capability had the command module landed uprange of the target point. Figure 12-1 shows the relative positions of the recovery ship, its aircraft, and the HC-130 aircraft prior to landing. The recovery forces assigned to the Apollo 17 mission are shown in table 12-I.

12.3.1 Command Module Location and Retrieval

Radar contact with the command module was reported by the Ticonderoga at 1915 G.m.t. December 19. Visual sighting of the command module occurred at 1920 G.m.t. by the Ticonderoga and the Photo helicopter. The initial sighting occurred shortly after the main parachutes were deployed. Two-way voice communications were established between the Apollo 17 crew and the recovery forces approximately 2 minutes later.

TABLE 12-I.- APOLLO 17 RECOVERY SUPPORT

Type ship/ type aircraft	Number	Ship name/aircraft staging base	Responsibility
ARS	1	USS Recovery	Launch site recovery ship and sonic boom measurement platform.
MSO	1	USS Alacrity	Sonic boom measurement platform on launch ground track.
MSO	1	USS Fidelity	Sonic boom measurement platform on launch ground track.
MSO	1	USS Assurance	Sonic boom measurement platform on launch ground track.
MSO	1	USS Adroit	Sonic boom measurement platform on launch ground track.
LST	1	USS Saginaw	Sonic boom measurement platform on launch ground track.
AO	1	USS Camden	Refuel primary recovery ship.
CVS	1	USS Ticonderoga	Primary recovery ship.
HH-53C	2[a]	Patrick Air Force Base	Launch site area.
HU-1	2	Patrick Air Force Base	Launch site area.
HC-130 P/N	1[a]	Eglin Air Force Base	Launch abort area and west Atlantic recovery area.
HC-130	1[a]	RAF Woodbridge, England	Launch abort area and contingency landing support.
HC-130	1	Hickam Air Force Base	Mid-Pacific earth orbital recovery zone, deep space secondary landing area on mid-Pacific line, and primary end-of-mission landing area.
SH-3G	4	USS Ticonderoga	Deep space secondary landing area and primary end of-mission landing area.
E-1B	1	USS Ticonderoga	Communications relay for primary end-of-mission landing area.

[a] Plus one backup.

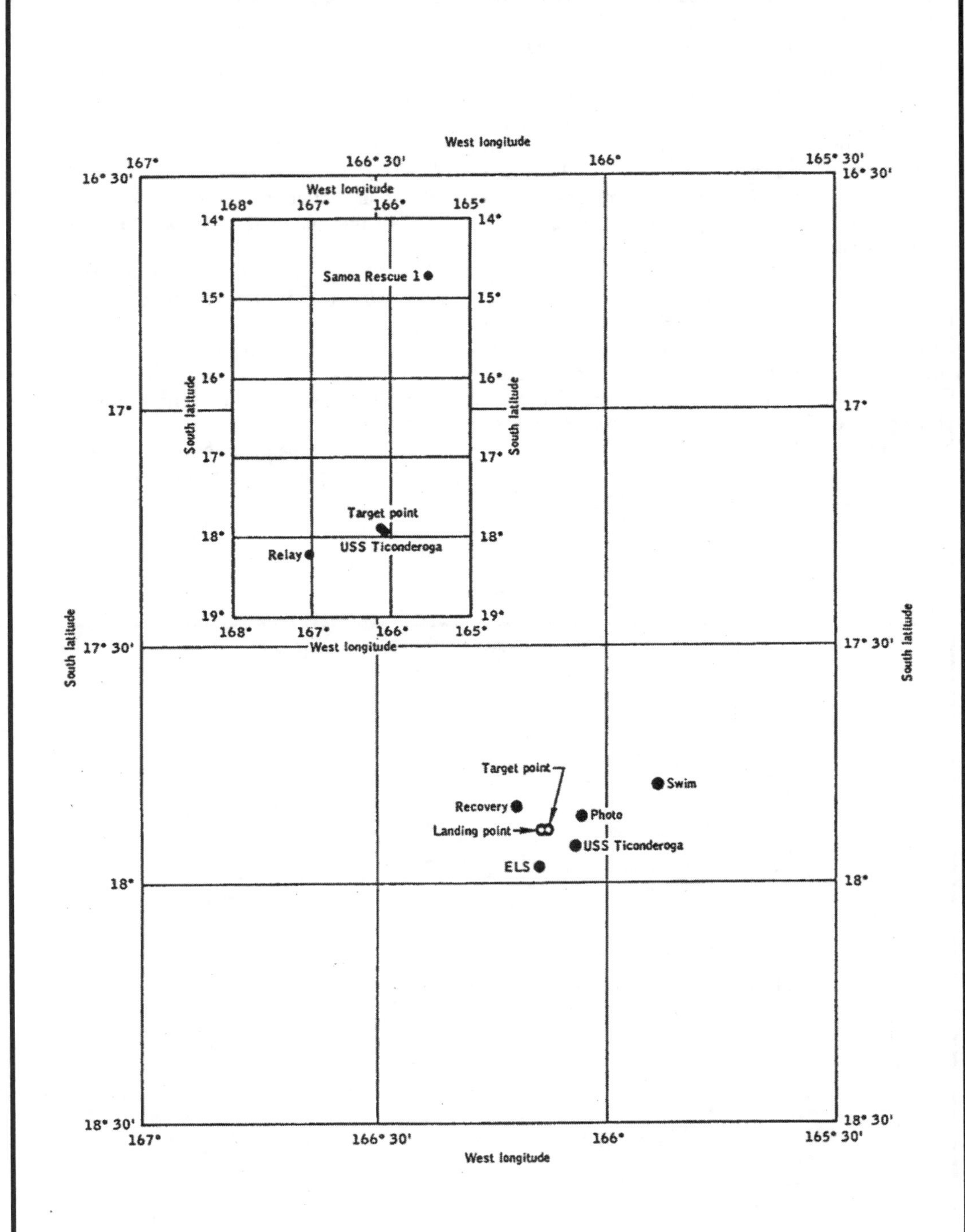

Figure 12-1.- End of mission recovery support.

The command module landed at 1925 G.m.t. December 19, 1973. Based upon a navigation satellite (SRN-9) fix obtained at 1818 G.m.t., the Ticonderoga's position at the time of landing was determined to be 17 degrees, 55 minutes south latitude and 166 degrees 04 minutes 18 seconds west longitude. Using this fix of the ship's position, plus visual bearings and radar ranges, the landing point coordinates of Apollo 17 were determined to be 17 degrees 53 minutes south latitude and 166 degrees 8 minutes 36 seconds west longitude.

The command module landed normally and remained in the stable I attitude. Swimmers were deployed to the command module at 1926 G.m.t. and the flotation collar was installed and inflated by 1940 G.m.t. The flight crew was delivered aboard the Ticonderoga by the recovery helicopter at 2017 G.m.t.

Command module retrieval was accomplished at 2128 G.m.t. In addition to the command module, all three main parachutes and the forward heat shield were recovered.

Table 12-II shows a chronological listing of recovery and post-recovery events.

12.3.2 Postrecovery Inspection

Visual inspection of the command module revealed the following discrepancies:

a. One square foot of the aft bulkhead ablator fell off when the command module was placed on the dolly.
b. Each of the command module windows was fogged.
c. There was 0.35 inch of water on the aft bulkhead.
d. The rock bag stowed in the bottom of the pressure garment assembly bag was damp.

TABLE 12-II.- SIGNIFICANT RECOVERY-POSTRECOVERY EVENTS

Extent	Time, G.m.t. December 19, 1972	Time relative to landing day:hr:min
Radar contact by Ticonderoga	1914	-0:00:11
Visual contact	1920	-0:00:05
VHF recovery beacon contact by Ticonderoga	1921	-0:00:04
Voice contact with Apollo 17 crew	1922	-0:00:03
Command module landing	1925	0:00:00
Swimmers deployed to command module	1935	0:00:10
Flotation collar installed and inflated	1941	0:00:16
Hatch opened for crew egress	1954	0:00:29
Flight crew aboard helicopter	2006	0:00:41
Flight crew aboard Ticonderoga	2017	0:00:52
Command module aboard Ticonderoga	2128	0:02:03
	December 20, 1972	
First sample flight departed ship	1725	0:22:00
	December 21, 1972	
First sample flight arrived Hawaii	0000	1:04:35
Flight crew departed ship	0038	1:05:13
First sample flight departed Hawaii	0310	1:07:45
First sample flight arrived Houston	1016	1:14:51
Flight crew arrived Houston	1550	1:20:25
	December 27, 1972	
Command module arrived North Island, San Diego, California	1930	8:00:05
	December 20, 1972	
Command module deactivated	2200	11:02:35
	January 2, 1973	
Command module departed San Diego	1900	13:23:35
Command module arrived Downey	2200	14:02:35

13.0 ASSESSMENT OF MISSION OBJECTIVES

The three primary mission objectives were assigned to the Apollo 17 mission (ref. 4) were:

a. Perform selenological inspection, survey, and sampling of materials and surface features in a pre-selected area of the Taurus-Littrow region.
b. Emplace and activate surface experiments.
c. Conduct inflight experiments and photographic tasks.

Table 13-I includes the eight detailed objectives, which were derived from the primary objectives, plus nineteen experiments and one inflight demonstration (ref. 5) which were conducted. Preliminary indications are that adequate data were obtained to successfully complete all objectives except as noted in the table.

The Department of Defense and the Kennedy Space Center performed eight other tests which are as follows:

a. Chapel Bell (Classified Department of Defense test)
b. Radar skin tracking
c. Ionospheric disturbance from missiles
d. Acoustic measurement of missile exhaust noise
e. Army acoustic test
f. Long-focal-length optical system
g. Sonic boom measurement

TABLE 13-I.- DETAILED OBJECTIVES AND EXPERIMENTS

Description	Completed
Detailed objectives	
Service module orbital photographic tasks	Yes
Visual light flash phenomenon	Yes
Command module photographic tasks	Yes
Visual observations from lunar orbit	Yes
Skylab contamination study	Yes
Food compatibility assessment	Yes
Protective pressure garment	Yes
Experiments	
Heat flow (S-037)	Yes
Lunar ejecta and meteorites (S-202)	[a] Partial
Lunar seismic profiling (S-203)	Yes
Lunar atmospheric composition (S-205)	Yes
Lunar surface gravimeter (S-207)	[b] Partial
Lunar geology investigation (S-059)	Yes
S-band transponder (command and service module/ lunar module) (5-164)	Yes
Far ultraviolet spectrometer (S-169)	Yes
Infrared scanning radiometer (S-171)	Yes
Traverse gravimeter (S-199)	Yes
Surface electrical properties (S-204)	Yes
Lunar sounder (S-209)	Yes
Lunar neutron probe (S-299)	Yes
Cosmic ray detector (sheets) (S-152)	Yes
Biostack IIA (M-211)	Yes
Biocore (M-212)	Yes
Gamma-ray spectrometer (S-160)	Yes
Apollo window meteoroid (S-176)	Yes
Soil mechanics (S-200)	Yes
Inflight demonstrations	
Heat flow and convection	Yes
Passive objectives	
Long term lunar surface exposure	Yes

[a] Operation restricted during lunar day due to overheating.
[b] Partial (obtaining data in the seismic and free oscillation channels only).

h. Skylab Medical Mobile Laboratory.

14.0 LAUNCH PHASE SUMMARY

14.1 WEATHER CONDITIONS

The broken high clouds were sufficiently thin and the low scattered clouds were sufficiently sparse in the launch area to permit a spectacular view of the launch. A gentle north wind was blowing at the surface level and the temperature was mild. The warm moist air mass over Florida was separated from an extremely cold air mass over the rest of the south by a cold front, oriented northeast-southwest and passing through the Florida Panhandle.

A flurry of isolated, very light, and short-lived rainshowers began soon after sundown in the middle part of the state and moved toward Cape Kennedy at about 10 knots. A few showers reached the west bank of the Indian River, but none occurred east of the Indian River (Launch Complex area). The maximum cloud tops reached 12 000 feet altitude, and had the showers occurred in the launch area they would have, at most, required only a short delay of launch.

The maximum wind speed observed in the troposphere was 90 knots; from 310 degrees and at an altitude of about 39 000 feet.

14.2 LAUNCH VEHICLE PERFORMANCE

The performance of the launch vehicle was satisfactory and all Marshall Space Flight Center mandatory and desirable objectives were accomplished except the precise determination of the S-IVB/instrumentation unit lunar impact point. Trajectory assessments indicate that the final impact solution will satisfy the mission objective.

The ground systems supporting the countdown and launch performed satisfactorily with the exception of the terminal countdown sequencer malfunction. This malfunction resulted in a 2-hour 40-minute unscheduled hold. The hold was caused when the terminal countdown sequencer failed to command pressurization of the S-IVB liquid oxygen tank. This command closes the liquid oxygen tank vent; opens the liquid oxygen tank pressurization valve; and arms the S-IVB liquid oxygen tank pressurized interlock. The tank was pressurized manually, thus satisfying the first two items, but the absence of the third item prevented actuation of the interlock in the S-IVB ready-to-launch logic train. The result was automatic cutoff at T-30 seconds. The launch was accomplished with the interlock bypassed by a jumper. Investigation indicates cause of failure to be a defective diode on a printed circuit card in the terminal countdown sequencer.

The vehicle was launched on an azimuth 90 degrees east of north. A roll maneuver was initiated at 13.0 seconds that placed the vehicle on a flight azimuth of 91.504 degrees east of north. Earth parking orbit insertion conditions were achieved 4.08 seconds earlier than nominal.

In accordance with preflight targeting objectives, the translunar injection maneuver shortened the translunar coast period by 2 hours and 40 minutes-to compensate for the launch delay so that the lunar landing could be made with the same lighting conditions as originally planned. Translunar injection conditions were achieved 2.11 seconds later than predicted with an altitude 313 miles greater than predicted and velocity 16.7 ft/sec less than predicted.

All S-IC propulsion systems performed satisfactorily, with the propulsion performance very close to the predicted nominal. The S-II propulsion systems performed satisfactorily throughout the flight with all parameters near predicted values. Engine mainstage performance was satisfactory throughout flight, and engine thrust buildup and cutoff transients were within the predicted envelopes.

The S-IVB propulsion system performed satisfactorily throughout the operational phase of first and second firings and had normal start and cutoff transients. S-IVB first firing time was 138.8 seconds. This difference is composed of minus 4.1 seconds due to the higher than expected S-II/S-IVB separation velocity and plus 0.4 second due to lower than predicted S-IVB performance. Engine restart conditions were within specified limits. S-IVB second firing time was 351.0 seconds. This difference is primarily due to the lower S-IVB performance and heavier vehicle mass during the second firing.

The S-IC thrust cutoff transients experienced by the Apollo 17 vehicle were similar to those of previous flights. The maximum longitudinal dynamic responses at the instrument unit were ±0.20 g and +0.27 g at S-IC center engine cutoff and outboard engine cutoff, respectively. The magnitudes of the thrust cutoff responses were normal. During the S-IC stage boost, 4- to 5-hertz oscillations were detected beginning at approximately 100 seconds. The maximum amplitude measured at the instrument unit was ±0.06 g. Oscillations in the 4- to 5-hertz range have been observed on previous flights and are considered to be normal vehicle response to the flight environment. Longitudinal oscillations did not occur during S-IC or S-II stage boost.

The navigation and guidance system successfully supported the accomplishment of all mission objectives with no discrepancies in performance of the hardware. The end conditions at parking orbit insertion and translunar. injection were attained with insignificant navigational errors.

The S-IVB/instrument unit lunar impact mission objectives were to impact the stage within 350 kilometers of the target, determine the impact time within 1 second, and determine the impact point within 5 kilometers. The first two objectives were met, but further analysis is required to satisfy the third objective. Based on present analysis, the S-IVB/instrument unit impacted the moon December 10, 1972 at 20:32:40.99 G.m.t. at 4 degrees 12 minutes south latitude and 12 degrees 18 minutes west longitude. This location is 155 kilometers (84 miles) from the target of 7 degrees south latitude and 8 degrees west longitude. The velocity of the S-IVB/instrument unit at impact, relative to the lunar surface, was 8346 ft/sec. The incoming heading angle was 83.0 degrees west of north and the angle relative to the local vertical was 35.0 degrees. The total mass impacting the moon was approximately 13 900 kilograms (approximately 30 700 lb).

15.0 ANOMALY SUMMARY

This section contains a discussion of the significant anomalies that occurred during the Apollo 17 mission. The discussion is divided into five major sections: command and service modules, lunar module, government-furnished equipment, lunar surface experiments, and orbital experiments.

15.1 COMMAND AND SERVICE MODULE ANOMALIES

15.1.1 Spurious Master Alarms

Several spurious master alarms without the accompanying caution and warning lights were reported after earth orbit insertion. The alarms could be initiated by tapping on panel 2, indicating a short circuit to ground in that panel.

The main bus A under-voltage warning light flickered coincident with one alarm. Several times, the alarm occurred coincident with the crew switching of the cryogenic pressure indicator switch from position surge/3 to position 1/2. However, no alarms occurred when the switch was transferred from position 1/2 to position surge/3.

The spurious master alarms are indicative of an intermittent grounding of the circuit which is possible in various locations in the system. An intermittent short-to-ground in the main bus A under-voltage warning circuit (fig. 15-1) would cause the alarm and the main bus A under-voltage warning light to flicker. The short could be in the wiring or could be conductive contamination in the main A reset switch. Postflight inspection of all wiring on panel 2 revealed no problems. Also, the main A reset switch located on panel 3 was disassembled and no contamination was found.

An intermittent ground in either the cryogenic pressure indicator switch, or in the wiring from the switch, or in the cryogenic oxygen pressure meter (fig. 15-2) would cause the alarms that occurred when the switch was placed to position 1/2. The switch and the meter were disassembled, but no contamination or problems were found. The wiring was also inspected for shorts to ground without success.

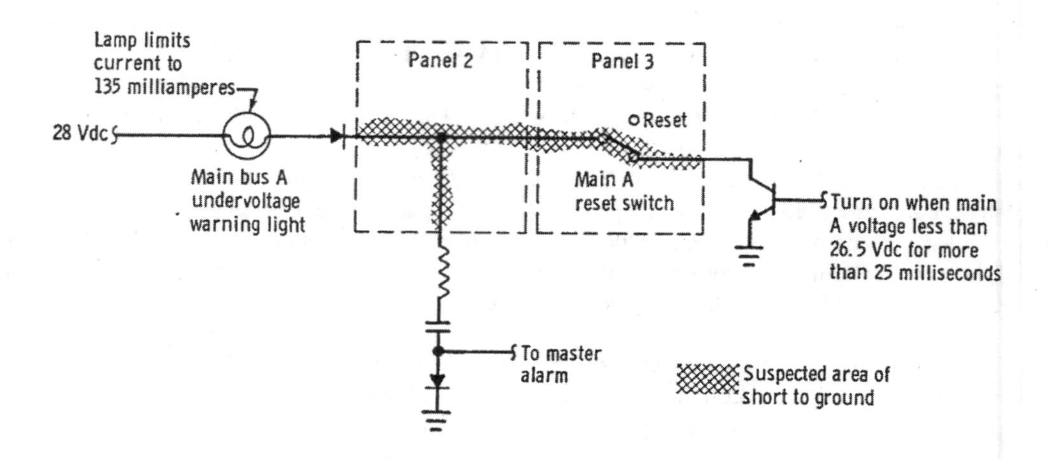

Figure 15-1.- Main bus A undervoltage warning circuit.

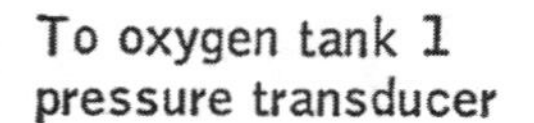

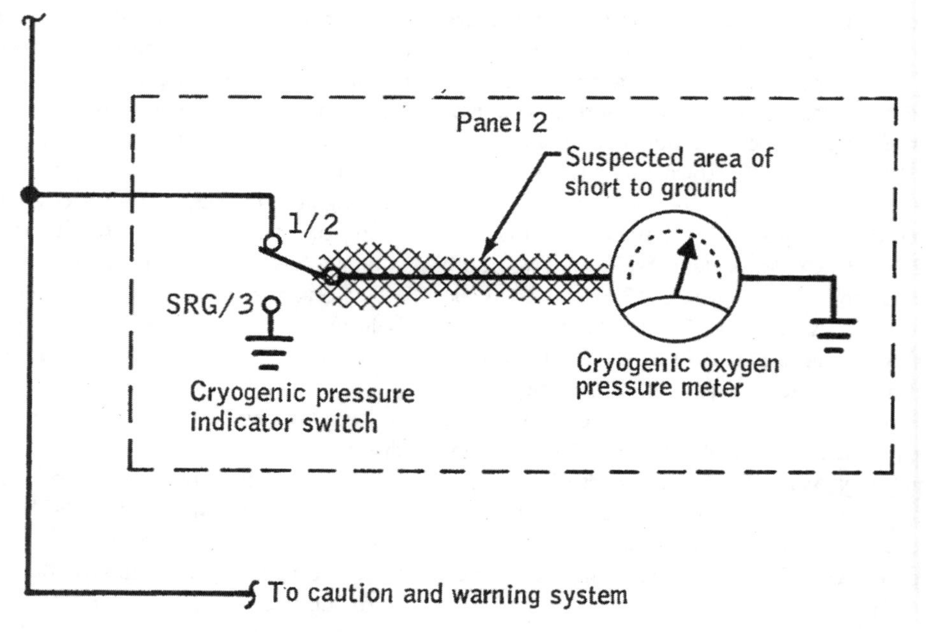

Figure 15-2.- Cryogenic oxygen pressure monitoring circuit.

The alarms must have been caused by some intermittent ground on panel 2 which depended upon zero-g conditions to occur. The spurious warnings were only a nuisance and valid warnings could not be inhibited by this type of grounding problem. There is not a safety hazard associated with grounding the circuit because of the inherent current-limiting characteristics of the lamp (see fig. 15-1). Corrective action is not necessary for the Skylab or Apollo-Soyuz spacecraft.

This anomaly is closed.

15.1.2 Mission Timer Slow

The mission timer in the lower equipment bay was 15 seconds slow at 1 hour and 58 minutes after lift-off. The timer was reset and worked properly for the remainder of the mission.

Postflight testing has been performed on the timer and it operated normally. The circuitry has been analyzed to determine possible causes of the time loss with the result that the most probable cause was an intermittent in one of the integrated circuits within the timer. This condition may have been caused by mechanical imperfections within the construction of the circuit.

There is another mission timer on panel 2 and event timers in the lower equipment bay as well as on panel 1. Consequently, the loss of one timer is not critical for Skylab or the Apollo-Soyuz missions, and corrective action is not necessary.

This anomaly is closed.

15.1.3 Retract Limit Switch On Lunar Sounder HF Antenna Boom 1 Did Not Actuate

The retract limit sensing switch on HF antenna boom 1 failed to operate, resulting in telemetry and command module display data that indicated the boom had not fully retracted.

The boom was one of two HF antennas, each utilizing two nested 0.008-inch by 4.00-inch steel elements to form an extendable /retractable tubular antenna 410 inches in length. Two limit switches were used, one to sense extension, and the other to sense retraction. Each microswitch was operated by a cam which was rotated by a spring-loaded cam follower arm (see, fig. 15-3). One element has a slot at the tip and another near the root. Each slot was aligned with one of the two arms so that at the appropriate, extension length, the follower arm dropped into the slot and rotated the cam to actuate the limit switch.

Current data of the extend/retract motor indicated that the antenna did fully retract; i.e., that the motor stalled after retracting for a nominal period. This verified that the antenna motor/gear box/retractor mechanism was operating normally and confined the problem to the retract limit switch system. Several likely problems with the limit switch system are:

a. Abnormal friction on the limit switch cam follower arm. (Fig. 15-3 shows the torque available in the spring and the torque required to rotate the cam. At the nominal position angle for capture by the element, there is a positive margin of 8 to 1.) b. Defective spring.

c. Misalignment of the hole in the element.

d. Internal limit switch assembly malfunction.

All of these possible problem areas have been investigated without a most-likely cause being established. Mechanical mechanisms requiring multiple operation and that are exposed directly to the space environment have, on several occasions within a relatively short time, exhibited slow operation or have not;operated. Examples are the mapping camera mechanism, the laser altimeter plume protection door, and the high frequency antenna booms. Since this device is not to be flown on future missions, no further action is warranted.

This anomaly is closed.

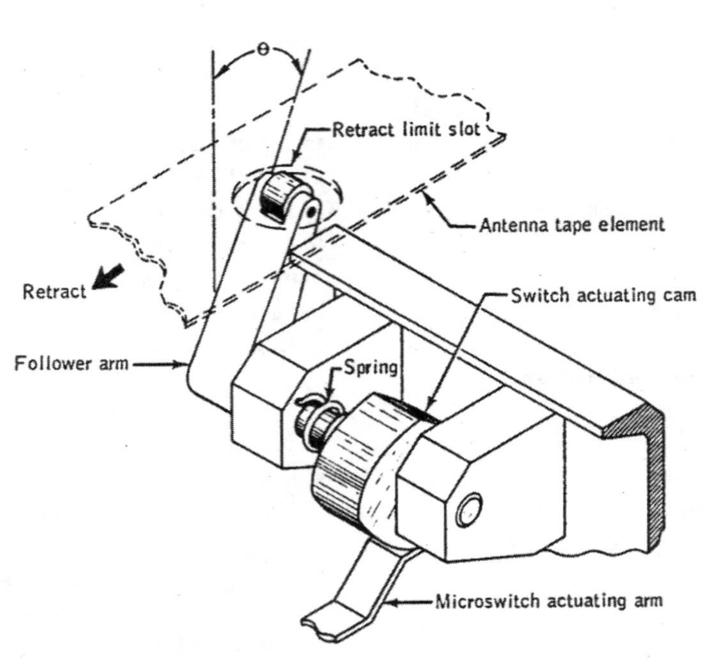

(a) Limit switch actuating mechanism.

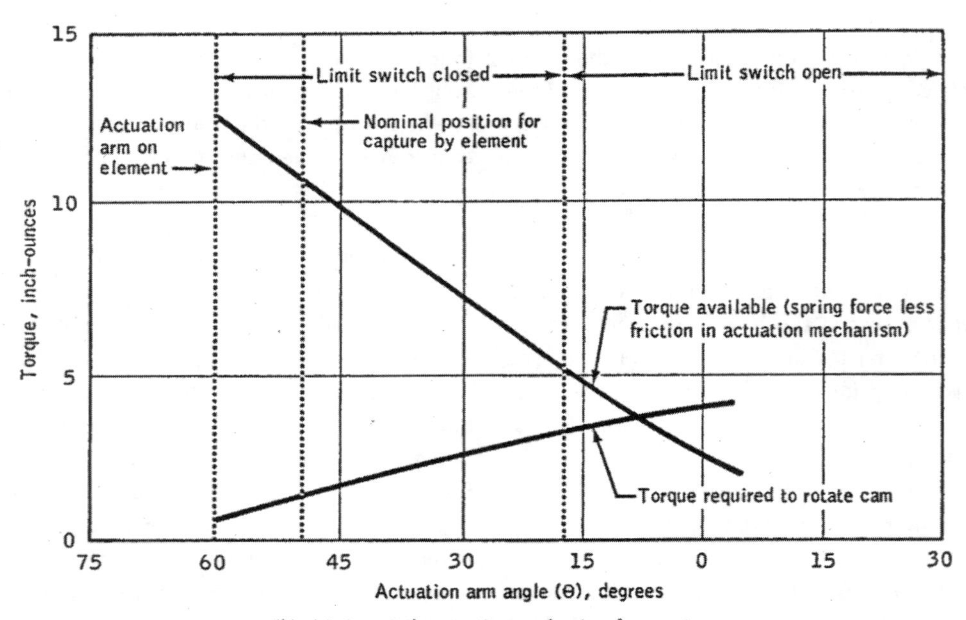

(b) Limit switch actuation mechanism forces at laboratory ambient conditions.

Figure 15-3.- Limit switch.

15.1.4 Lunar Sounder HF Antenna Boom 2 Deployment Was Slower Than Expected

Extension of the lunar sounder HF antenna boom 2 at approximately 191:38 required more time than expected. The motor stalled after approximately one-half deployment. Motion was re-established by cycling the power switch until the unit fully deployed. Subsequent extensions repeated the condition.

The two halves of the "bi-stem" antenna were stored on two spools which were simultaneously driven to extend or retract the elements through the guide/antenna forming assembly (see fig. 15-4). The guide/antenna forming assembly is normally the predominant friction source in the mechanism, and components in the mechanism are believed to be sensitive to a thermal/vacuum environment in an as yet unidentified manner. In any event, current data shows that the load on the motor was greater than expected.

Increased and varying loads in a thermal/vacuum environment are a peculiarity experienced in mechanisms several times on Apollo (refs. 6 and 7). The assembly is not planned to be used for any future mission. Consequently, further analysis on this specific mechanism will not be performed.

This anomaly is closed.

15.1.5 Entry Monitor System Accelerometer Bias Shift

The entry monitor system accelerometer null bias shifted prior to the transearth midcourse correction maneuver. Throughout the mission, the bias was measured to be plus 0.01 ft/sec^2.

Prior to the midcourse correction, the bias was plus 0.66 ft/sec^2 and after the maneuver, it had shifted to plus 0.20 ft/ sec^2.

During step 1 of the system test prior to entry, the 0.05g light illuminated erroneously, indicating an acceleration level greater than 0.05g. A large accelerometer null bias, as indicated in the

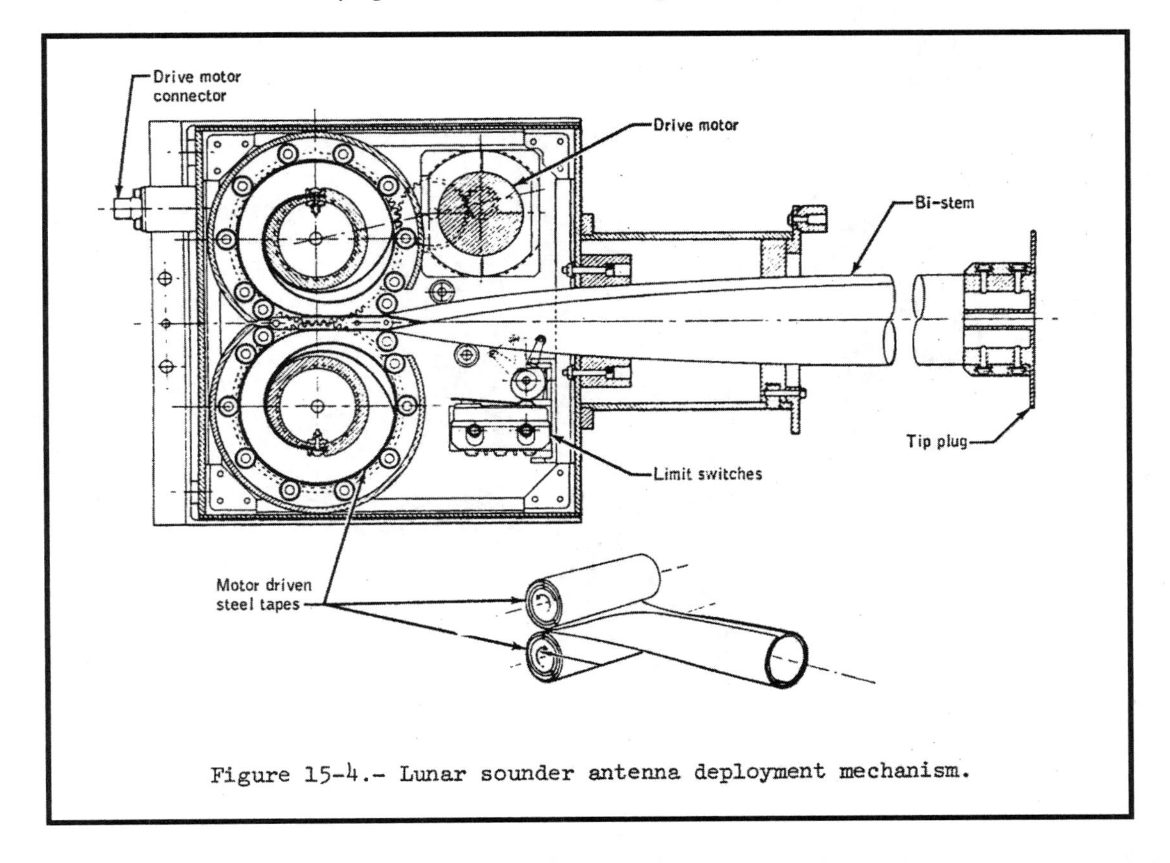

Figure 15-4.- Lunar sounder antenna deployment mechanism.

earlier bias tests, can account for the sensed 0.05-g level. The entry procedures were revised to circumvent the premature 0.05-g indication during entry. Post-flight, the system was removed from the spacecraft and tested. The system parameters and characteristics were normal.

The accelerometer consists of a cylindrical coil bobbin suspended in a magnetic field as shown in figure 15-5. Acceleration forces along the bobbin axis move the bobbin. As the bobbin moves, a ferrite plate attached to the bobbin changes the air gap, and consequently, the inductance of a coil which is part of an oscillator circuit shown in figure 15-6. The resulting oscillator frequency change i s detected and converted to a voltage that is applied to the rebalance coil wound on the bobbin.

Thus, the bobbin motion is stopped and the bobbin is forced to a new balance position. The voltage applied to the rebalance coil is, therefore, proportional to the sensed acceleration.

For a self test of the system, a voltage is applied to a second coil wound on the bobbin and this voltage causes a displacement force on the bobbin by motor action. The rebalance circuit again forces the bobbin to a new null position.

Three possible conditions that could cause a bias shift at one time, and then disappear are schematically shown in figures 15-5 and 15-6.

These possible causes are:

a. A contaminant particle or bubble in the damping fluid in which the accelerometer is immersed. The bubble or particle could migrate into the magnetic gap or become trapped in the balance hair springs shown in figure 15-5. This would cause an offset force on the bobbin. Acceptance tests are designed to detect this condition, but these tests were rerun on the accelerometer during postflight tests and the accelerometer was normal.

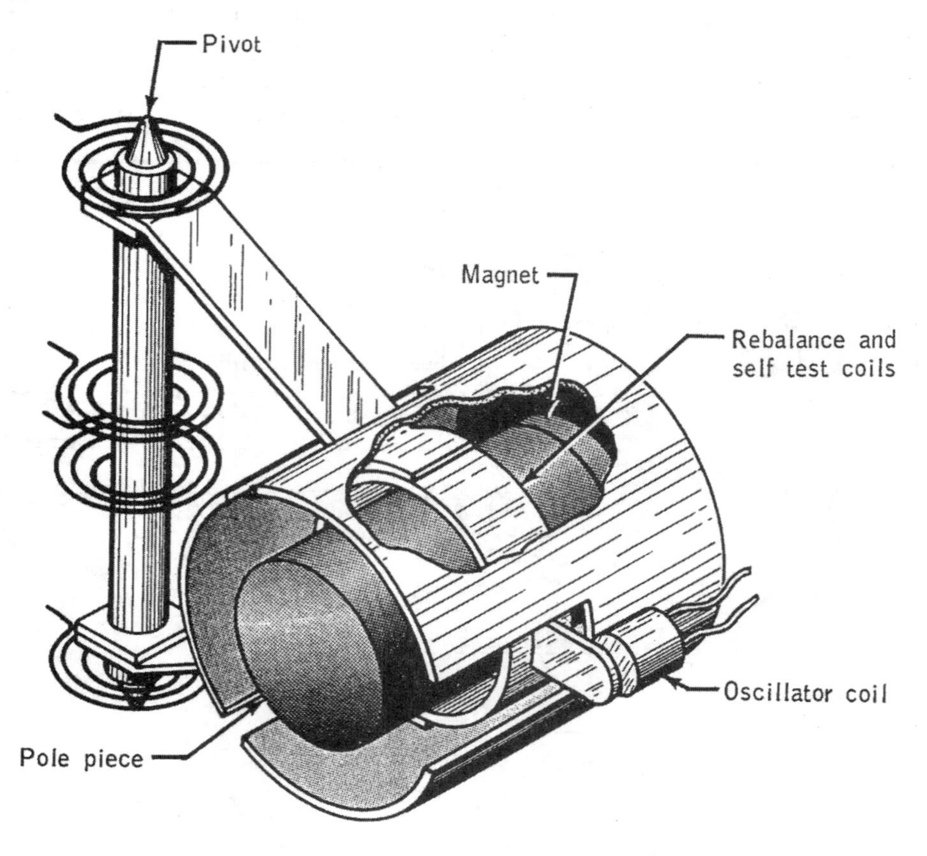

Figure 15-5.- Entry monitor system accelerometer.

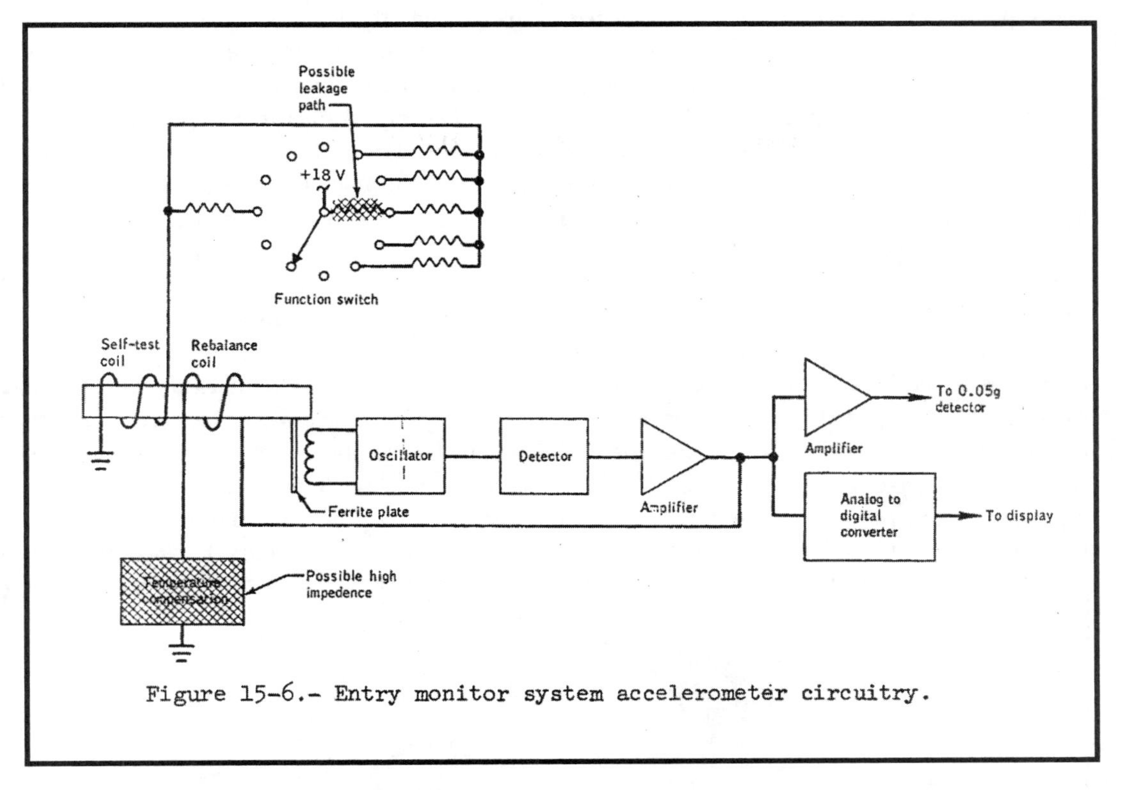

Figure 15-6.- Entry monitor system accelerometer circuitry.

b. An intermittent open in the temperature compensation circuit of the accelerometer rebalance circuit (fig. 15-6) could have existed; however, circuit analysis showed that only half of the bias shift observed in flight would result.

c. An intermittent short between contacts of the 12-position rotary function switch (fig. 15-6) could supply a small current to the accelerometer self-test torquer coil. During postflight testing, however, the switch was normal in all respects.

No further action will be taken since, with a similar problem:

a. The accuracy of the velocity counter will be reduced during service propulsion system maneuvers, but is acceptable.

b. The accuracy of the velocity counter will be reduced to a point where the counter will be meaningless during 2- or 4-thruster plus- or minus-X axis translation.

c. The Skylab entry profile can tolerate the inaccuracy from this condition. Additionally, the entry monitor system is a backup to the primary guidance system.

This anomaly is closed.

15.1.6 Chlorine Ampule Leaked

Leakage occurred in and around the casing assembly of the injector mechanism. Difficulty was also noted occasionally in rotating the knob which advances the injector piston.

Thirteen chlorine ampules and thirteen buffer ampules were used during the mission. Six of the chlorine ampules and five of the buffer ampules had bladder tears around the periphery of the adhesive bond between the bladder and the end cap, or the bladder and the front surface of the ampule bore, as shown in figure 15-7.

The process used to bond the bladder to the front face of the ampule and to the end cap was changed for this mission. During previous missions, the bonds failed in many cases and the end caps came loose. As a result of the change, the adhesive bond no longer fails. However, a high-stress tension-to-shear transition region occurs in the bladder at the periphery of the adhesive bond when the ampule is pressurized while the end cap is free, i.e. the piston is not against the end cap. This occurs because the bladder balloons as shown in figure 15-8. In a test of two ampules that were bonded using the new process, the bladders tore around the periphery of the adhesive bond at 11 and 17 psid, respectively. The tears were similar to those experienced during flight. Two other ampules bonded using the previous process, sustained 27 and 29 psid, respectively, before failure. The end caps came loose before the bladders failed, and the tears that occurred were tension tears.

During injection, the ampule is placed in the injector and the injector piston is advanced until the crewman can feel the piston contact the ampule end cap. The piston knob is then rotated an additional 1/8 turn to compress the bladder. This procedure will in some cases allow a small amount of bladder ballooning to remain. When this occurs, the bladder should tear when the injector assembly is mated to the needle assembly, since the water system supplies a 20 psid back pressure to the ampule.

The injector needle assembly outer needle was bent. As a result, when the injector assembly was mated to the needle assembly, the needle entered the ampule gland off center and tore a hole through the gland septums (fig. 15-9) In several cases, when rotating the knob to advance

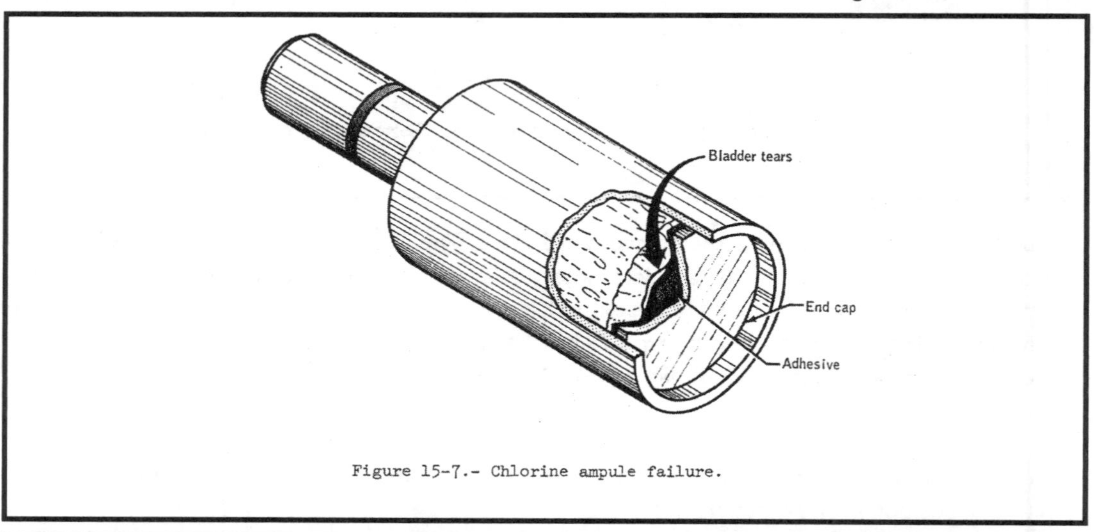

Figure 15-7.- Chlorine ampule failure.

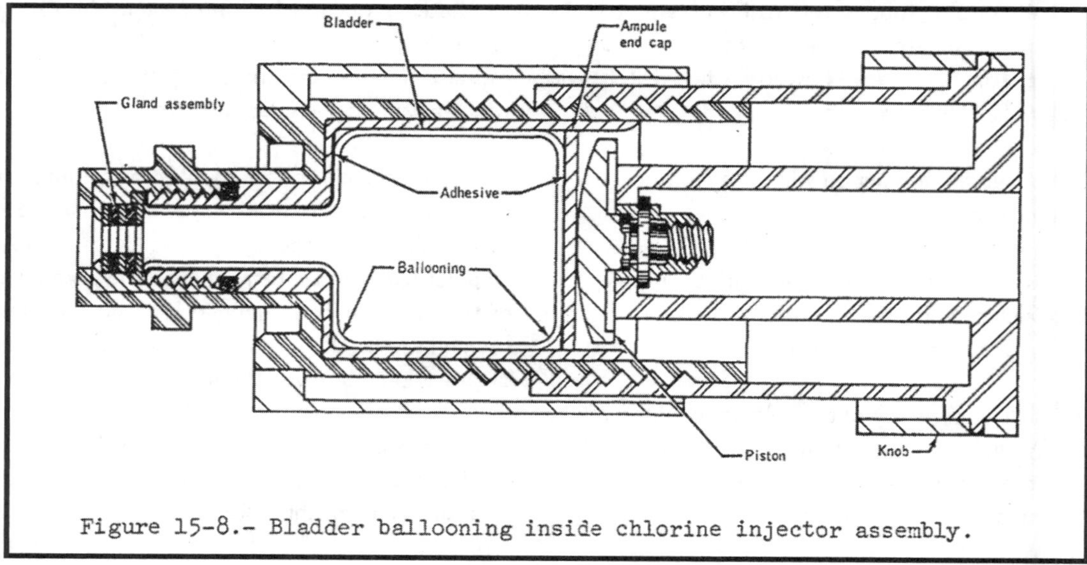

Figure 15-8.- Bladder ballooning inside chlorine injector assembly.

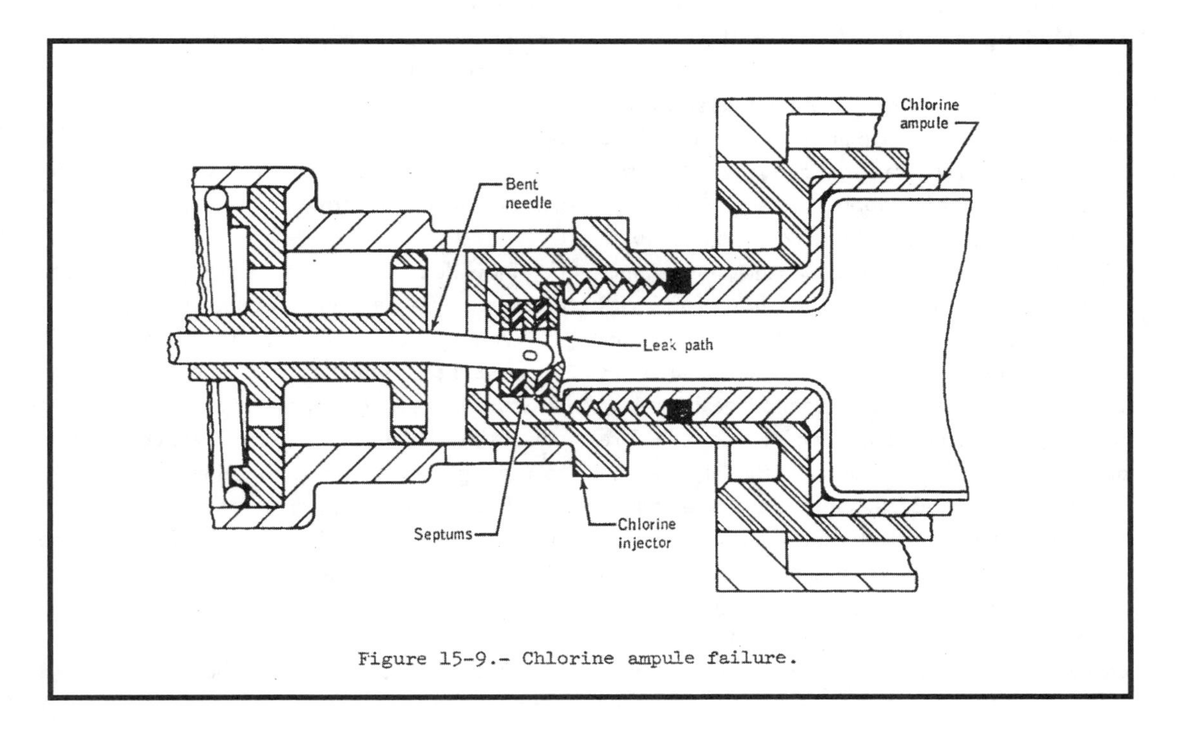

Figure 15-9.- Chlorine ampule failure.

the piston was difficult, the crew demated the injector from the needle assembly, removed the ampule from the injector, and then repeated the complete injection procedure. In one case, this resulted in two needle holes torn through the gland septums. Septum tears, however, could only result in minor leakage.

For the Skylab and Apollo-Soyuz missions, the crew will be instructed to screw the injector piston down far enough to remove all bladder ballooning before connecting the injector to the needle assembly. In addition, the needle assembly will be inspected after the last ground chlorination to assure that the needle is straight. In any event, if ampule leakage recurs, chlorination can still be accomplished since spare ampules are carried aboard the spacecraft.

This anomaly is closed.

15.1.7 Mapping Camera Reaction Control System Plume Shield Door Failed to Close

Photographic data taken from the lunar module during rendezvous shows the mapping camera reaction control system plume shield door in the open position with the mapping camera apparently retracted.

The door is attached to the spacecraft with a hinge that includes a torsion spring to close the door when the camera is retracted (fig. 15-10). The door is pushed open by the camera when it is deployed.The photographs show the camera to be fully retracted which indicates that the problem is with the door itself.

Other mechanical systems anomalies have occurred on Apollo 15, 16 and 17 which suggests a common factor which is as yet unknown. In all cases, some degree of sliding between metal surfaces was required, which is the case with the door hinge. It is therefore conceivable that the friction between these surfaces may have been significantly increased by the effects of the space environment, but these effects have not been duplicated in ground testing.

Since this was the last mission for this device, no further action will be taken.

This anomaly is closed.

15.1.8 Pressure Operating Deadband For Hydrogen Tanks 1 And 2 Shifted

Cryogenic hydrogen system operation was normal until approximately 12 hours. The operating pressure range (deadband) after this time decreased from about 16 psi (238-254 psi) to approximately 3 psi (244-247 psi) by 13 hours (fig. 15-11) and continued in this mode until automatic heater operation for tanks 1 and 2 was terminated at about 15 hours. Automatic heater operation was resumed in tanks 1 and 2 at about 24 hours. Only about five automatic heater cycles (all normal) occurred in tanks 1 and 2 before approximately 71 hours because the fans were in the automatic mode in tank 3, and the tank 3 pressure switch controlled at a higher pressure than tanks 1 and 2. At that time, the tank 3 fans were turned off and the operating pressure deadband again decayed to 3 psi (248-251 psi) as shown in figure 15-11. This deadband was maintained until automatic heater operation was terminated in tanks 1 and 2 at 87 hours. Hydrogen system pressure was maintained for the remainder ofthe mission by use of manual fan operation in tanks 1 and 2 and automatic fan operation in tank 3. Hydrogen system operation was normal in this mode.

Individual pressure switches monitor pressures in tanks 1 and 2 as shown in figure 15-12. When automatic heater or fan operation is selected, both switches must be in the low-pressure position to apply power to the heaters or fans in each tank. When either switch transfers to the high pressure position, the motor switch transfers and removes power from the heaters and fans. Figure 15-13 is a cutaway sketch of one of the pressure switches in the low-pressure position.

Decreasing or restricting the full downward

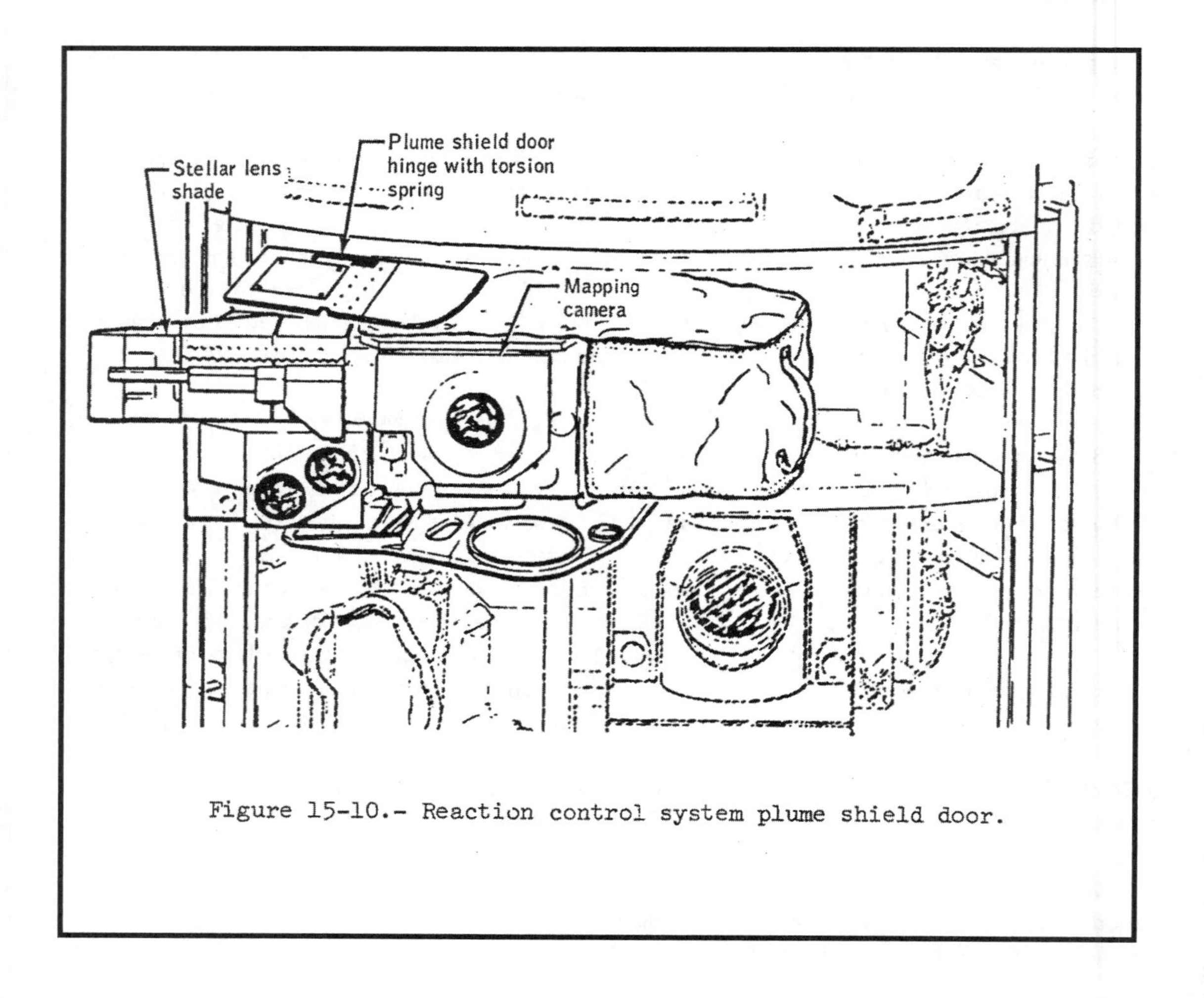

Figure 15-10.- Reaction control system plume shield door.

travel of the stop will increase the pressure at which the shaft transfers to the upper position. Decreasing or restricting the full upward travel of the stop will decrease the pressure at which the shaft transfers to the lower position. Consequently, hard particles floating in the stop mechanism area shown in figure 15-13 can have a variable effect which can appear and disappear depending on the location of the floating particles.

The size of the particles required to change the pressure settings is relatively small. The full travel of the stop may be as small as 0.007 inch. The pressure change to move from one stop to the other is about 15 psi. Consequently, the upper limit can be reduced to about 7 psi with a piece of metal about 0.004-inch in size between the stop on the shaft and the upper case stop.

The toggle plates are pin-joined to the shaft at one end and to a horseshoe-shaped spring at the other end as shown in figure 15-13. Grooves are formed in the opposite edges of the toggle plates by a coining operation. Figure 15-14 shows the toggle plate taken from a switch which was disassembled for inspection. In this instance, burr's were found that were about to break off. The particles could then move around and become trapped in the stops under zero-gravity conditions.

Other possibilities were investigated, however, only contamination satisfies all of the peculiar flight characteristics and data.

Further inspection of assembled pressure. switches is not necessary independent of the practicality of performing such an inspection, because for Skylab and Apollo-Soyuz Test. Program a problem of this nature can be handled by manual control of the fans and heater cycles. This anomaly is closed.

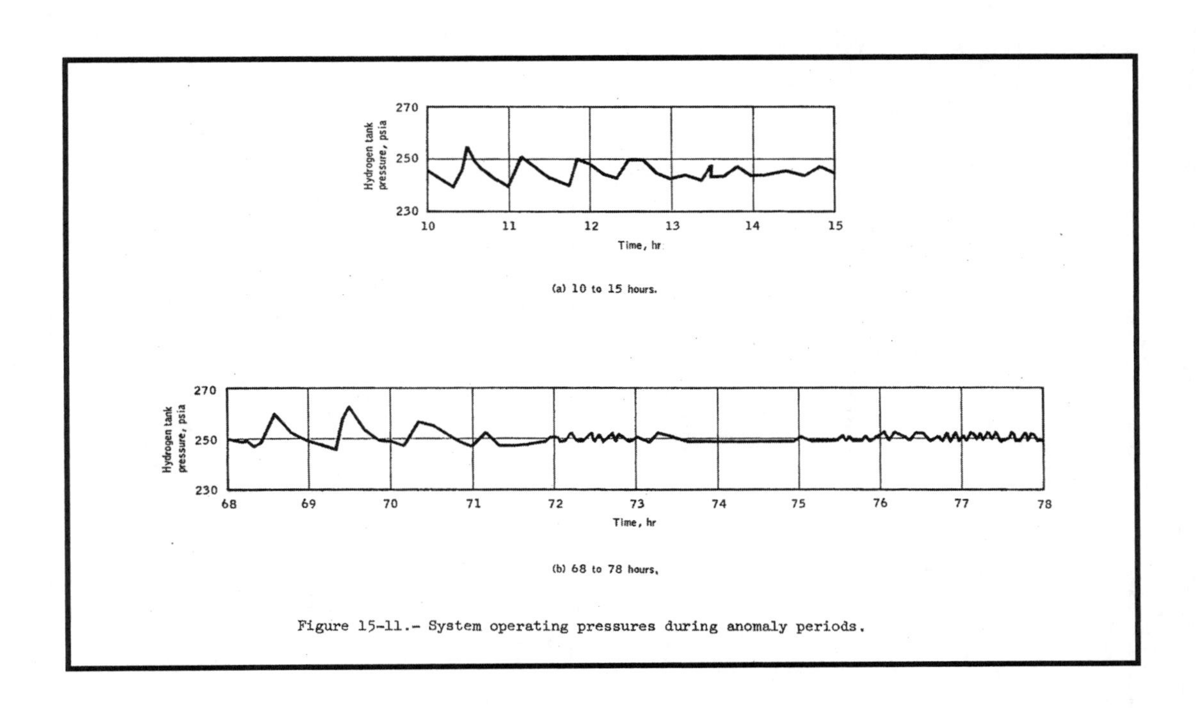

Figure 15-11.- System operating pressures during anomaly periods.

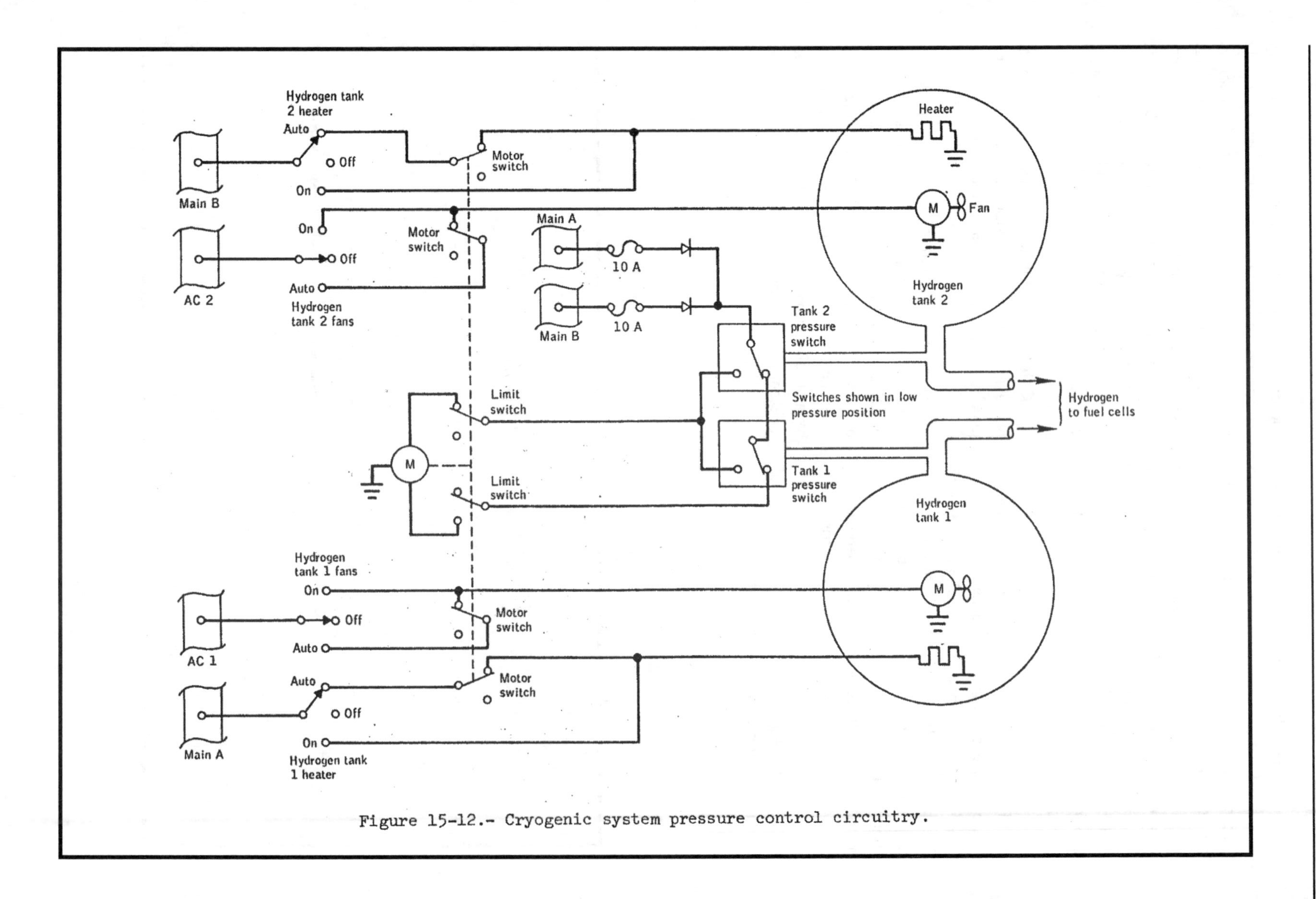

Figure 15-12.- Cryogenic system pressure control circuitry.

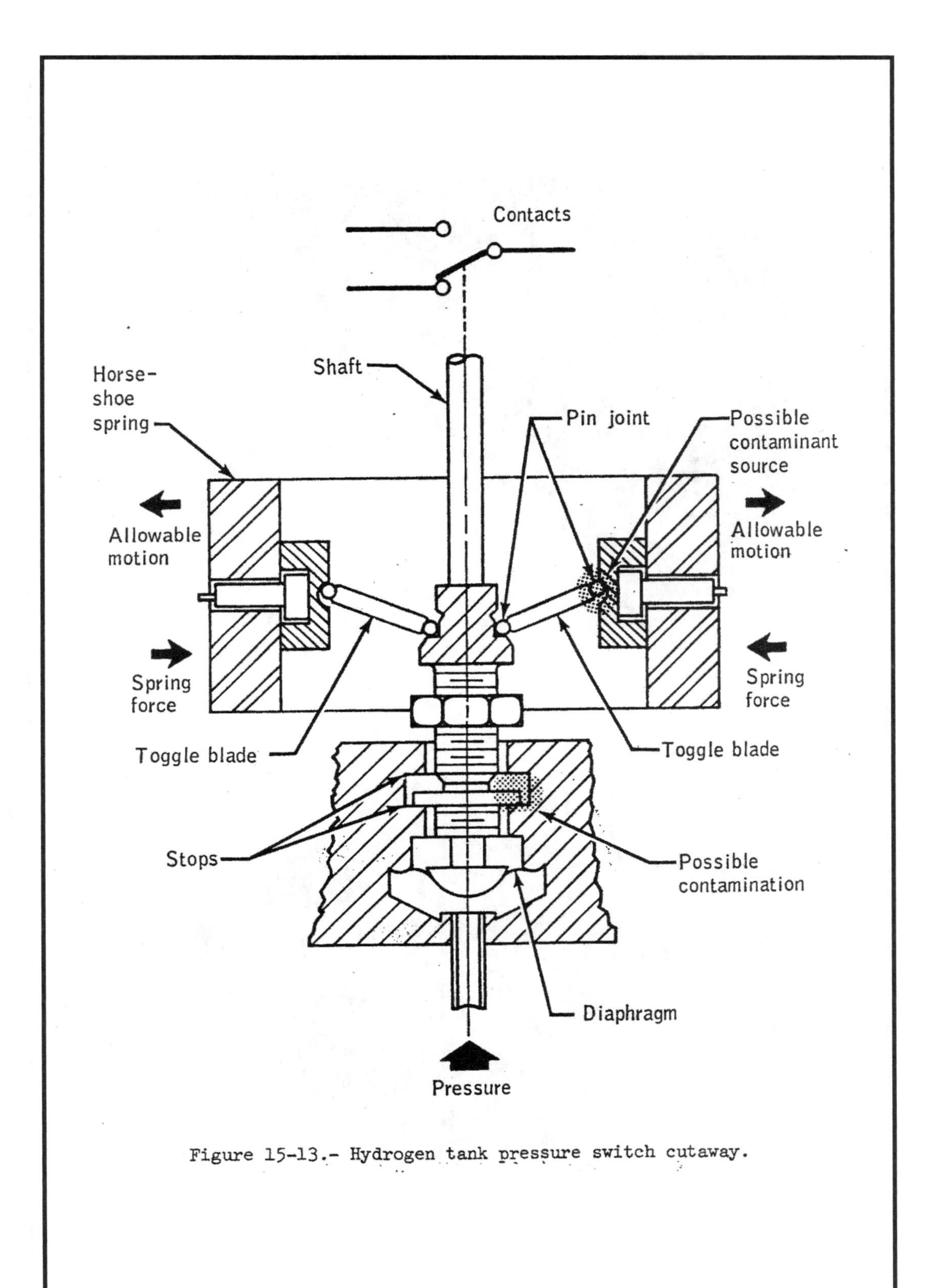

Figure 15-13.- Hydrogen tank pressure switch cutaway.

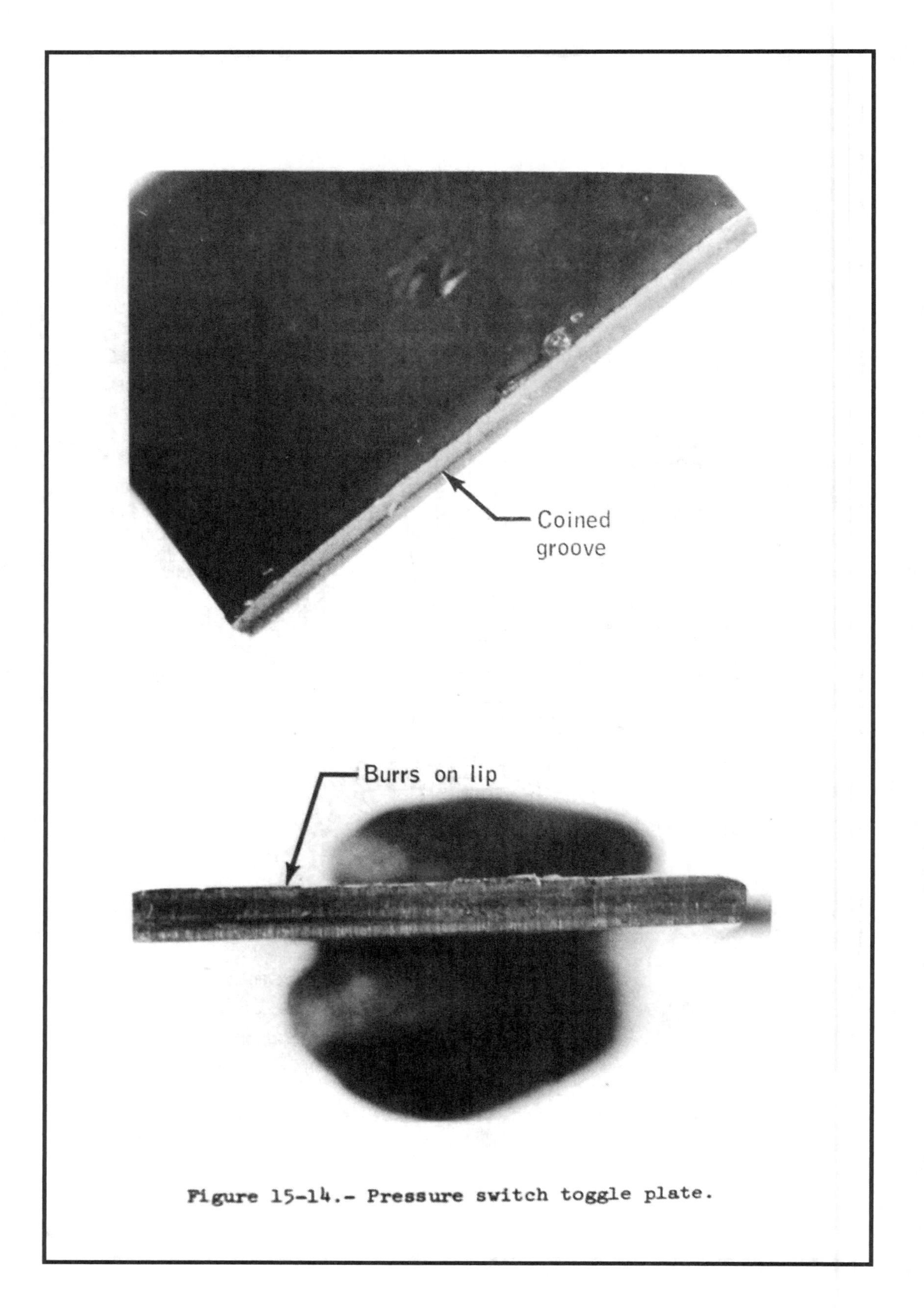

Figure 15-14.- Pressure switch toggle plate.

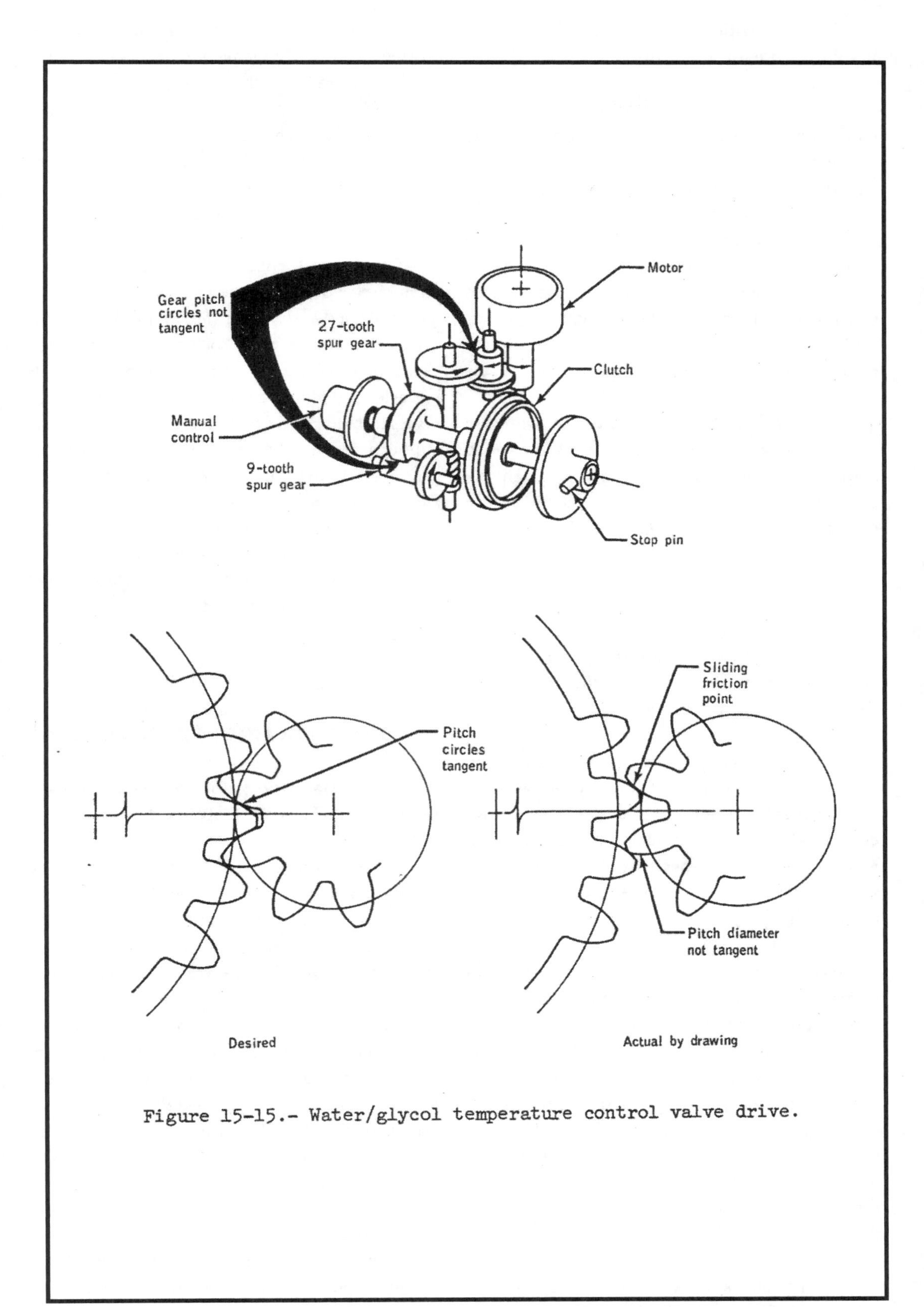

Figure 15-15.- Water/glycol temperature control valve drive.

15.1.9 Temporary Dropout Of Several Instrumentation Parameters

Erroneous readings occurred on several command and service module measurements between 191:40:39 and 191:42:32. Data received at two different ground stations indicated the problem existed on the downlink signal. The measurements affected were limited to the one sample per second channels. Postflight tests were performed on the spacecraft, and also on the bench where the pulse code modulation unit was subjected to shock and vibration. The system operated normally during both tests.

Data shows the problem to be with the timing pulse network which controls the word matrices. Consideration is being given to providing a workaround in the event a similar condition occurs on Skylab or Apollo-Soyuz missions.

This anomaly is open.

15.1.10 Water/Gylcol Temperature Control Valve Failed To Maintain The Evaporator Temperature

During four early lunar orbit revolutions, the water/glycol temperature control valve failed to open properly as the radiator outlet temperature decreased. The mixed coolant temperature momentarily fell as much as 4° F below the specification control band of 42° to 48° F during mixing startup.

No corrective action was taken and initiation of mixing was proper during all subsequent lunar orbits and during transearth coast. The system behavior indicates that some slight sticking occurred in the valve gear train after having been driven hard against the full closed stop during the hot portions of the lunar orbits. A larger than normal temperature error and modulated electrical signal to the valve were required to free the gear train so that the valve would start to drive open.

During checkout on previous spacecraft, these valves have occasionally stuck when driven hard against either the full open or full closed stop. Manual disengagement of the valve clutch mechanism relieves the back pressure on the gears and frees the gear train. This occurred during checkout on the Apollo 17 valve and was accepted for flight because of the manual operation capability and demonstration of manual operation during the Apollo 16 mission.

Dimensional checks have shown that the pitch circles of two of the gear pairs in the valve gear train are not tangent (fig. 15-15). This results in sliding rather than rolling action between contacting teeth. One valve which previously experienced the same sticking has been disassembled and the gears were normal, although the dry lubricant was scraped off the gear teeth in the area of tooth-to-tooth contact. Sliding friction between the unlubricated gear teeth may have been high enough to cause the sticking experienced. Since manual valve operation is satisfactory for the Skylab and Apollo-Soyuz missions, no further action is necessary.

This anomaly is closed.

15.2 LUNAR MODULE ANOMALIES

15.2.1 Battery 4 Voltage Reading Was Lower Than That Of Battery 3

At. approximately 108:06, the lunar module battery 3 and battery 4 voltage data indicated a voltage difference of approximately 0.5 volt, with battery 3 voltage greater than that of battery 4. Previously, battery 3 and 4 voltages were equal.

Batteries 3 and 4 are parallel and should reflect equal voltages. Reverse current from battery 3 to battery 4 would have been indicated by the reverse current triggering the battery malfunction light. This alarm is triggered by reverse current of more than 10 amperes for 4 to 6 seconds which was exceeded by battery 4 current readouts. The battery current measurement does not indicate current direction. Since no alarm occurred, battery 4 was delivering positive current to the bus, indicating that the battery 4 voltage readout was not a true reading. This 0.5-volt difference also appeared on the battery open circuit voltages.

The battery 4 voltage measurement is conditioned by a signal conditioner which consists of an input resistive divider network feeding a dc-to-dc converter (fig. 15-16). An increase of resistance of 1 percent in this network would account for the observed change. Another possibility of failure exists in the zero-adjust network. In any event, no further action is required because this circuitry will not be flown on any planned future mission.

This anomaly is closed.

15.2.2 Oxygen Demand Regulation Leakage

While using demand regulator A for cabin depressurization in preparation for the third extravehicular activity, the data show the suit loop to be leaking at a 0.04 lb/hr rate. The crew switched to regulator B for the remainder of the mission. The suit loop leakage could have been caused by either contamination of the regulator A seat or the crew could have inadvertently bumped the regulator handle during cabin activities, although the latter is not considered likely. Specificially, the leakage could have occurred if a particle of 2.5 microns became lodged between the regulator (ball) poppet and the seat. The upstream contamination level is 25 microns, which was considered adequate due to the low probability of particle entrapment in a ball-type seat. One previous instance of regulator leakage caused by a particle trapped in the seat. In that case, the particle was built in to the regulator during manufacture since it was too large (0.04-inch) to pass through the filters.

Corrective action will not be required as this equipment is not to be used on future spacecraft.

This anomaly is closed.

15.3 GOVERNMENT FURNISHED EQUIPMENT ANOMALIES

15.3.1 No Extravehicular Activity Warning Tone In Command Module Pilot's Communications Carrier

At 253:44, during preparation and checkout for the transearth coast extravehicular activity,

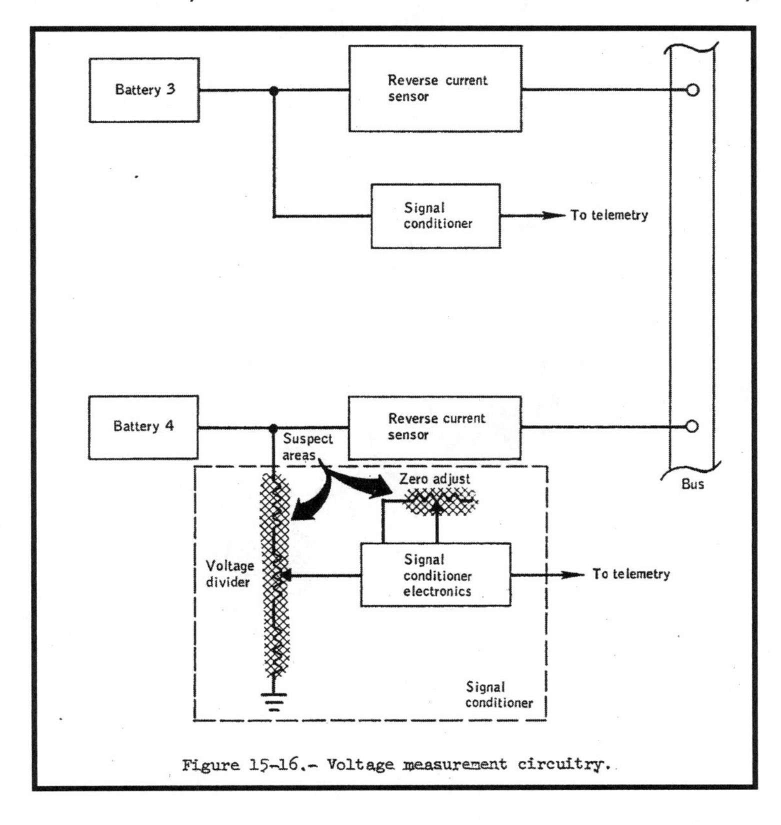

Figure 15-16.- Voltage measurement circuitry.

the Command Module Pilot could not hear the caution and warning tone in his communications carrier, although his transmission and reception were apparently normal. For the extravehicular activity, the Command Module Pilot used the Lunar Module Pilot's communications carrier which operated normally, including the caution and warning tones.

After the extravehicular activity, the Command Module Pilot partially removed the sleeve covering his communications carrier electrical pigtail and found that of the nine leads (twisted shielded pairs) making up the pigtails, two were broken (fig. 15-17).

Postflight discussions with the Command Module Pilot indicated that the pigtail lead became twisted and wound up to the point of knotting due to his twisting and turning while performing his normal cabin functions. Additionally, the Command Module Pilot did not wear the inflight coverall garment and the constant wear garment electrical harness was not restrained. In this configuration, twisting of the leads is possible and could lead to the failure observed. The communication carrier was designed for use primarily in the pressure garment where space restrictions make the small flat cable desirable and where ample support for the wire is provided. When used with the coverall or constant wear garment, it must be anchored to the garment to provide strain relief and control against twisting.

Failure analysis of the returned flight item has shown that the failed wires were broken about 1/4 inch below their exit point from the potting of the flex relief attached to the lower end of the communications carrier splice block. The break occurred where these wires lay across an adjacent set of wires in a braided configuration. The functions carried by these wires were caution/warning tone and left earphone circuits.

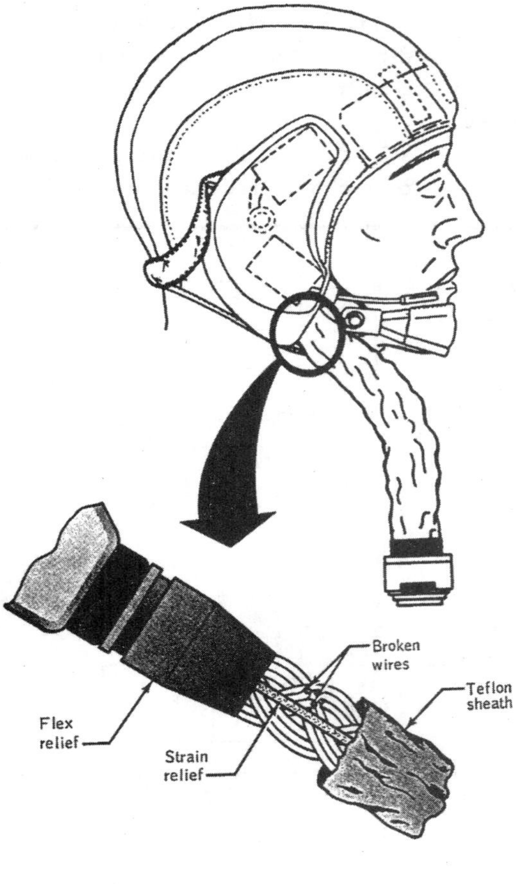

Repeated twisting and knotting of similar wire bundles to the extent described by the crewman has demonstrated that the point of breakage was inherently the most highly loaded due to the geometry of the pigtail, and thereby was subject to flex/tensile failure.

Skylab training and flight procedures are being revised to reflect a harness configuration that prevents the twisting of the wire harness. For normal Skylab operations, the training and flight procedures designate the lightweight headset, which is designed for this type of use. For occasions when it is necessary to use the communications carrier,

Figure 15-17.- Communications carrier wire failures.

when not wearing the pressure suit, both the carrier harness and crew garment have additional provisions for attachment.

This anomaly is closed.

15.4 LUNAR SURFACE EXPERIMENT EQUIPMENT ANOMALIES

15.4.1 Lunar Surface Gravimeter Sensor Beam Cannot Be Stabilized In the Null Position

Centering the sensor beam capacitor plate in the proper stable position between the fixed capacitor plates (fig. 15-18) could not be accomplished following the initial experiment turn-on. When the command was given to add any or all of the masses to the sensor beam assembly, the data indicated that the beam would not move away from the upper capacitor plate. The only way to bring the beam down was by caging it against the lower capacitor plate.

For normal operation, the sensor beam must be horizontal, with the capacitor end-plate centered between the two fixed capacitor plates. Adjustment of the suspension point of the sensor spring and the addition and removal of individual masses are provided to assist in centering the beam in its reference position, equidistant between the two capacitor plates. Small changes in the vertical component of local gravity tend to displace the beam upward or downward.

The displacement is sensed as a capacitance change between the center or beam plate and each of the two fixed outside plates. A voltage proportional to this displacement is generated by the capacitance change and integrated to supply a feedback voltage which electrostatically forces the beam plate back to the center of the gap.

Review of sensor records revealed that a mathematical error resulted in the sensor mass weights being about 2 percent lighter than the proper nominal weight for 1/6-g operation of the flight unit. The sensor mechanism allows up to only ±1.5 percent adjustment from the nominal by ground command for possible inaccuracies. The error was made in the conversion calculations from 1-g to 1/6-g mass for the flight unit by including an erroneous value in the calculations from the uncorrected calculations for the qualification unit.

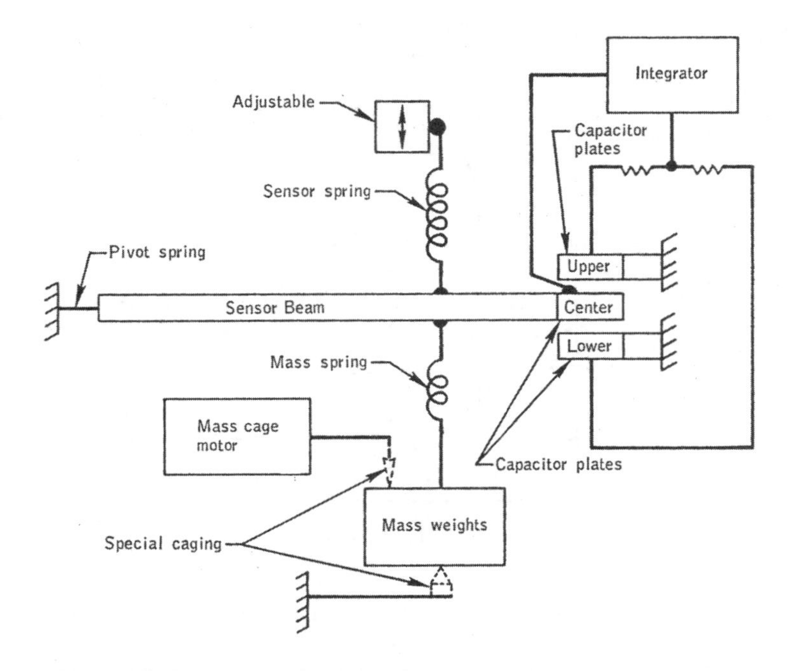

Figure 15-18.- Lunar surface gravimeter sensor beam centering.

Because the mass weights do not provide enough downward force, the beam has been balanced and centered by applying a load on the beam through the mass support springs. This is accomplished by special caging of the mass weight assembly with ground commands to the mass cage motor as shown in figure 15-18. In this configuration, the mass spring permits balance of the beam and function of the instrument. However, the spring constant of the mass spring considerably reduces the sensitivity of the system. This anomaly is closed.

15.4.2 Surface Electrical Properties Receiver Temperature Higher Than Predicted

The receiver temperature was about 5° F less than normal at the end of the first extravehicular activity as shown in figure 15-19. However, during the rest period between the first and second extravehicular activities, the temperature rose to 80° F instead of dropping to about 28° F as predicted. Between the second and third extravehicular activities, the temperature dropped about 8° F instead of the expected drop of about 50° to 60° F.

The receiver was protected by a multilayered aluminized Kapton thermal bag (fig. 15-20). The thermal bag had two flaps which protected the optical solar reflector (mirror) on top of the receiver from lunar dust accumulation. A dust film of about 10 percent on the mirror surface could result in the indicated degradation of thermal control and a filmof this amount may not be apparent to the crew.

Folding back one, or both, flaps during rest periods was to result in cooling of the receiver by radiation of heat energy to deep space.

With normal system efficiency, and the experiment turned off , opening the tab A cover (fig. 15-20) at the end of the first extravehicular activity should have resulted in the predicted temperature drop to about minus 14° F by the start of the second extravehicular activity. Opening both the A and B flaps was provided for contingencies requiring more rapid cooling. This procedure was used throughout the remainder of the mission when the lunar roving vehicle was not in motion.

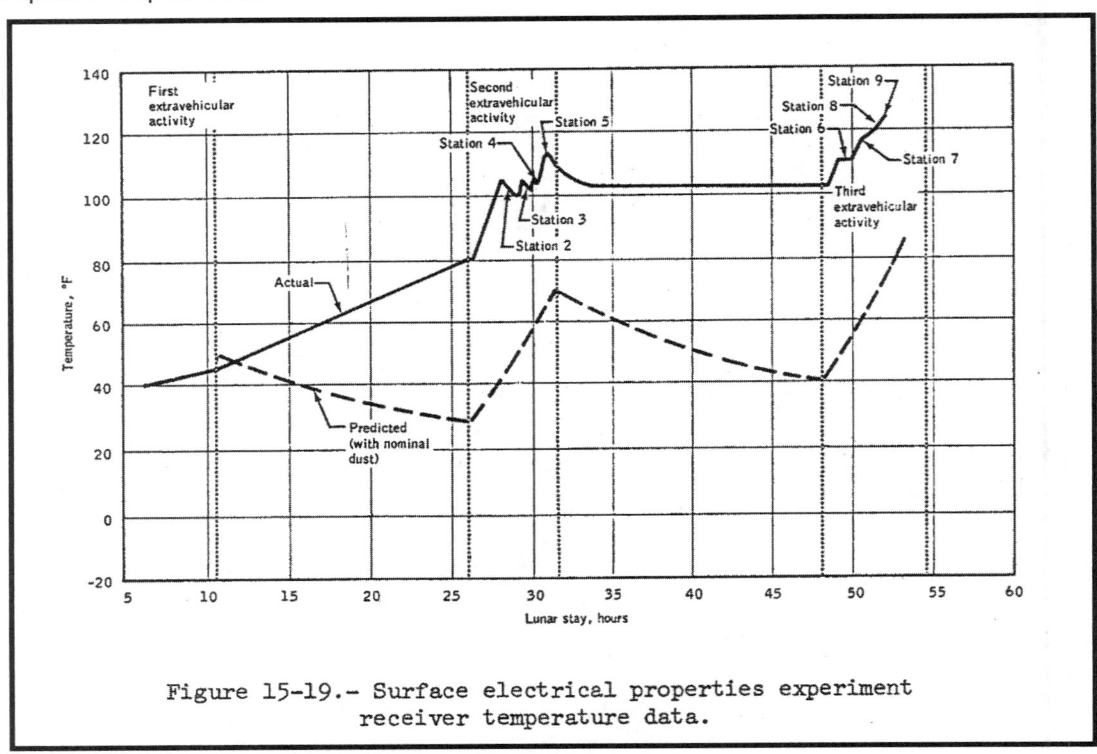

Figure 15-19.- Surface electrical properties experiment receiver temperature data.

Cover design depended upon Velcro straps and pads to hold the Kapton flaps tightly closed to keep out dust and sunlight (fig. 15-20). The Velcro pile pad was bonded to the Kapton bag and the Velcro hook strap was bonded to the Kapton flaps. The bond of the Velcro pads for both flaps had already failed before the Lunar Module Pilot configured the receiver at the end of the first extravehicular activity, thus resulting in dust accumulation on the mirror surface under both flaps. The bond of the Velcro pads to the Kapton failed, leaving no trace of the adhesive on the Kapton, and the pads remained attached to the straps. The polyurethane FR-127 A and B bonding material used was acceptable and recommended for bonding Velcro to Kapton. The failure most likely resulted from a weak bond caused by improper bonding preparation or procedure. The mixing and timing of the bonding application and mating are critical, as well as maintaining the surface free of contamination.

This experiment is not scheduled for a future mission; however, similar bonding configurations will require stringent quality control of the bonding process.

This anomaly is closed.

15.4.3 Lunar Ejecta And Meteorite Experiment Temperature Figh

The temperature of the lunar ejecta and meteorite experiment was higher than predicted during the first and second lunar days (fig. 15-21). The high temperatures occurred with all combinations of experiment modes: on, off, and standby, with all dust covers on, with only the sensor covers on, and with all covers off. Whenever the experiment was in the "operateon" mode, the science data indicated normal operation of the experiment. The maximum allowable temperature for survival of the electronic components has not been exceeded, however, it was necessary to command the experiment from "operate-on" to "off" at a sun angle of about 153 degrees during the first lunar day and at a sun angle of about 16 degrees during the second lunar day. Following sunrise of the second lunar day, the temperature rose from 0° F at 0° sun angle to about 168° F at 15° sun angle (fig. 15-21). The instrument was commanded to standby and then to off because the temperature continued to rise. In the off mode, with no power to the instrument, the temperature rise rate was lower.

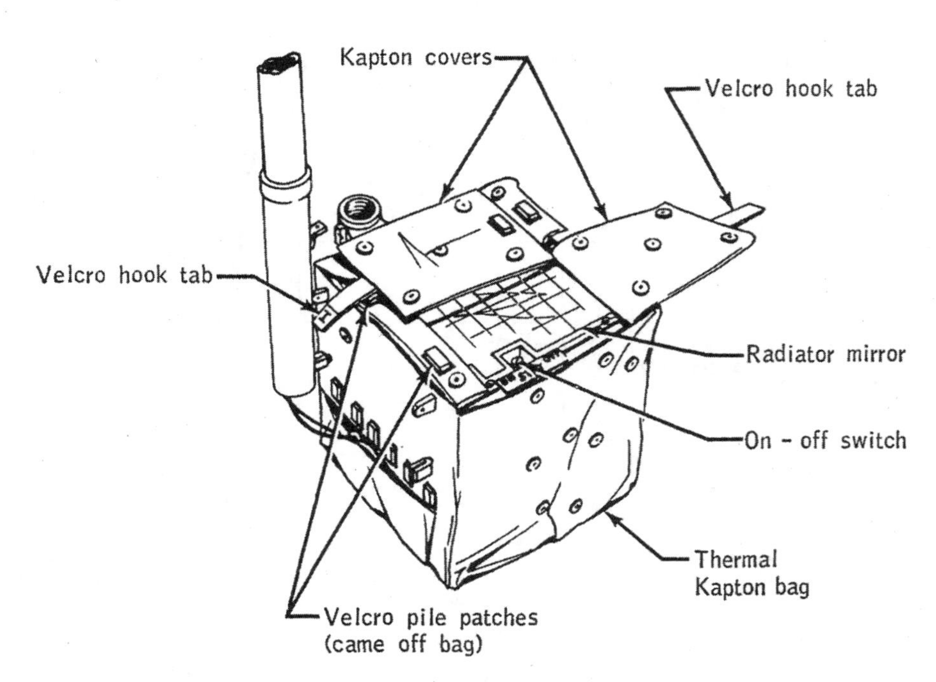

Figure 15-20.- Surface electrical properties experiment receiver.

The experiment temperature was cooler during the morning of the third lunar day as compared to the second. This could be attributed to the procedural change which turned the experiment off for 1 1/2 hours through sunrise and sunset. Data from the suprathermal ion detector and charged particle lunar environment experiments, deployed on previous Apollo missions, indicate that a flux of -100 to -750 volts can occur near the optical terminator (before optical sunrise and after optical sunset). During the lunar day, the surface is stable with photo electron layering at +10 to +20 volts. It is postulated that when the experiment is on (sensor film at -3 volts and suppressor grid at -7 volts), the charge differential observed at these times may result in an accretion of lunar dust on the east and west sensors. Based on this, the experiment was turned off each sunset and sunrise after the second lunar day. The presence of dust on the sensor film and grid would degrade the thermal control system and result in higher experiment temperatures during the lunar day.

The current thermal profile permits experiment operation for 100-percent of the nighttime and 30 percent during the lunar day. If the thermal profile does not improve, consideration will be given to thermal testing on the qualification unit to ascertain whether or not the temperature limits can be raised to permit additional daytime operation.

Preliminary results from an examination of the science data indicate that the instrument is operating properly.

Since the experiment is not scheduled for future missions, no corrective actions will be taken.

This anomaly is closed.

15.4.4 Cask Dome Removal Was Difficult

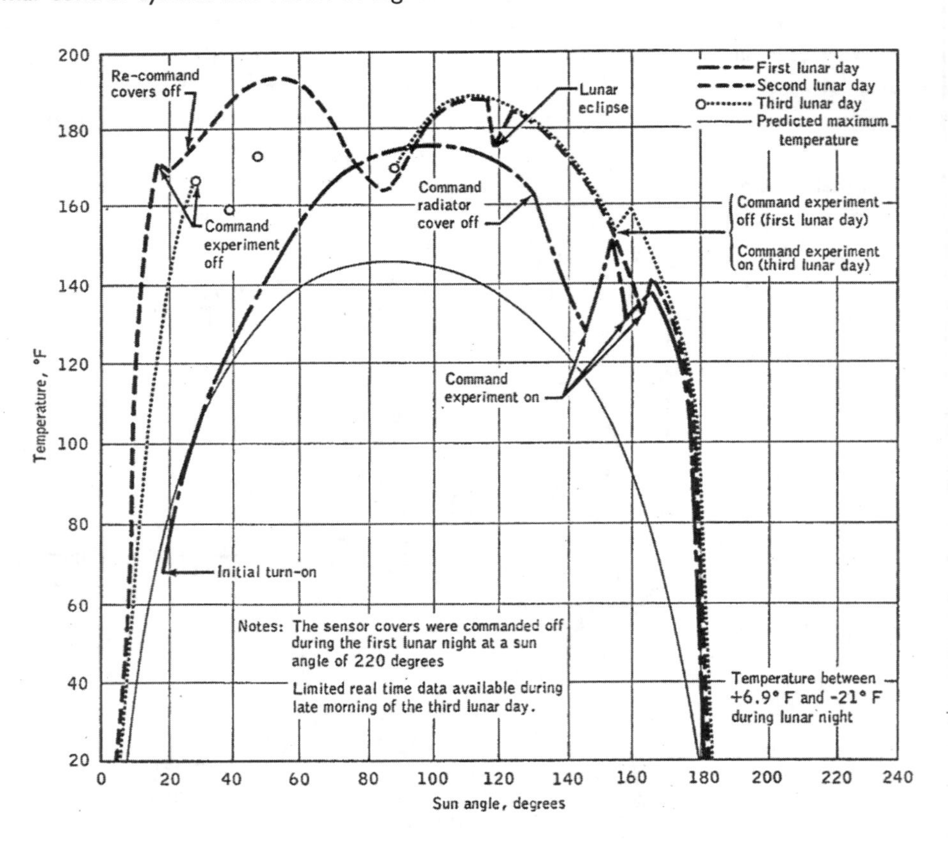

Figure 15-21.- Lunar ejecta and meteorites experiment temperature data.

The Lunar Module Pilot was not able to remove the cask dome with the removal tool.

The socket on the removal tool can engage the nut on the dome before the pins on the tool lock into the recess in the dome (fig. 15-22). The Lunar Module Pilot did not verify that the pins were locked. In this configuration, rotating the tool clockwise will rotate the nut on the dome. A 90-degree rotation of the nut releases the dome retaining straps, as noted by the crew. This release allows the dome to rotate when the tool is rotated another 60 degrees, thus disengaging the threaded dome/ cask interface. However, with the pins not locked into the dome recess, the dome could be cocked, but not withdrawn. The dome was easily wedged off the cask with the hammer. The sequence can be duplicated with either broken pins or by incomplete insertion and locking of the tool pins.

No further investigation will be performed since the cask will not be flown on future missions.

This anomaly is closed.

15.4.5 Background Noise in the Lunar Atmospheric Composition Experiment Data

A zero offset was noted in part of the lunar atmospheric composition experiment data on the mid-mass and low mass channels (fig. 15-23), and occasionally on the high mass channel. This offset was the result of background noise in the detector system. The condition is stable and has caused no loss of data. However, it will require additional processing during data reduction.

Analysis of the data and the sensor circuit indicates that the offset is'the result of electronic noise coupling between the unshielded high voltage wires (fig. 15-24) and the unshielded collectors in the sensor package. -Commands to either of the sensor ion sources or the sensor electronic multipliers does not affect a change in the offset; however, the presence of the offset is affected by the voltage level to the sensors (fig. 15-23). Since the data are usable and the offset is a stable condition, no corrective action i s necessary. This anomaly is closed.

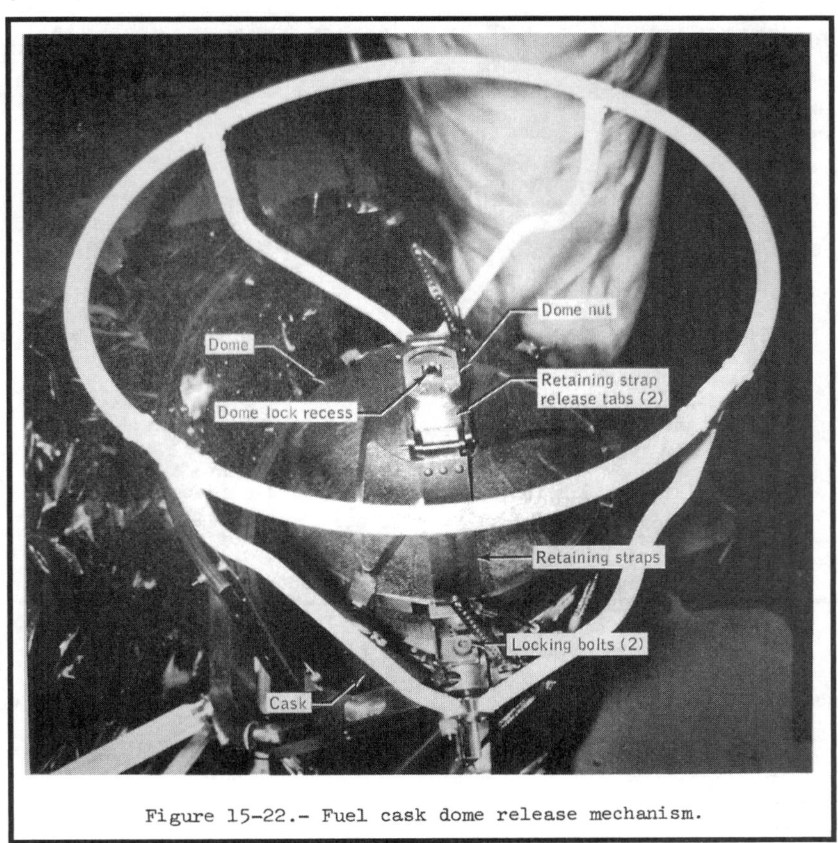

Figure 15-22.- Fuel cask dome release mechanism.

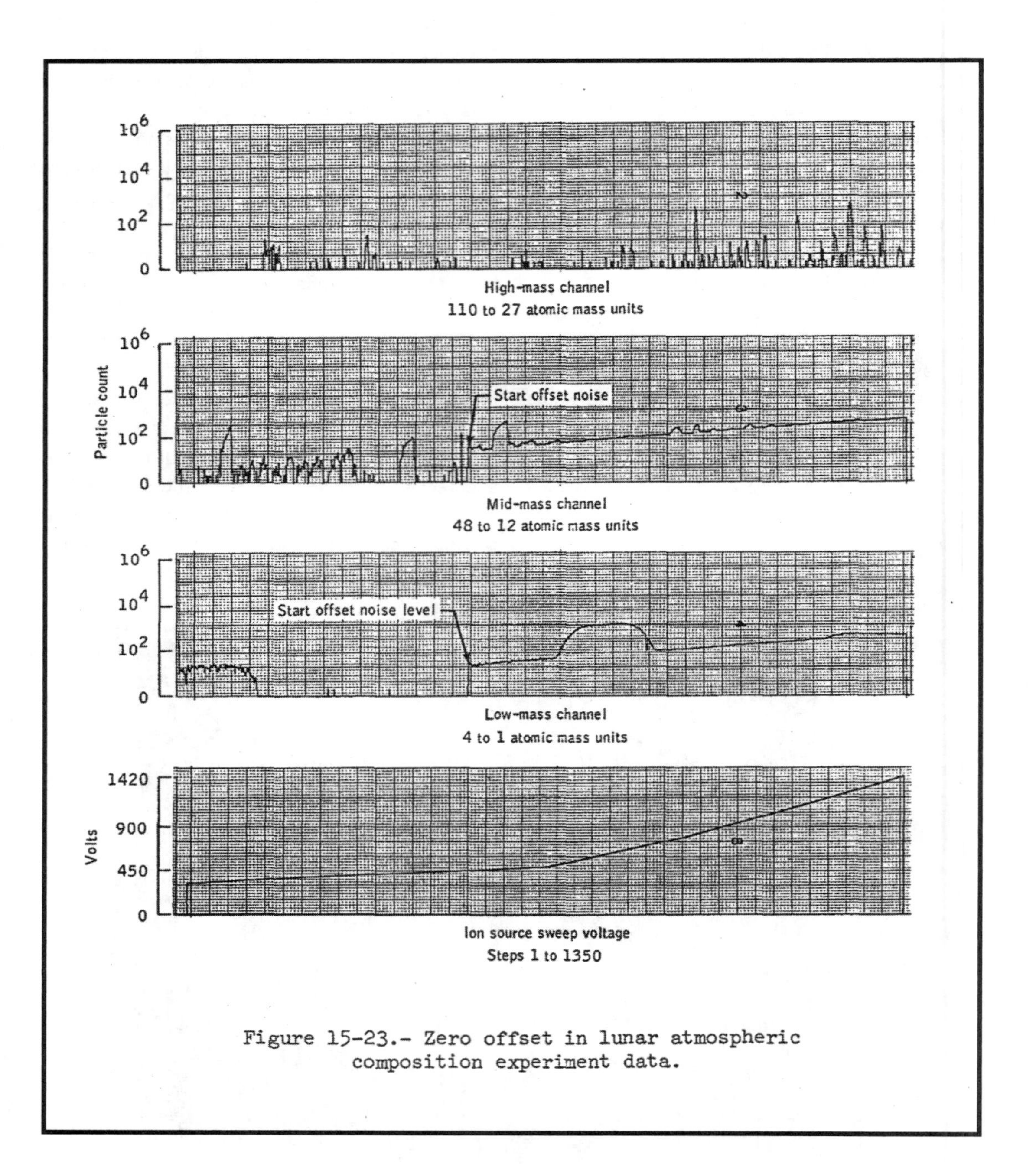

Figure 15-23.- Zero offset in lunar atmospheric composition experiment data.

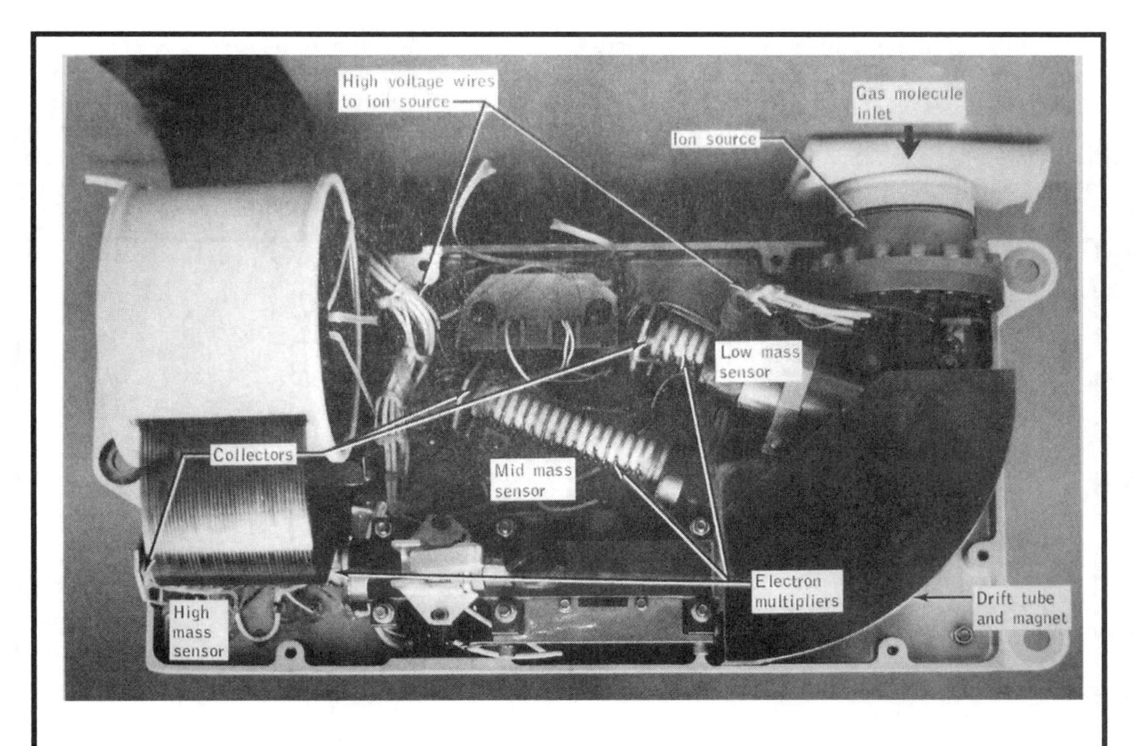

Figure 15-24.- Lunar atmospheric composition experiment detection system.

15.5 ORBITAL EXPERIMENT EQUIPMENT ANOMALIES

15.5.1 Panoramic Camera Velocity/Altitude Sensor Operated Erratically

Telemetry data from the second panoramic camera pass on revolution 13 indicated that the velocity/altitude sensor output was erratic. The velocity/altitude sensor measured the rate of travel of the spacecraft relative to the lunar surface. The output signal from the sensor (f i g . 15-25) controls the cycling rate of the camera and the forward motion compensation, as well as the exposure since the width of the exposure slit is dependent upon the scanning rate of the lens. The sensor operates in the range from 45 to 80 miles altitude. An override switch in the spacecraft provides the capability for locking out the sensor and substituting preset voltages which correspond to rates for the 55- and &-mile altitudes. In accordance with the flight plan, the velocity/altitude sensor was not used during the first panoramic camera operating period; instead, the override switch was used to substitute a preselected velocity /altitude value and thereby achieve the desired forward motion compensation. The sensor was used in the normal mode during the second camera operating period, but due to an indication that the normal mode was operating erratically, the override capability was used for the remainder of the mission.

Data analysis indicates that the erratic operation may have resulted from a downward shift in the scaling of the sensor output signal. If so, the shift was most likely caused by a component failure within the sensor circuitry. Because this was the last mission for the panoramic camera, no further investigation is necessary. This anomaly is closed.

15.5.2 Mapping Camera Exposure Pulse Absent At Low Light Levels

One of the mapping camera telemetry channels incorporates a pulse whose presence each time the film is exposed confirms that light has been transmitted through the metric lens and shutter to the film. This exposure pulse failed to appear at the lower light levels.

The pulse is generated by a light sensor and associated amplifier and a one-shot multivibrator (fig. 15-26). The light sensor consists of four photo diodes, with one near each edge of the 4-1/2 by 4-1/2 inch picture format (fig. 15-27). All four diodes are connected in series.

During Apollo 15 and 16, this pulse was absent only at the very low light levels, thereby indicating that the Apollo 17 discrepancy may have been due to a threshold shift in the light sensor circuitry. A 50-foot strip of metric photography from revolution 62 was processed in order to establish whether this was the case or some other problem was actually reducing the

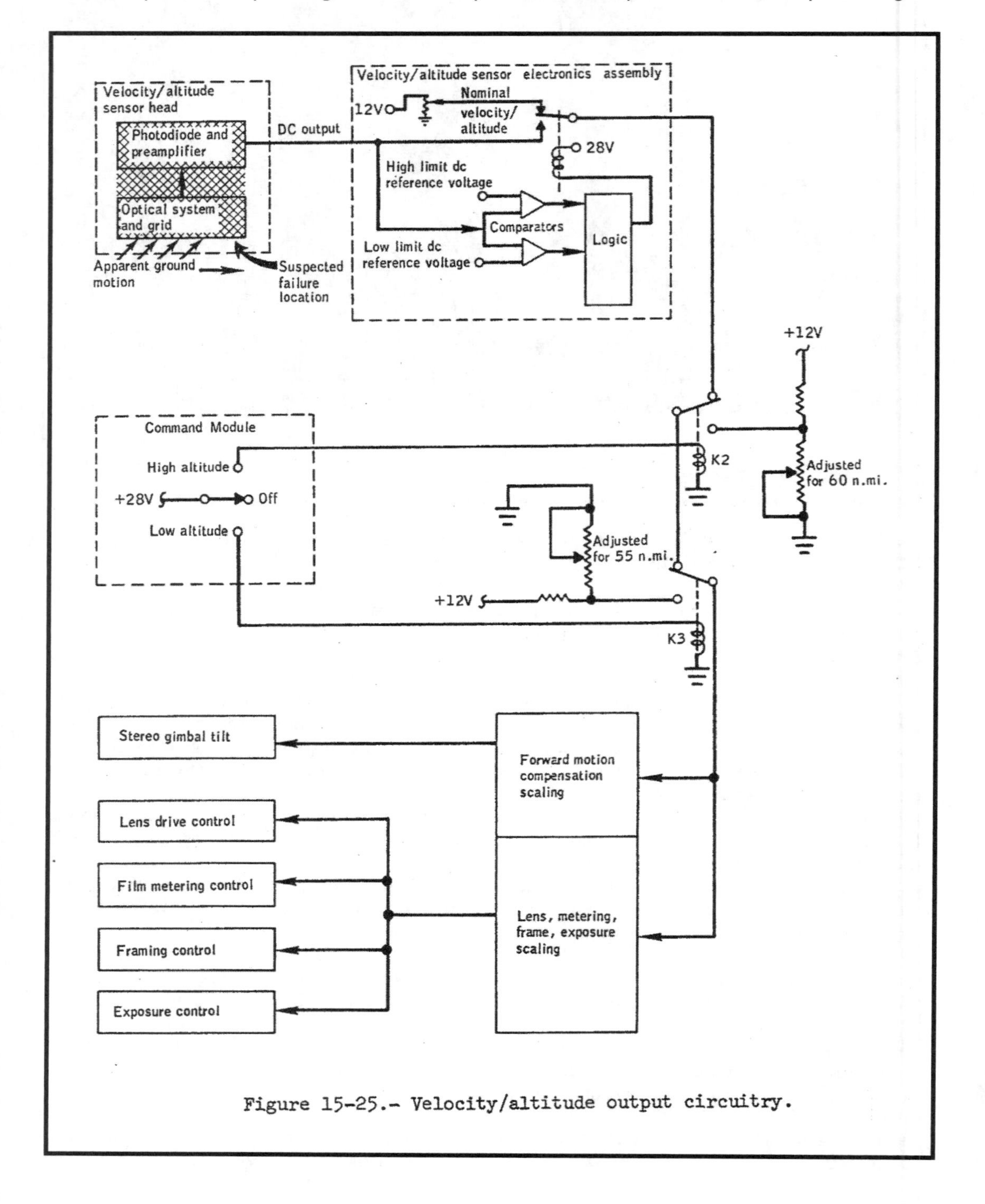

Figure 15-25.- Velocity/altitude output circuitry.

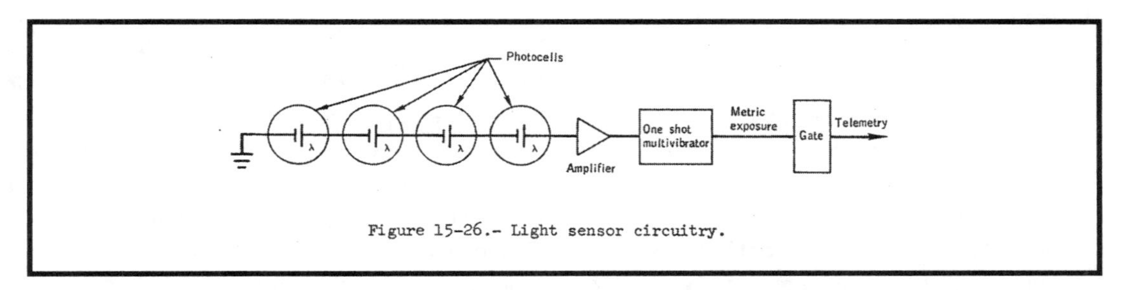

Figure 15-26.- Light sensor circuitry.

light being transmitted to the film.

The test strip indicated that the film had been properly exposed, thus indicating that the light sensor circuitry had indeed experienced a threshold shift. This shift could have been the result of a photo diode failure or a change in the performance of the amplifier, the one-shot multivibrator, or the circuitry between these two elements.

Since this was the last mission for this camera, no further analysis is necessary.

This anomaly is closed.

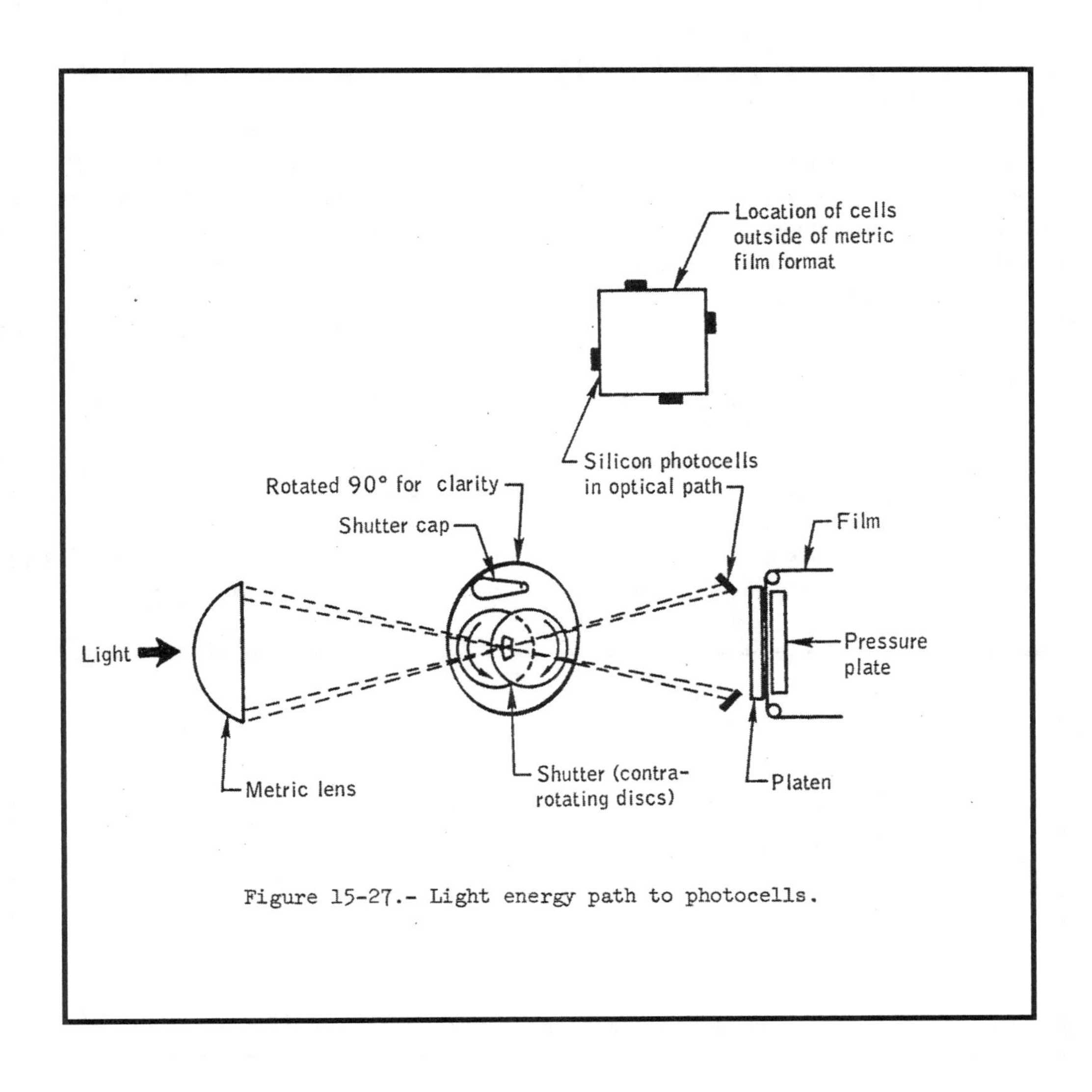

Figure 15-27.- Light energy path to photocells.

15.5.3 Panoramic Camera Gimbal Drive Failure

The panoramic camera lost its stereo capability 8 minutes prior to completion of its final photographic pass in lunar orbit.

Stereo photography is achieved by gimbaling the lens assembly fore and aft between exposures so that one picture of a stereo pair is taken with the lens viewing ahead of the nadir and the other picture is taken with the lens pointing behind the nadir. Two independent telemetry points indicated that the lens assembly had ceased to gimbal (fig. 15-28). These were the forward motion compensation tachometer voltage and the automatic exposure command voltage. Monographic photography continued, but it was degraded somewhat by the associated loss of forward motion compensation that is achieved by slowly gimbaling the lens during film exposure.

This inability to gimbal the lens system was apparently the result of a failure in either the gimbal drive motor or the electrical circuitry that provides power to the motor. Most likely, the motor failed since the motor brushes are limited-life items.

Since this was the last mission for this camera, no further analysis will be performed.

This anomaly is closed.

15.5.4 Ultraviolet Spectrometer Temperature Measurement Failures

Both ultraviolet spectrometer internal temperature measurements (electronics and motor) failed simultaneously near the end of the mission at approximately 282:20. At the same time, the input current to the spectrometer increased about 10 milliamps (approximately 4 percent). Since these temperatures were for housekeeping purposes and their circuitry is independent of the scientific data circuitry, their failure did not affect the performance of the ultraviolet spectrometer science mission. An external temperature measurement attached to the instrument case continued to provide thermal status after the loss of the internal measurements.

A block diagram of the temperature sensing circuits is shown in figure 15-29. Each thermistor temperature sensor is connected to its own bridge network which in turn drives a d-c operational amplifier whose 0 to 5-volt output is then telemetered. Both temperature circuits are identical and independent except that: (1) they share common returns, (2) they share common ±15 volt power supplies, and (3) each bridge network is biased with a common plus 1-volt precision stable reference. The plus 1-volt reference is derived via a temperature-stabilized zener diode and an inverting d-c amplifier powered from the minus 15-volt supply.

At the time of failure, the temperature indications changed dramatically in a 1-second interval. Such a change was not consistent with the thermal characteristics of the ultraviolet spectrometer. Since the primary indications were

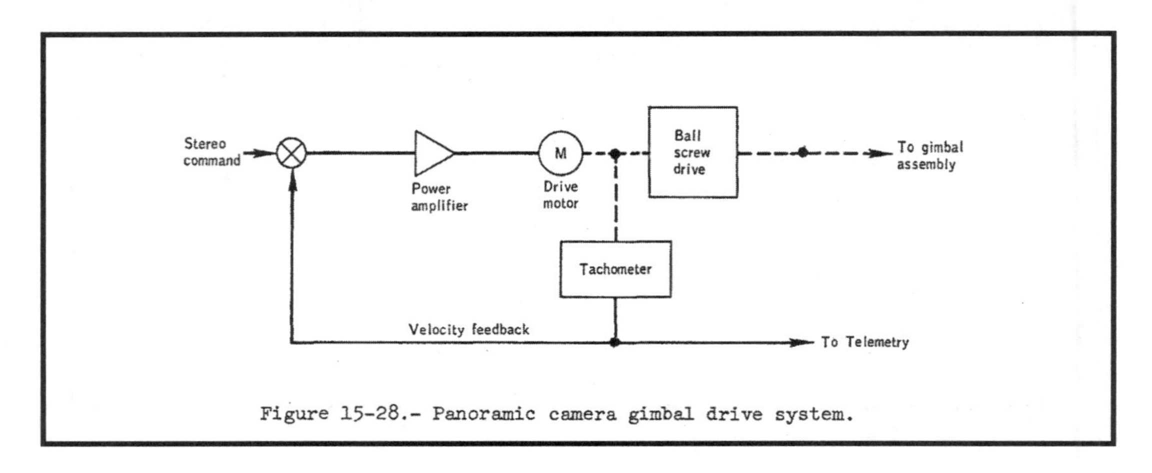

Figure 15-28.- Panoramic camera gimbal drive system.

the failed temperature channels coupled with the slight increase in input current, the failure most likely occurred in one of the circuits common to the temperature signal conditioners.

Further analyses and testing with the prototype ultraviolet spectrometer indicate that the anomaly was the result of a failure of a component or a short in the circuit wiring in the 1-volt reference voltage or the minus 15-volt supply voltage.

Since this was the only mission for this instrument, no further analysis will be accomplished.

This anomaly is closed.

15.5.5 Mapping Camera Deploy/Retract Times Were Excessive

The first and third extensions of the mapping camera were nominal, but all other deploy and retract times were longer than expected. The behavior was quite similar to that during the Apollo 15 and 16 missions.

The cause of the Apollo 15 and 16 anomalies has not been identified, but as a result of extensive investigations, several precautionary fixes had been implemented, such as removal of the no-back device, improved contamination covers, and a change in the lubricant on the drive screw - all apparently to no avail. Dynamic testing in support of the Apollo 15 and 16 investigations included thermal-vacuum and simulated zero-g tests. These tests provided no clues to explain the sluggish behavior of the deployment mechanism. Either these tests were not capable of sufficiently simulating the scientific instrument module bay environment or another, as yet unknown, factor was involved.

Other anomalies have occurred with. scientific instrument module bay equipments, thus suggesting a common factor. The mapping camera

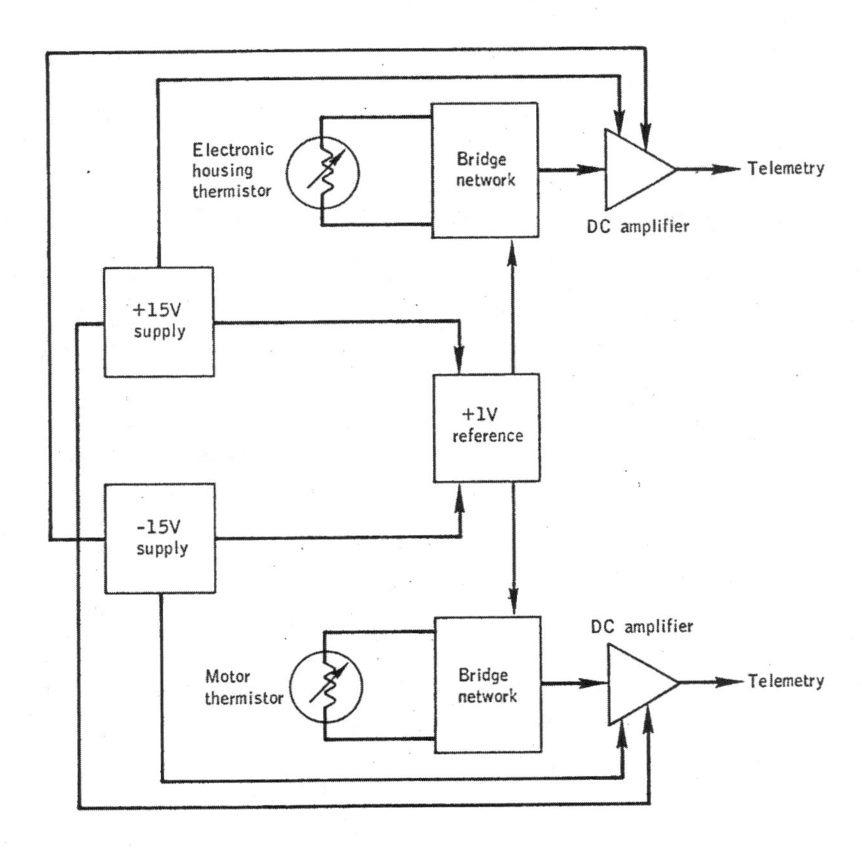

Figure 15-29.- Block diagram of ultraviolet spectrometer temperature sensing circuits.

reaction control system plume shield did not properly close on Apollo 15 and 17, and the mapping camera stellar glare shade failed to retract on Apollo 16. The lunar sounder high frequency antenna exhibited sluggish deployment on Apollo 17. In all cases, some degree of sliding between metal surfaces was required. It is, therefore, conceivable that the friction between those surfaces may have been significantly increased by the effects of a hard vacuum that is unobtainable in ground testing.

Since this was the last mission for this device, no further analysis will be performed on this specific device. However, the general problem area is being studied for Skylab and future missions and a treatment of the subject will be covered in a separate report. This anomaly is closed.

16.0 CONCLUSIONS

All facets of the Apollo 17 mission were conducted with skill, precision, and relative ease because of experienced personnel and excellent performance of equipment. The following conclusions are drawn from the information contained in this report.

1. The Apollo 17 mission was the most productive and trouble-free manned mission. This represents the culmination of continual advancements in hardware, procedures, training, planning, operations, and scientific experiments.

2. The Apollo 17 flight demonstrated the practicality of training scientists to become qualified astronauts and yet retain their expertise and knowledge in the scientific field.

3. Stars and the horizon are not visible during night launches, therefore out-of-the-window alignment techniques cannot be used for attitude reference.

4. The dynamic environment within the cabin during the early phases of launch is such that system troubleshooting or corrective actions by the crew are not practical. Therefore, either the ground control or automation should be relied upon for system troubleshooting and, in some cases, corrective actions.

5. As a result of problems on this and other missions, further research is needed to increase the dependability of mechanisms used to repeatedly extend and retract equipment in the space environment.

One Apollo program record and four World class records were exceeded on the Apollo 17 mission. The collected samples of 110.40 kilograms (243 pounds) established a new Apollo program record for total weight of returned lunar samples. Pending approval of the formal application, the following items will constitute new World records:

1. The total time of 21 hours 49 minutes and 24 seconds for one crewman outside the spacecraft on a single mission.

2. The total time of 147 hours 41 minutes and 13 seconds in lunar orbit.

3. The maximum radial distance of 7370 meters traveled away from a spacecraft on the lunar surface.

4. The elapsed time of 301 hours 51 minutes and 59 seconds for the total duration of a lunar mission.

APPENDIX A - VEHICLE AND EQUIPMENT DESCRIPTION

This appendix contains the configuration changes to the spacecraft and the extravehicular systems since the Apollo 16 mission. In addition, the scientific experiment equipment and equipment flown for the first time in the Apollo program is described.

The Apollo 17 command and service module (CSM-114) was of the block II configuration, but was modified to essentially the Apollo 15 configuration to carry out a greater range of lunar orbital science activities than had been programmed on missions prior to Apollo 15. The launch escape system and spacecraft/launch vehicle adapter were unchanged. The lunar module (LM-12) was modified, as were the Apollo 15 and 16 lunar modules, to increase the lunar surface stay time and return a larger scientific payload. The Saturn V launch vehicle used for this mission was AS-512, and the significant configuration changes for that vehicle are given in reference 2.

Many minor changes were made because of problems which occurred during the Apollo 16 mission. Not all of the details of these changes are discussed in this section of the report as the anomaly summary of the Apollo 16 mission report (reference 7) contains detailed discussions of these modifications.] This reference should be used if the detailed information is required.

A.1 COMMAND AND SERVICE MODULES

Four pressure transducers were added to the instrumentation system to provide additional data during prelaunch checkout and lunar orbit insertion operations. Two of these transducers were added in the lines between the helium check valves and propellant tanks of the command module reaction control system. These transducer provided data for determining differential pressure across the tank bladders during prelaunch checkout.

The remaining two pressure transducers were added to the service propulsion system manifold to provide redundant pressure readings during the lunar orbit insertion firing.

An automatic relief valve was added in series with the existing manual vent valve to the entry battery manifold overboard dump line as a result of the high battery manifold pressures experienced on the Apollo 16 mission. The manual vent valve was left open during the mission.

An orthostatic countermeasure garment was provided for use by the Command Module Pilot during entry and recovery operations. The garment's intended use was to evaluate its effectiveness in preventing or lessening the effects of postflight orthostatic problems that may result from extended space flight.

Additionally, the length of the pressure garment assembly stowage bag in the command module was increased 9 inches to ease suit handling when placing them in the bag. Also, the extra length will reduce the possibility of suit damage when placing the suits in the bag.

A.2 LUNAR MODULE

The lower edge of the thermal shields on the aft equipment rack was redesigned to prevent exhaust gases from entering the cavity behind the shields, and additional vents were added to the thermal blankets to increase venting during earth launch. This modification resulted from the loose thermal shields noted in the Apollo 16 mission.

A.3 LUNAR SURFACE MOBILITY SYSTEMS

A.3.1 Extravehicular Mobility Unit

The extravehicular mobility unit was modified to improve its operational capability. Significant changes were as follows:

a. The zipper restraint patches on the pressure garment assemblies were strengthened to preclude tearing, as noted during postflight inspection of the Apollo 16 suits.

b. A piece of Velcro was added to the feedport on the inside of the Commander's helmet to act as a nose scratcher.

c. The lanyard on the purge valve was shortened and stiffened to preclude inadvertent removal as occurred on the Apollo 16 mission. As an added precaution, the internal spring, which forces the valve open upon actuation, was removed.

d. A removable cap was added to the in-suit drink-bag port and the tilt valve stem was reclocked to prevent spillage of liquid as noted on the Apollo 16 mission.

e. Dust covers were added to the wrist rings on the pressure garment assemblies.

f: A hook was added to the pressure garment assembly zipper as a donning aid.

g. One spare oxygen purge system antenna was packed in the buddy secondary life support system bag on the lunar roving vehicle. This additional antenna will enable the antenna to be changed, if breakage such as experienced on Apollo 16 should occur.

h. Spring-loaded clips were added to the top of the sample container bag holder and a spring-loaded hook to the bottom. Also, the pull force required to actuate the quick-release shackle has been increased.

i. One of the two sets of spacers, which were subject to galling from dust, were removed from the lunar extravehicular visor assembly.
j. The cover gloves that are worn over the extravehicular gloves were modified to provide better hand mobility.

A.3.2 Lunar Roving Vehicle

The third lunar roving vehicle was essentially unchanged from the lunar roving vehicle (2) flown on Apollo 16. Fender extension stops were added to each fender to prevent loss of the fender extension, and a signal cable was added to provide navigation information from the lunar roving vehicle navigation system to the surface electrical property experiment tape recorder. In addition, a decal was added to the aft chassis to aid the crew in locating the proper hole in which to place the pallet stop tether.

Extensive changes were made to the experiment pallet carried on the lunar roving vehicle to accomodate the Apollo 17 unique experiments. Also, an index ring was added to the low-gain antenna alignment dial to facilitate azimuth pointing.

A.4 EXPERIMENT EQUIPMENT

A.4.1 Lunar Surface Science Equipment

The lunar surface science complement contained eleven experiments. Seven of these experiments were flown for the first time on Apollo 17. These were the lunar seismic profiling experiment, the lunar surface gravimeter experiment, the lunar atmospheric composition experiment, the lunar ejects and meteorites experiment, the surface electrical properties experiment, the lunar neutron probe experiment, and the traverse gravimeter experiment.

The four remaining Apollo 17 lunar surface experiments plus the lunar surface tools were flown previously.

Six of the experiments were performed by the crew while on the lunar surface. The other five experiments contained in the Apollo lunar surface experiments package will remain on the lunar surface in an operative state to collect data for about two years. These latter five experiments were deployed and configured for activation by ground command.

Lunar geology and soil mechanics experiments.- The hardware used to perform the lunar geology and soil mechanics experiments was stowed in a vacuum-tight container to protect the contents from the earth environment prior to launch and on return to earth. A tabulation of the hardware used in performing these experiments and a brief description of each is contained in the following paragraphs.

Apollo lunar sample return container: The Apollo lunar sample return container was made of aluminum with exterior dimensions of 19.0 inches in length, 11.5 inches in width, and 8.0 inches in height (fig. A-1). The interior volume was approximately 1000 cubic inches. The major components of the container were the handle and latch pins, seals, seal protectors, York-mesh liner, and strap-latch system. The two latch pins operated from a central lever which also served as a carrying handle for the container. The pins and linkage system supported the container in the lunar module and command module stowage compartments under all vibration and g-force conditions. A triple-seal arrangement maintained a vacuum during translunar and transearth flight. Seal protectors were also provided to prevent lunar dust from getting on the seals. The strap-latch system consisted of four straps and two cam latches. When closing the container, the crewman engaged the cam latches, thus tightening the straps over the lid.

Sample collection bags: The sample collection bags (fig. A-1) were constructed of vulcanized Teflon sheets having a consistency of heavy oil cloth. Each bag had a full-opening flip-type cover for insertion of bulk samples and a diagonal slit in the cover for insertion of small singular samples. Three interior pockets were incorporated for holding the drive tubes. Exterior pockets were provided to hold the special environmental sample container and a drive tube cap dispenser. Two straps on the front of the bag facilitated handling the bag. Three restraint points were provided on the back of the bag for attachment on the side of the portable life support system and the Apollo lunar hand tool carrier or lunar roving vehicle pallet.

Special environmental sample container: The special environmental sample container fig. A-1) was a thin-walled stainless-steel can with a knife edge machined into the top rim. A grip assist was attached to the bottom of the container body to aid in generating sufficient torque for sealing. A three-legged press assembly was attached to the lid and used to exert a force between the lid and the body. A machined groove in the lid was filled with an Indium-Silver alloy gasket into which the knife edge cut to effect a seal for earth return. During stowage, both the lid and the knife edge were protected by Teflon seal protectors. Each of these protectors had a tab sticking out to facilitate removal upon completion of the sample collection.

Core sample vacuum container: The core sample vacuum container (fig. A-1) consisted of the top portion of the special environmental sample container that was tapered to accomodate a drive tube without the adapter plug. The container had an insert that held the knurled section of drive tube and provided lateral and longitudinal restraint.

Documented sample bag dispenser: The documented sample bag dispenser (fig. A-1) consisted of 20 Teflon bags and a mounting bracket. Each bag was 7.5 inches wide and 8.0 inches high and was designed to contain a 4.5-inch diameter rock sample. The bags were pre-numbered to identify the sample and had two Teflon tabs. One tab was attached to the dispenser and was torn when the flat bag was pulled open. The other was used by the crewmen to pull open the bag. After sample collection, the rim of the bag was rolled down and Z-crimped to retain the samples.

Drive tubes: The drive tubes (fig. A-1) consisted of three major components; the drive tube, drive tube tool, and cap dispenser. The drive tube was a hollow aluminum tube 16 inches long and 1.75 inches in diameter with an integral coring bit. The tube could be attached to the tool extension to facilitate sampling. A deeper core sample was obtainable by joining the tubes in series. The drive tube tool was used to position the drive tube keeper against the core sample to preserve sample integrity. A cap dispenser, which mounts on the Apollo lunar hand tool carrier, contains three Teflon caps to seal the tube after sample collection.

Organic sampler: The organic sampler (fig. A-2) consisted of six rolls of York mesh packing material in a Teflon bag which served as a quantitative collector of hydrocarbons and other organic compounds during the mission. There were two identical samplers used. One was analyzed and sealed as a quantitative measure of contamination before flight. The second sampler was placed in the sample return container and handled like other sample bags, in that it was taken out on the lunar surface, sealed, and returned to the container. The sampler bag was sealed when the container was first opened on the lunar surface to prevent lunar dust from getting on the six rolls.

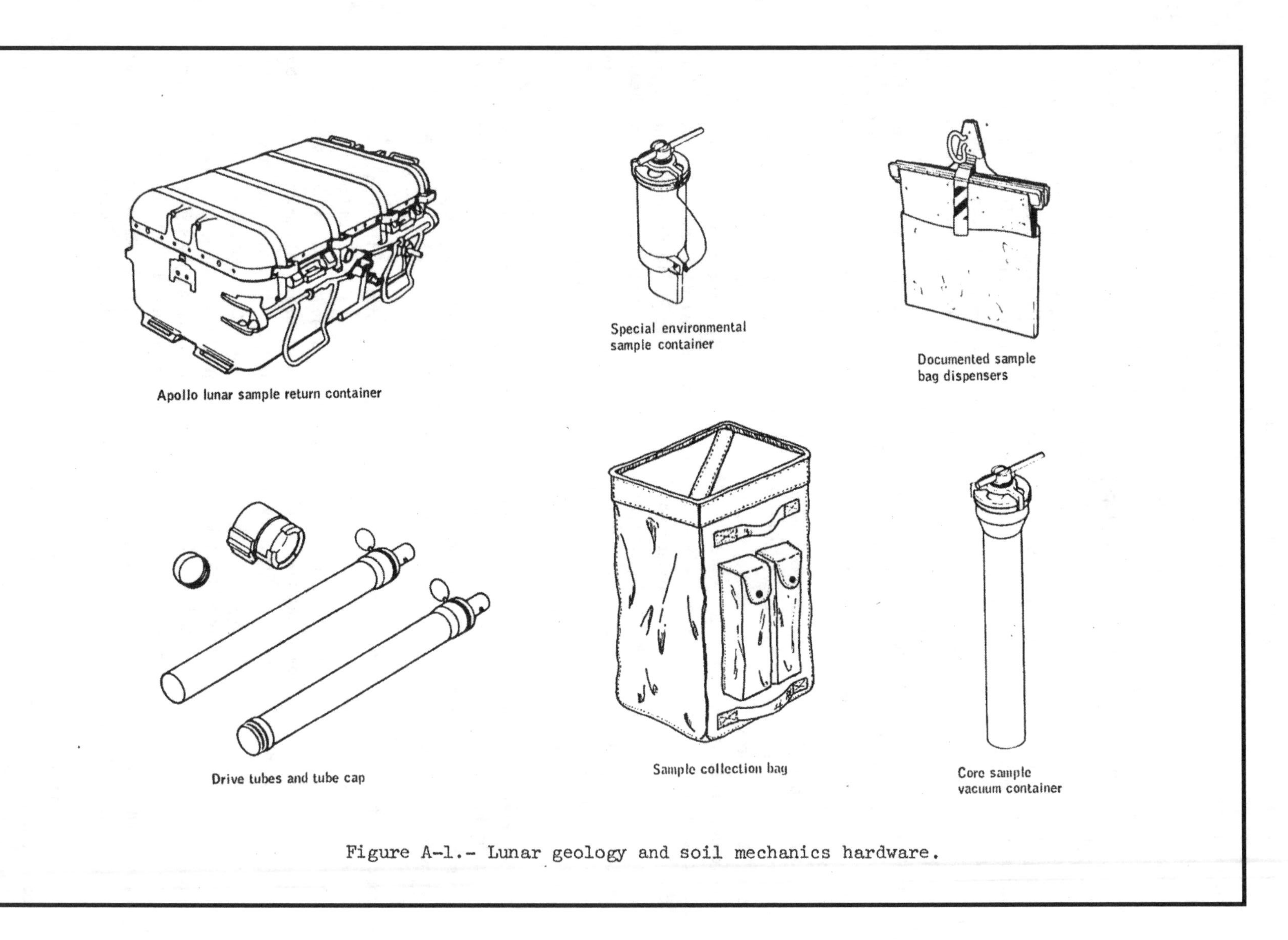

Figure A-1.- Lunar geology and soil mechanics hardware.

Round documented sample bag: Each round documented sample bag (cup) (fig. A-3) was 3.25 inches in diameter and 5.25 inches deep with 12 bags to a group and 4 groups or 48 bags total. The individual bags were packed in a telescope manner with a Teflon spacer between each group. Each bag had an aluminum supported rim to facilitate sealing after the sample had been collected from the lunar surface.

Lunar geology and soil mechanics tools: The geological tools used to sample or probe the lunar surface are listed. A notation of the specific design characteristics of each is contained in the following paragraphs.

Hammer: The hammer (fig. A-3) head was made of tool steel suitable for impact use and coated with vacuum-deposited aluminum to minimize solar heating. The handle was made of aluminum. The end of the hammer head opposite the striking surface was shaped for use as a pick or chisel, and when attached to the tool extension, as a hoe for trenching or digging. Lunar rake: The rake (fig. A-3) was adjustable for stowage and sample collection. The stainless-steel tines were formed in the shape of a scoop. An aluminum handle, approximately 10 inches long, was attached to the tool extension for sample collection.

Sample scale: The sample scale (fig. A-3) had graduated markings in increments of 5 pounds. Maximum weight capacity of the sample scale was 80 pounds. The sample scale was stowed in the lunar module ascent stage where it was used to weigh lunar samples.

Adjustable sample scoop: The adjustable sample scoop (fig. A-3) was about 11-3/4 inches long. The pan of the scoop had a flat bottom, flanged on both sides with a partial cover on the top. It was used to retrieve lunar samples too small for the tongs. The pan was adjustable from 0° (horizontal) to 55° for scooping and to 90° for trenching. The handle had a tool extension adapter. The pan and adjusting mechanism were made of stainless steel and the handle was made of aluminum.

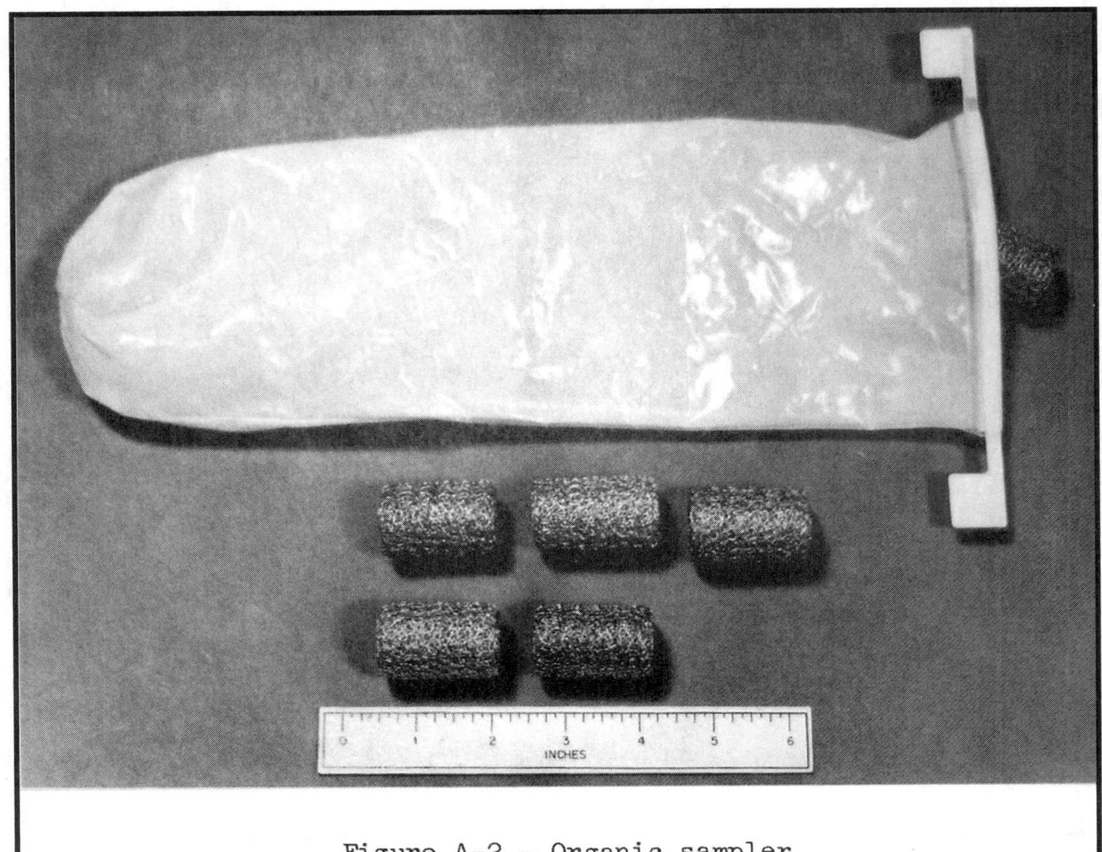

Figure A-2.- Organic sampler.

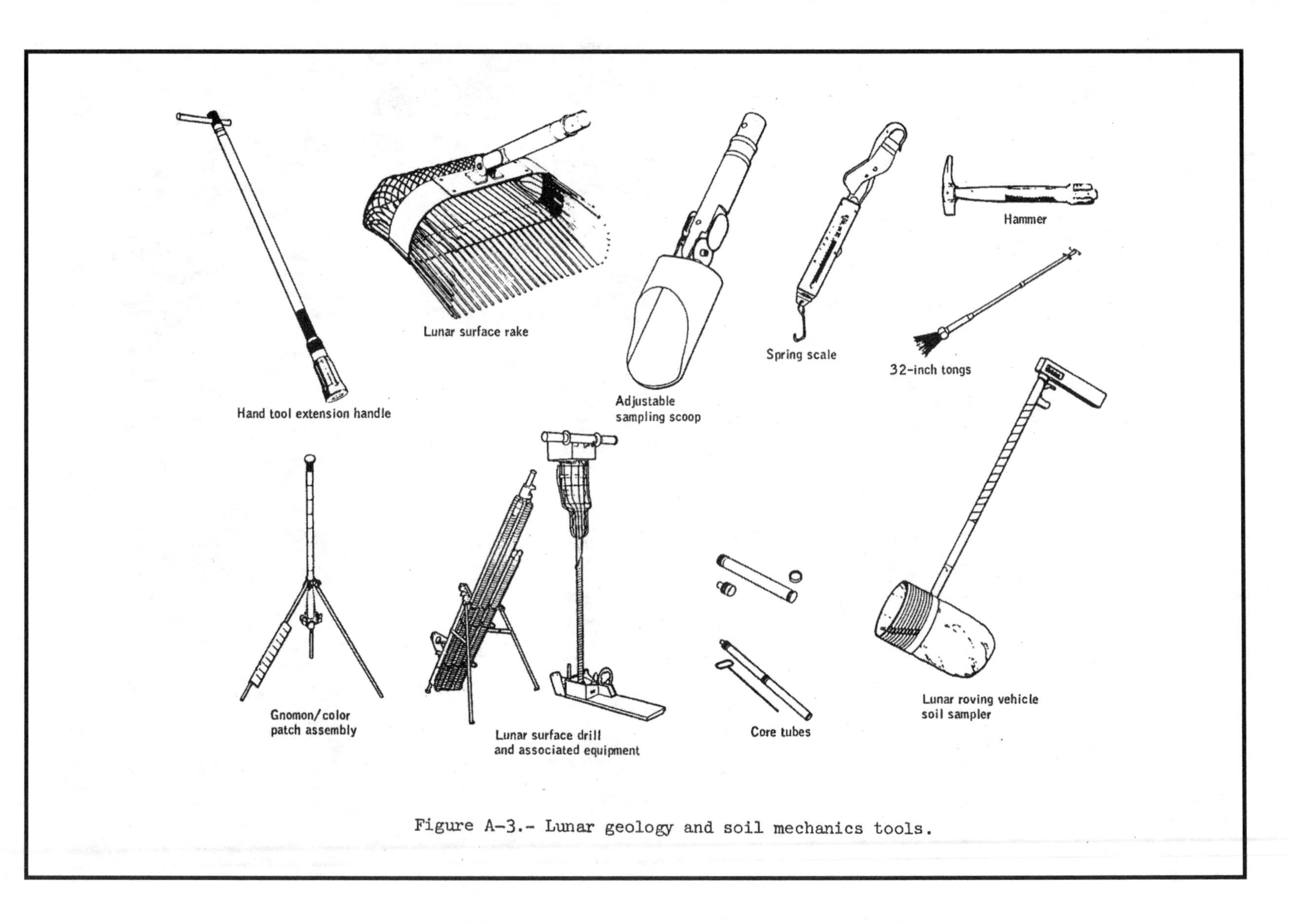

Figure A-3.- Lunar geology and soil mechanics tools.

Tongs: There were two pair of tongs (fig. A-3) and each consisted of a set of opposing spring-loaded fingers attached to a 28-inch handle. The tines were made of stainless steel and the handle of aluminum. The tongs operated by squeezing the handles to actuate the cable that opened the fingers.

Tool extensions: The tool extension (fig. A-3) provided 30 inches of length to the handles of various tools. The lower end of the tool extension had a quick-disconnect mount and lock for attaching the scoop, hammer, drive tubes, and rake. The upper end was fitted with a sliding "T" bar to assist with torquing operations. There were two tool extensions.

Gnomon: The gnomon (fig. A-3) was a stadia rod mounted on a tripod. It was constructed such that the rod righted itself and pointed to the vertical when the legs were placed on the lunar surface. The gnomon indicated the gravitational vector, and provided accurate vertical reference and calibrated length for determining the size and position of objects in nearfield photographs. It was painted in shades of gray ranging in reflectivity from 5 to 35 percent in 5-percent increments, and a color scale of blue, orange, and green. The color scale provided a means for accurately determining colors in color photographs. The rod was 18 inches long, and the tripod base folded for compact stowage.

Lunar roving vehicle soil sampler: The lunar roving vehicle soil sampler fig. A-3) consisted of a ring that holds twelve plastic cups. The ring could be attached to the universal handling tool. The soil sampler, with the universal handling tool attached on a handle, allowed the crewmen to take samples from the lunar surface without getting off the lunar roving vehicle. As each sample was taken, the cup containing the sample was removed from the stack of twelve individually sealable cups.

Surface electrical properties (S-204).- The equipment used to conduct the surface electrical properties experiment consisted of two units - a fixed-location transmitter, and a transportable receiver and recorder capable of being mounted on the lunar roving vehicle. The experiment was performed by sequentially transmitting (from a fixed location on the lunar surface) a series of multiple wavelength signals at each of six different transmitting frequencies, and alternately rotating the energy plane 90° by means of two standard dipole transmitting antennas that were lying at right angles to each other on the lunar surface. The receiver, mounted on the lunar roving vehicle, received two radiated energy patterns - one from the ground wave transmitted along the lunar surface, the other from the reflection of the transmitted energy from any subsurface layering. An operational configuration of this experiment is shown in figure A-4. Any moisture present was easily detected by this experiment because even minute amounts of water in rocks or the subsoil can change the electrical conductivity by several orders of magnitude. The receiver (at any given point in the traverse) received, through three orthogonally arranged loop antennas, energy at varying phase relationships, amplitudes, and densities. Using six different channels (corresponding to the six transmitting frequencies), the receiver translated these varying signal levels into a 300- to 3000-hertz audio tone that varies in accordance with the signal strength. The resultant audio output was recorded along with the location of the lunar roving vehicle on magnetic tape mounted in the receiver unit. The lunar roving vehicle's navigation system provided position data to the data storage assembly. At the end of the experiment, the recorder/storage unit was removed from the receiver housing and returned to earth for subsequent data analysis.

Lunar neutron probe experiment (S-229).- The lunar neutron probe experiment consisted of a cylindrical probe shown in figure A-5 that was inserted in the core stem hole on the lunar surface. The instrument action within the probe provided data from which the rate of neutron capture in lunar materials can be calculated, and variations of the capture rate with the depth beneath the surface could be determined, as well as information on the lunar neutron energy spectrum.

The 2-centimeter diameter, 2.3-meter long probe was made up of two sections, which when

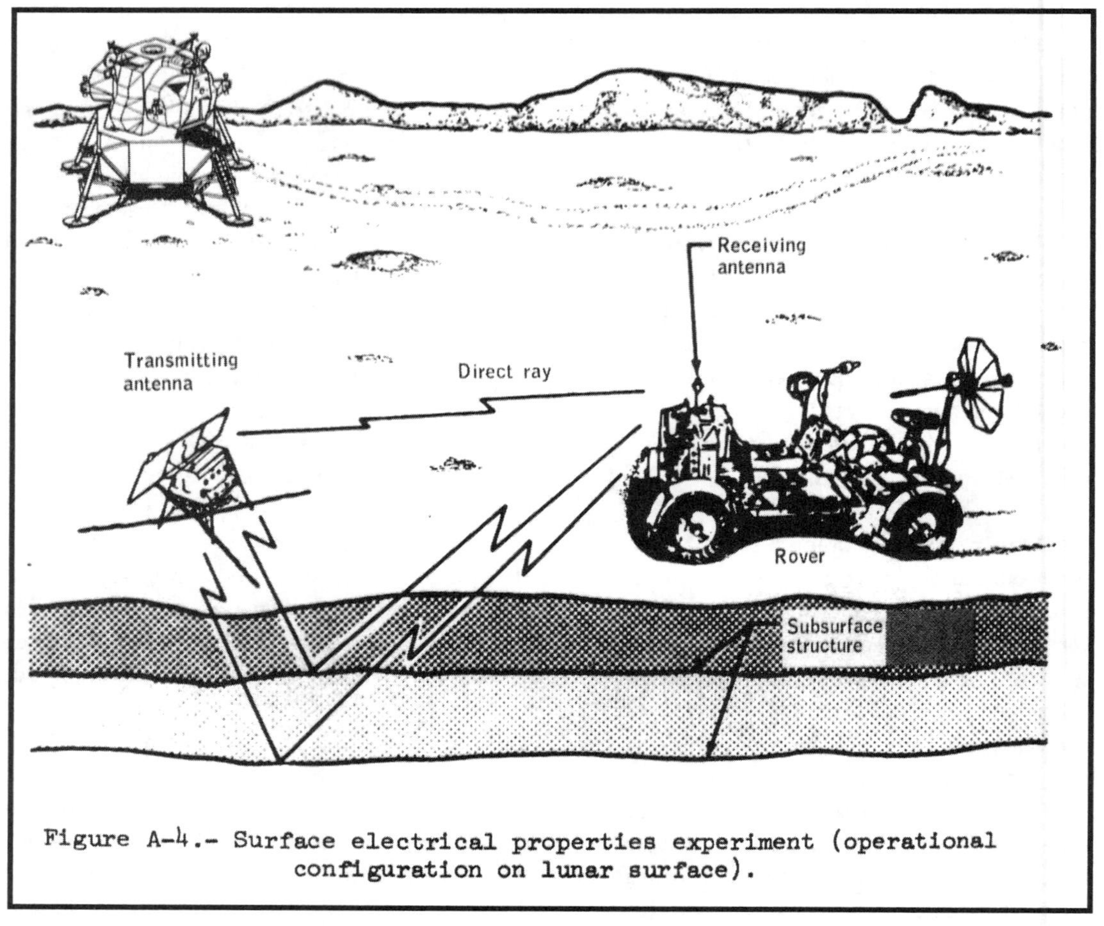

Figure A-4.- Surface electrical properties experiment (operational configuration on lunar surface).

assembled by the crew, were lowered into the core stem hole.

The lower section had a point at the bottom end (fig. A-5). The cap was removed from the top of the lower section and inverted and used as a tool to rotate the exposed end of the inner rod for activation or deactivation of the section. The upper section had a handle at the top which could be used for the same purpose.

Each rod section had a 12-station inner rod, a 12-station rib cage cylinder over the inner rod with windows at each station, and a protective outer cylinder.

Boron-10-coated semi-cylinders of tantalum were bonded to 23 of the stations on the inner rods starting at the bottom of the lower section. Cellulose triacetate film covered all 24 stations on the rib-cage cylinders to record the alpha particle tracks resulting from the neutron bombardment of the boron targets when the boron was aligned with the cage windows in the active position (fig. A-5).

Two mica sheets were bonded at each of eight stations on the inner rod on the opposite side from the boron 10 targets. Uranium 235 foil, bonded to the cage cylinders at these stations, covered the windows. The windows were aligned with the mica when the inner rod was rotated to the active position (fig. A-5).

The boron 10 cellulose triacetate provided the most accuracy in track densities in the range above 10 3/cm2. However, degradation of the data occurs above 70° C, so four temperature sensors were mounted on the inner rod, one on the opposite side from the boron targets, one at the bottom, and one at the middle and top of the probe.

The mica-uranium 235 provided more accurate data for track densities in the range of 102 to 10 3/cm2 because the fission fragment data tracks in the mica are easier to identify and count.

Three uranium 238 wagers, mounted at three stations on the same side of the inner rod as the boron targets, provided an alpha particle reference (fig. A-5).

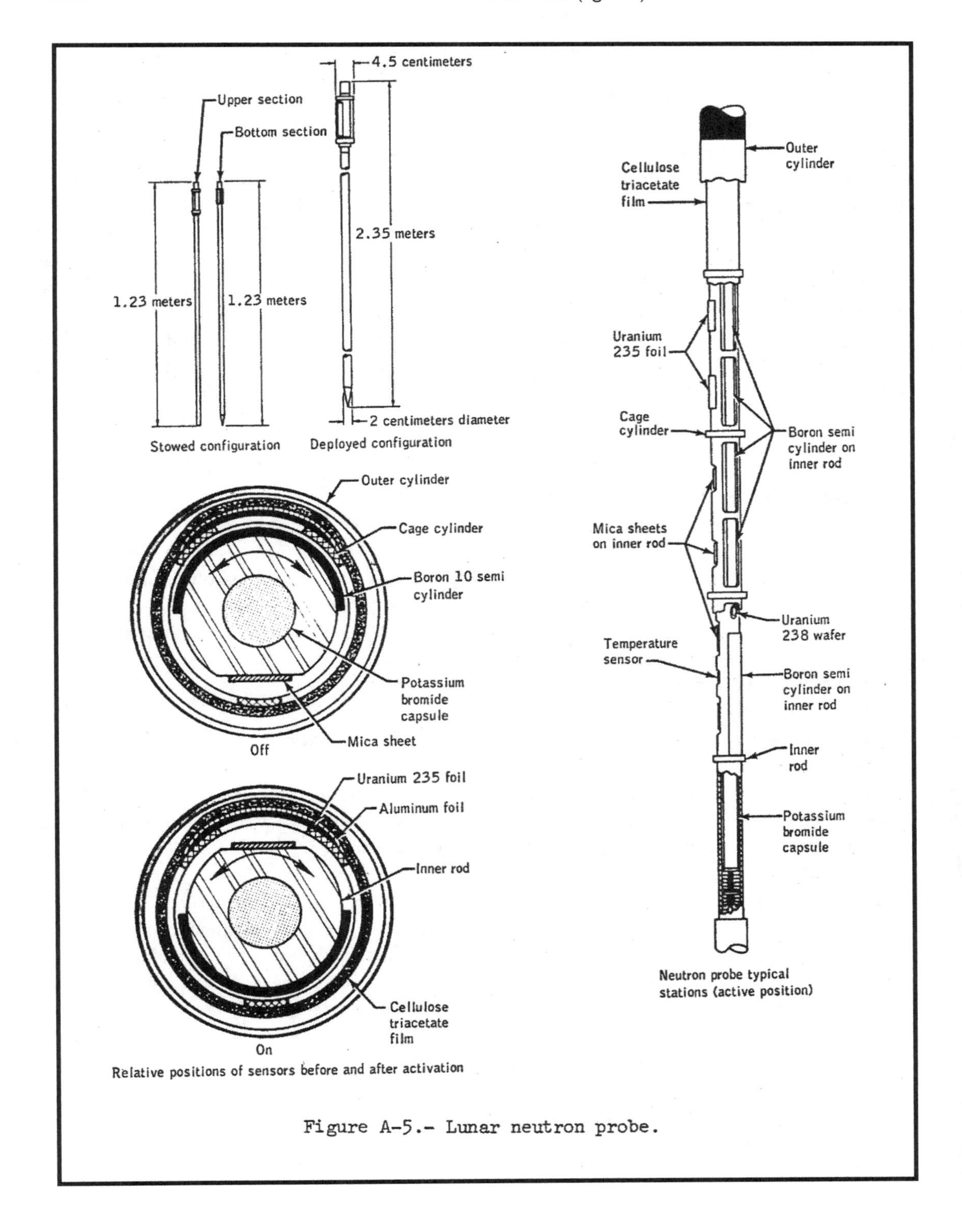

Figure A-5.- Lunar neutron probe.

Three nickel capsules, containing potassium bromide, were mounted in the bottom end of the inner rod of the lower section, in the bottom end of the inner rod of the upper section, and in the handle. These capsules provided a measure of the neutron flux in the 30 to 300 eV range by analysis of the krypton produced by the reaction of the neutrons with the potassium bromide; whereas, the boron 10 targets respond to neutrons below 10 eV. The capsule in the handle served as a control sample to correct for krypton produced by neutrons from the radioisotope thermoelectric generator and the lunar module.

The lower section of the experiment was activated by rotating the inner rod 180° from off to on, after which the two sections were mated and the upper section was activated by rotating the inner rod 180° from off to on. Following activation the probe was lowered into the core stem hole.

To retrieve the experiment, the assembly is withdrawn from the hole, the two sections separated, and each deactivated.

Traverse gravimeter experiment (S-199).- The traverse gravimeter experiment (fig. A-6) was a portable self-contained gravimeter that was also lightweight and operated automatically, once activated on the lunar surface. Power for experiment operation was provided by a 7.5-volt battery located inside the experiment package. The experiment provided measurements of local gravity at various stations along the traverse and in the immediate area of the lunar module. All gravity readings were made by the crew and relayed by voice communications to the earth-bound scientists.

Cosmic ray detector (sheets) experiment.- The cosmic ray detector experiment consisted of a thin aluminum box with a sliding, removable cover. Four particle detector sheets, consisting of platinum, aluminum, glasses, and mica were attached to the bottom of the box. Two platinum strips were mounted on the sides of the box. Four particle detector sheets consisting of platinum, glasses, plastic, and mica were attached to the inner surface of the cover. Pull rings mounted on the opposite ends of the cover and box were utilized to slide the cover from the box. Figure A-7 shows the inner surfaces of the box and cover.

The experiment was stowed in the lunar module and deployed during the first extravehicular activity period. The crew removed the cover and using the Velcro strap, hung the box on a strut of the landing gear, oriented so that the detector sheets faced directly into the sun with minimum obstruction to the field of view that was shaded from the sun at all times and oriented so that the exposed detector sheets faced away from the sun with a view of the dark sky the cover was hung in a location and with minimum obstruction to the field of view. The experiment was retrieved and placed in a protective bag for transport back to earth for analysis.

Apollo lunar surface experiments package.

The Apollo lunar surface experiments package was divided into two subpackages which contained five experiments and the central station. A description of the central station and experiments is contained in the following paragraphs.

Central station.

The central station (fig. A-8) was essentially sub-package no. 1 without the experiments mounted on it. The central station consisted of the data subsystem, helicel antenna, power conditioning unit, and experiment electronics. There were provisions for thermal control of the electronics, for alignment of the antenna, for electrical connections to the experiments and the radioisotope thermoelectric generator, and for the activation switches. A more detailed description of this subpackage is in appendix A of reference

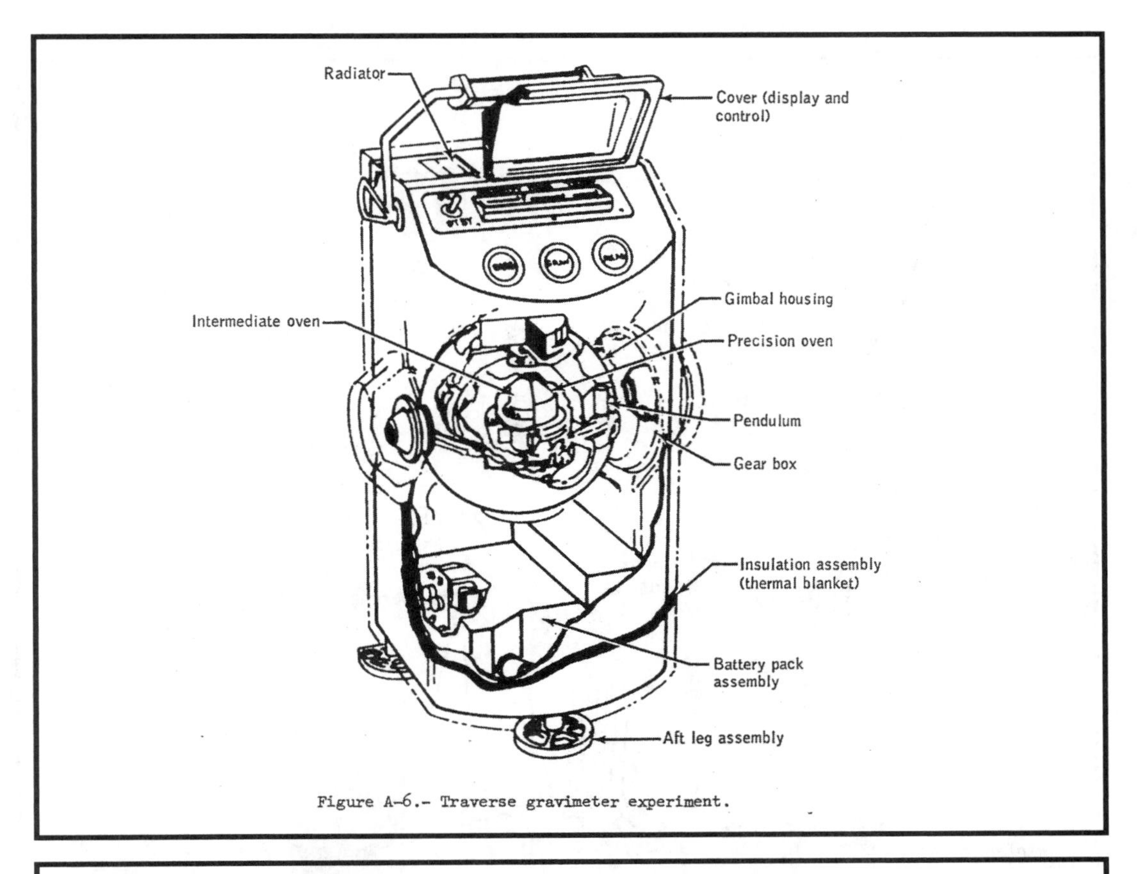

Figure A-6.- Traverse gravimeter experiment.

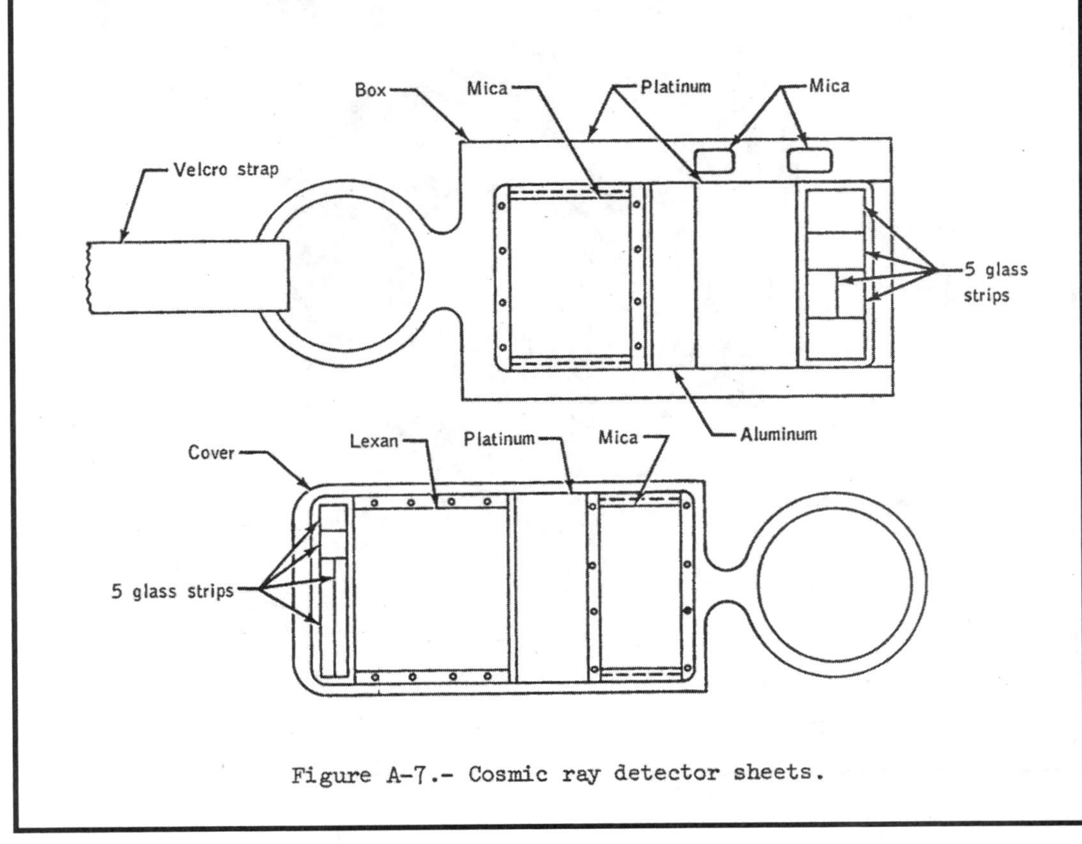

Figure A-7.- Cosmic ray detector sheets.

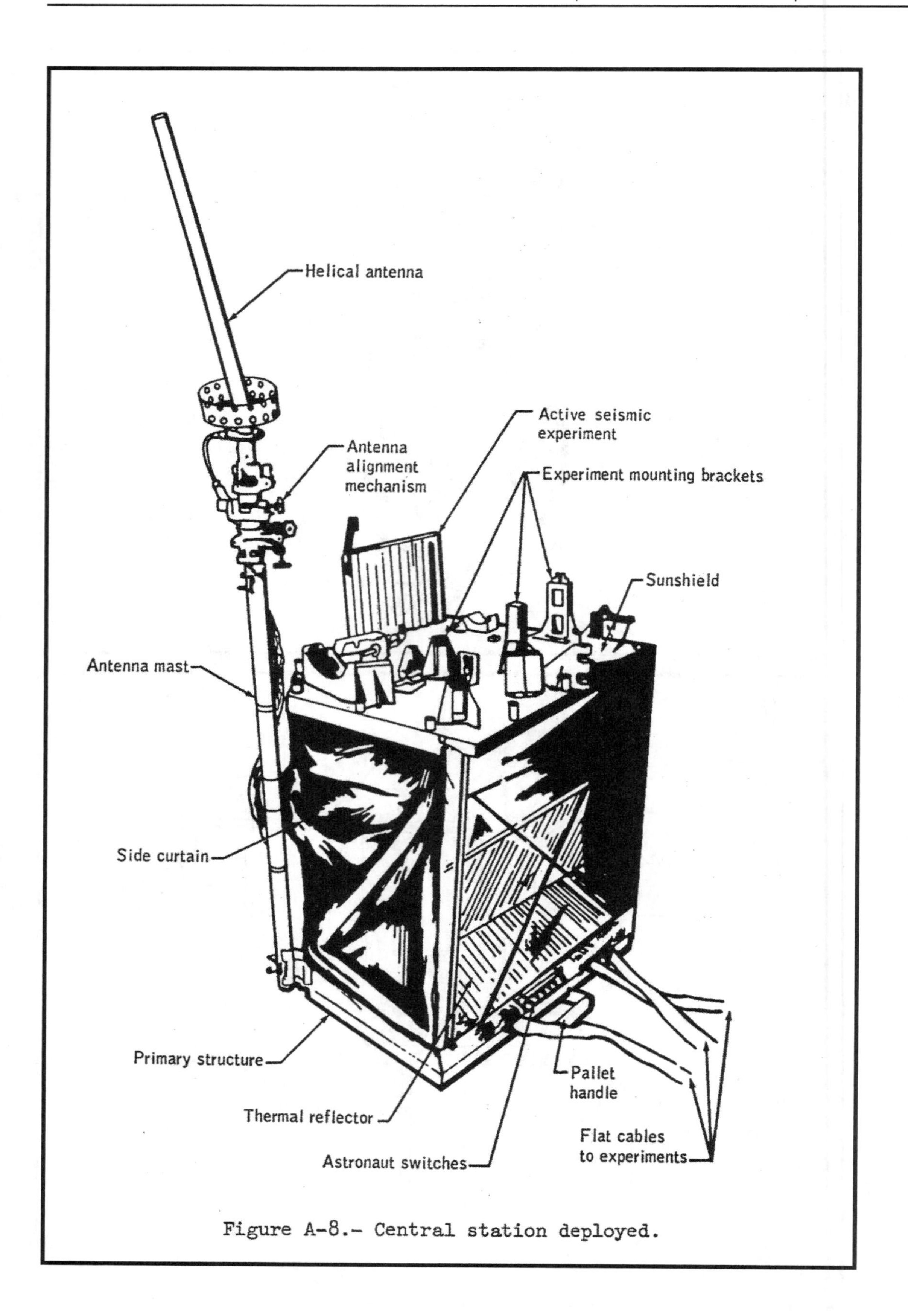

Figure A-8.- Central station deployed.

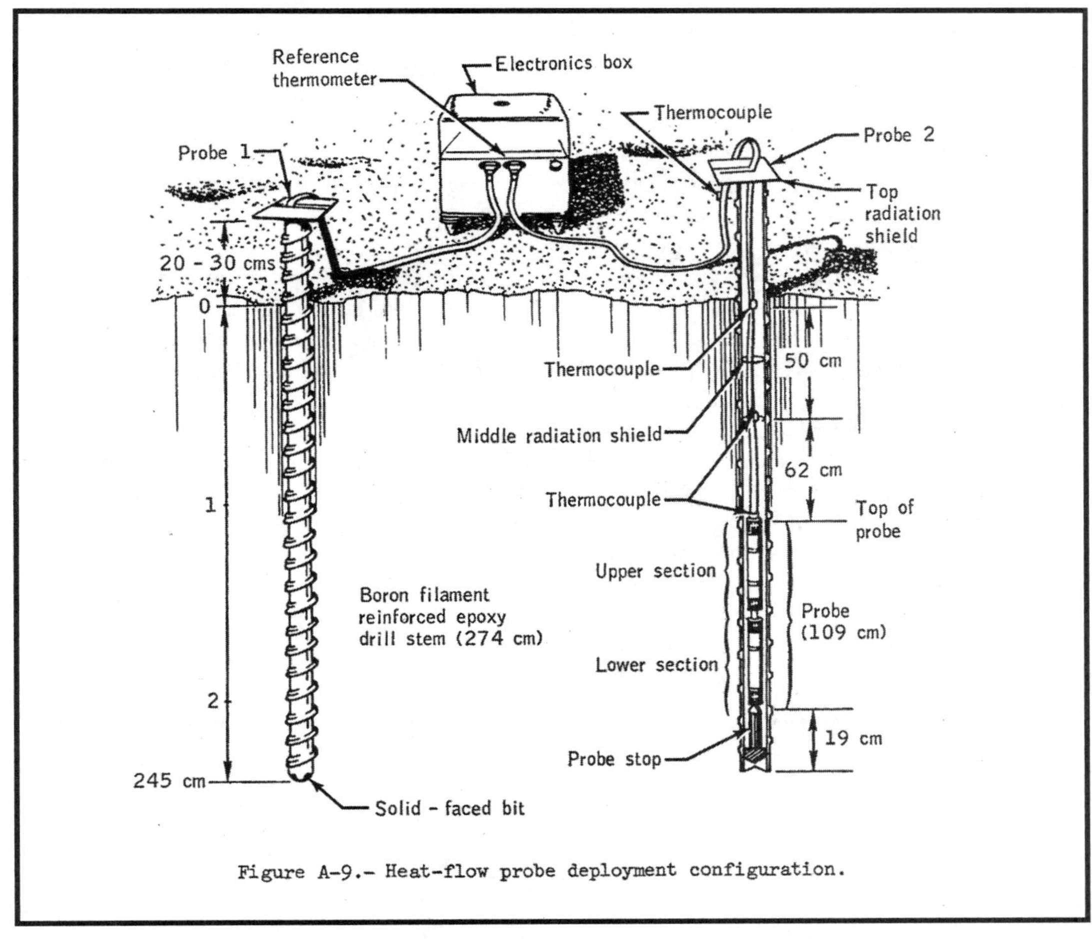

Figure A-9.- Heat-flow probe deployment configuration.

Heat flow experiment (S-037):

The major components of the heat flow experiment (fig . A-9) were two heat flow a probes and an electronics package. The probes were fiberglass tubular structures which supported the temperature sensors, heaters, and associated electrical wiring. Each probe section contained two gradient sensors , one located at each end, that were wired in a differential bridge configuration.

Each of the gradient sensors was surrounded by a 1000-ohm heater coil that could be activated on command from earth. In addition, two smaller ring sensors for measuring thermal conductivity were located 10 centimenters from each end of the probe section. Four thermocouples, located in the cable connecting the electronics to the probes, measure the absolute temperature as a function of time in the upper portion of the bore hole.

Placement of the heat flow electronics package required about 9 meters separation from the Apollo lunar surface experiments package central station, and alignment in an east-west direction. The separation required for the two bore holes for the heat flow probes was a minimum of 9 meters and no less than 5 meters from the electronics package.

Lunar seismic profiling experiment (S-203): The lunar seismic profiling experiment hardware (figs. A-10 and A-11) consisted of four geophones, marker flags, a geophone module with marker flag, and an electronics package in the Apollo lunar surface experiment package central station, a transmitter antenna, and eight explosive packages.

The four geophones sensed the seismic signals as each of the eight explosive packages were detonated, and the signals were telemetered back to earth through the central station along with timing data.

All experiment components with the exception of the explosive packages were deployed during ;the first extravehicular activity. The explosive packages consisted of a receiving antenna, a receiver, an explosive train, a signal processor, and a firing pulse generator. These were deployed at designated sites during the three lunar surface traverses. The packages were detonated by earth command at a designated time shortly after the crew left the lunar surface.

Lunar surface gravimeter (S-027): The lunar surface gravimeter (fig. A-12) experiment was a spring-mass type of the gravity sensor that required fine leveling for proper operation. Also the experiment temperature was maintained within very close limits. The gravimeter was mounted in three containers, one inside another. The three containers were an inner heater box, a pressurized intermediate instrument housing container, and an outer case. The intermediate box was suspended in the sealed instrument housing that was evacuated and back filled with nitrogen to a pressure of 10 torr. This provided the damping necessary for operating of the sensor and servo systems. The electronics package was mounted on top of the instrument housing.

The total instrument housing and electronics assembly was suspended from a gimbal for self-leveling. For fine adjustment, the center of gravity of the gimbal-suspended mass was adjusted by driving small, weighted motors attached to the heater box along a screw to a new position. Clearance between the instrument housing and the outer container permitted a swing of slightly more than 3 degrees in all directions.

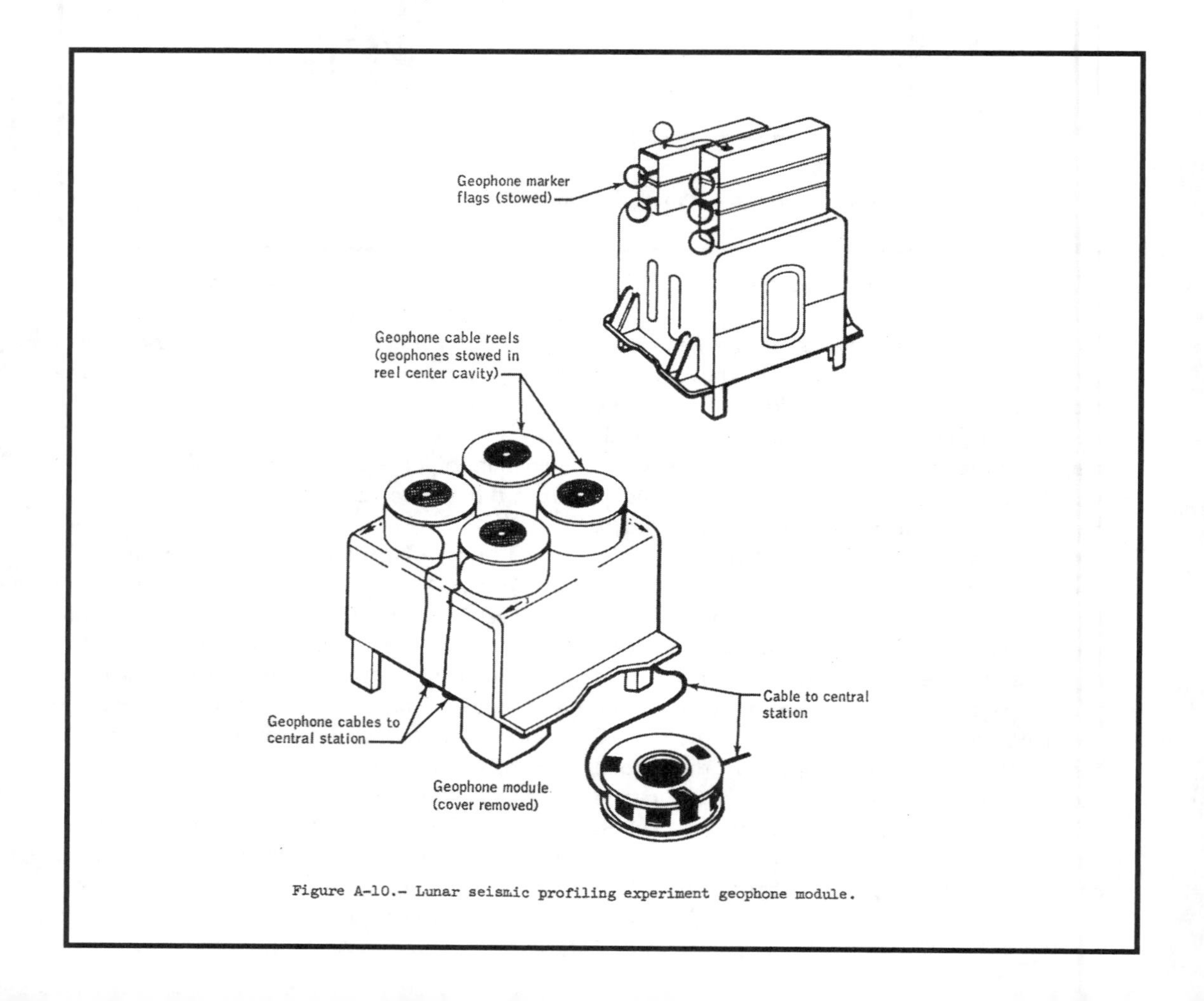

Figure A-10.- Lunar seismic profiling experiment geophone module.

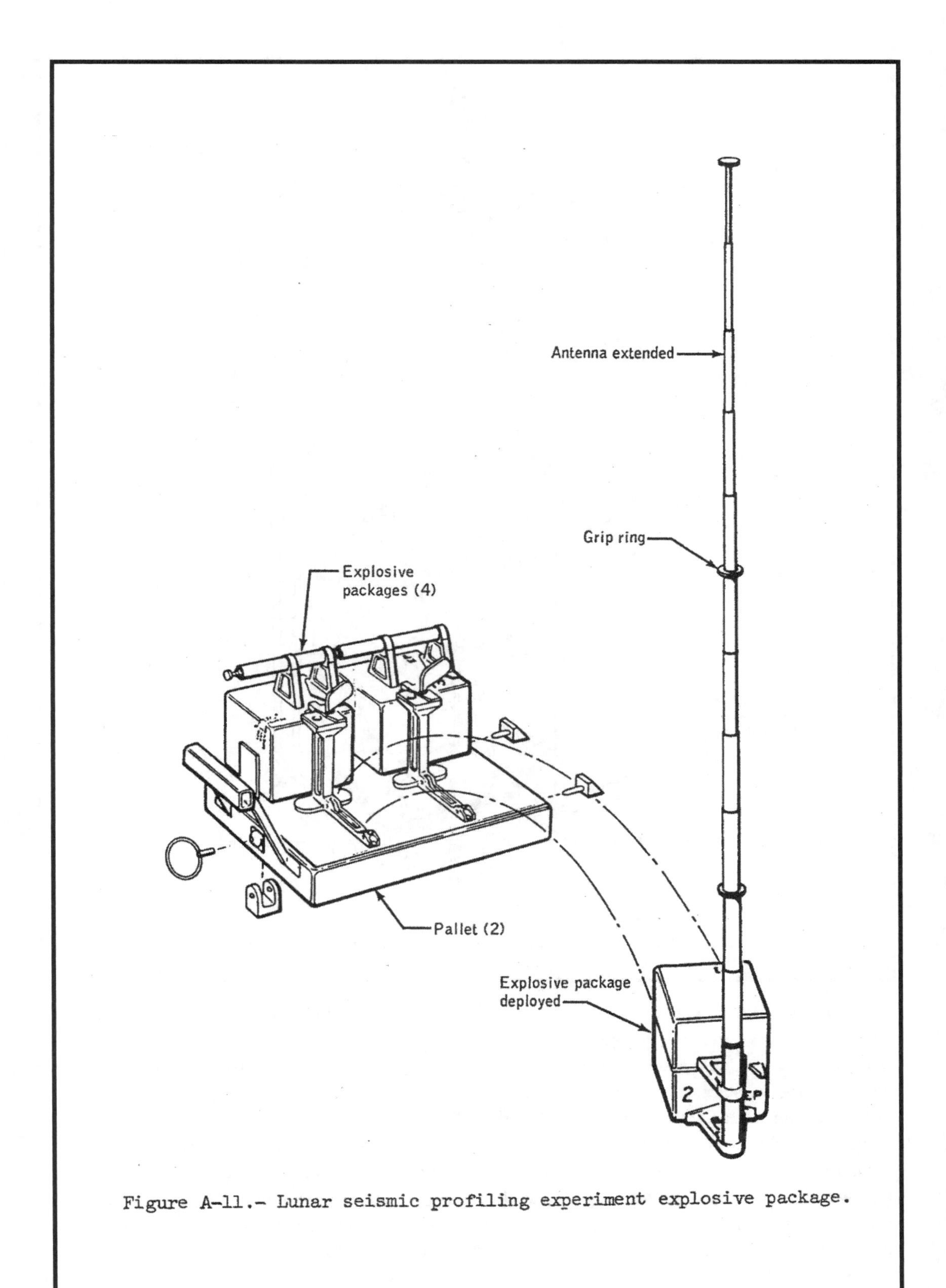

Figure A-11.- Lunar seismic profiling experiment explosive package.

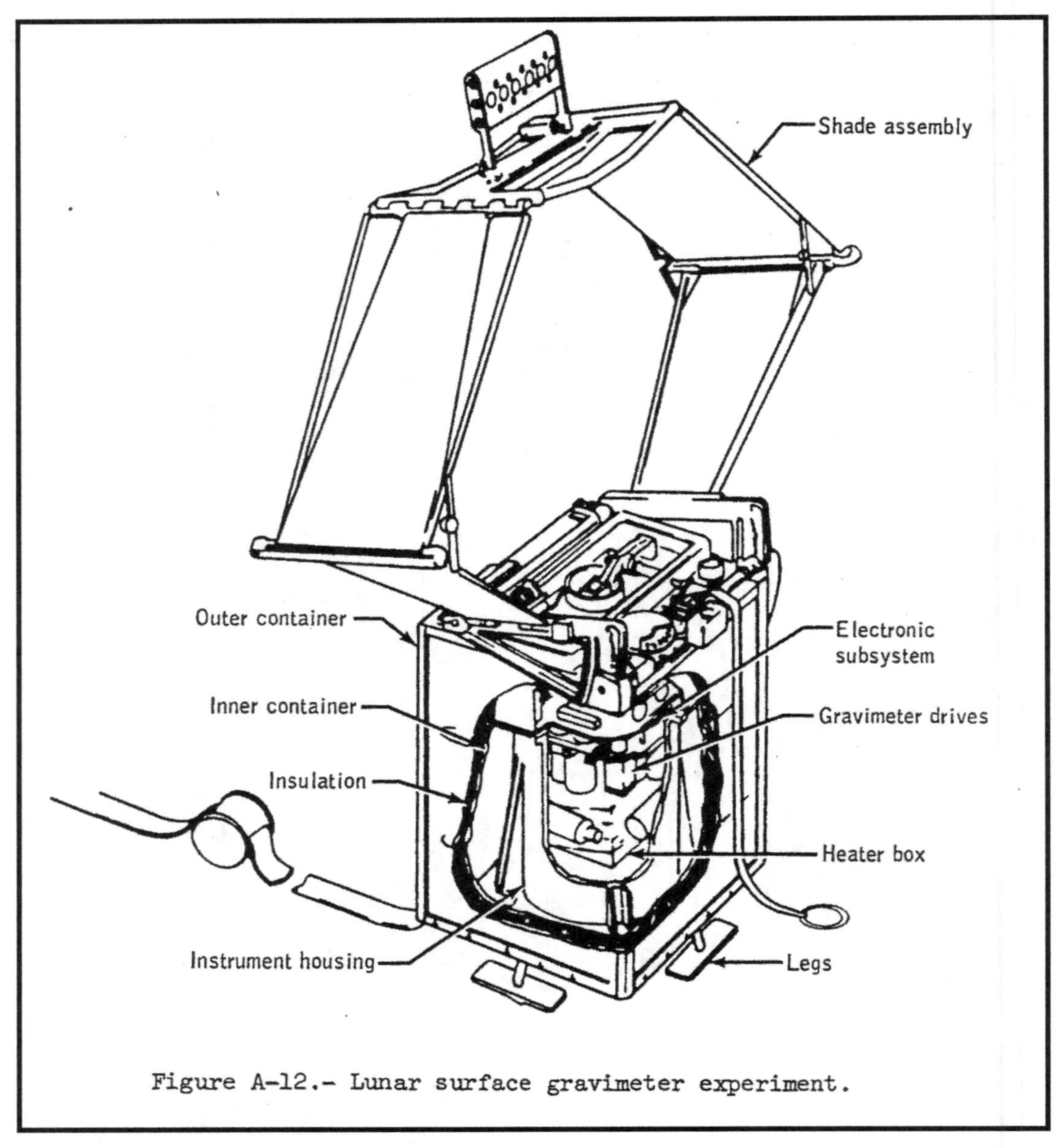

Figure A-12.- Lunar surface gravimeter experiment.

The container that enclosed the entire suspended mass was composed of insulation between two aluminum shells. Four legs, used for lunar surface emplacement, projected from the bottom of the container. The top had a cavity that contained the thermal radiator and the gimbal-actuator mechanism. Located on the top of the unit was a bubble level, the handling tool socket, the sun shield with its tilt mechanism, the tilt indicators, and detents for locking in a tilted position. A ribbon cable interconnected the unit with the central station. The unit required alignment to within 23 degrees of the sun line using the sunshield shadow.

Lunar atmospheric composition experiment (S-205):

The lunar atmospheric composition experiment equipment (fig. A-13) was composed of two primary systems - the gas analyzer and the electronics package with heaters. The electronics system was contained in a thermal bag, covered by a second-surface-mirror radiator plate to provide thermal control and maintain the electronics system temperature between 0° F (lunar night) and 150° F (lunar day). The gas analyzer system contained no thermal control and was subject to the lunar temperature range of -300° F to +250° F.

Gas molecules entered the gas analyzer and were ionized in the ion source, drawn out of the source, focused into an ion beam, and accelerated through a slit assembly leading to the drift tube. In the drift tube, the ions were deflected by a permanent magnet and collected by three electron multipliers (Heir-typ 90-degree magnetic-sector field-mass analyzers) at three exit-slit locations. The mass spectrum was scanned by varying the ion accelerating voltage in the ion source in a stepwise manner and counting the number of ions impinging on each detector at each voltage step. The magnitude of the count determined the concentration of each constituent of the gas sample in the ion source, and the voltage step number was calibrated to identify that constituent.

A vent valve was provided to allow the escape of the Krypton gas with which the experiment was filled on earth. The experiment required leveling within ±15 degrees and power was received from the central station.

The dust cover was closed on deployment for release by earth command at an appropriate time after lunar module ascent stage lift-off.

Lunar ejects and meteorites experiment S-202):

The experiment hardware fig. A-14) consisted of a deployable unit with detector plates electrically connected to the central station. Approximately 20 percent of the exposed outer surface of the unit was opened to admit microparticles through the top east and west sensors. Two of the three sensors had dualfilm assemblies to measure the time history of the particle flight. The third has a rear array only. One of the dual-film sensors faced up when deployed on the lunar surface and the other faced eastward. The singlefilm sensor faced westward. When a particle struck the front film of any sensor with sufficient momentum, it penetrated the very thin film and continued on toward the rear impact plate. As the metallized thin film was penetrated, an ionized plasma was produced and the particle lost some of its kinetic energy. The positive ions and electrons in the plasma were collected by the electrically biased films and grids, respectively, producing two coincident pulses. These were individually amplified and processed in the experiment electronics to identify the area of impact, to initiate time-of-flight measurement, and to analyze the pulse height of the positive film signal (which is related to the particle energy loss). Some of the particles that penetrate the front film intercepted the rear film also,

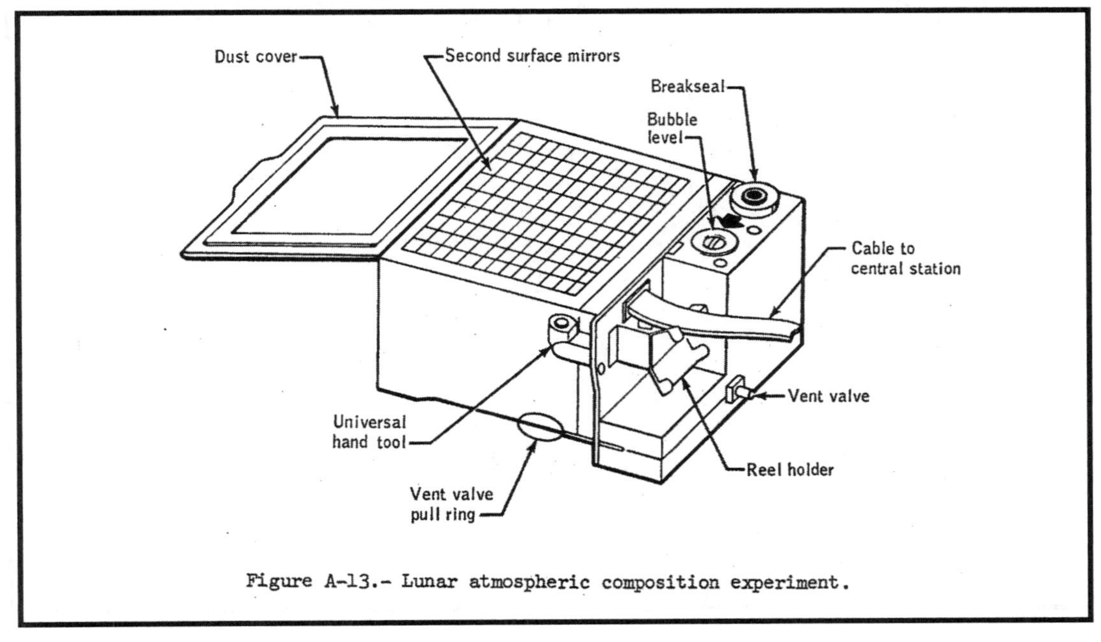

Figure A-13.- Lunar atmospheric composition experiment.

depending on their trajectory angle, and generated a plasma and a microphone pulse. These coincident signals were individually amplified and processed in the experiment electronics for transmittal through the central station back to earth.

The experiment required positioning so that no objects were above a 10-degree azimuth from a horizontal plane through the detector plates. Leveling and alignment of the unit was performed with a bubble level, shadowgraph, and gnomon. The detector plates were shielded by a dust cover that was opened by earth command after lunar lift-off.

A.4.2 Medical Experiments

Two medical experiments were flown. One was flown previously on Apollo 16 and the other was flown for the first time on Apollo 17. Biostack experiment (M-211).- The biostack experiment was flown on Apollo 1 and no change was made in the experiment hardware for Apollo 17 however, two biological specimens and one radiation detector were added. The biological

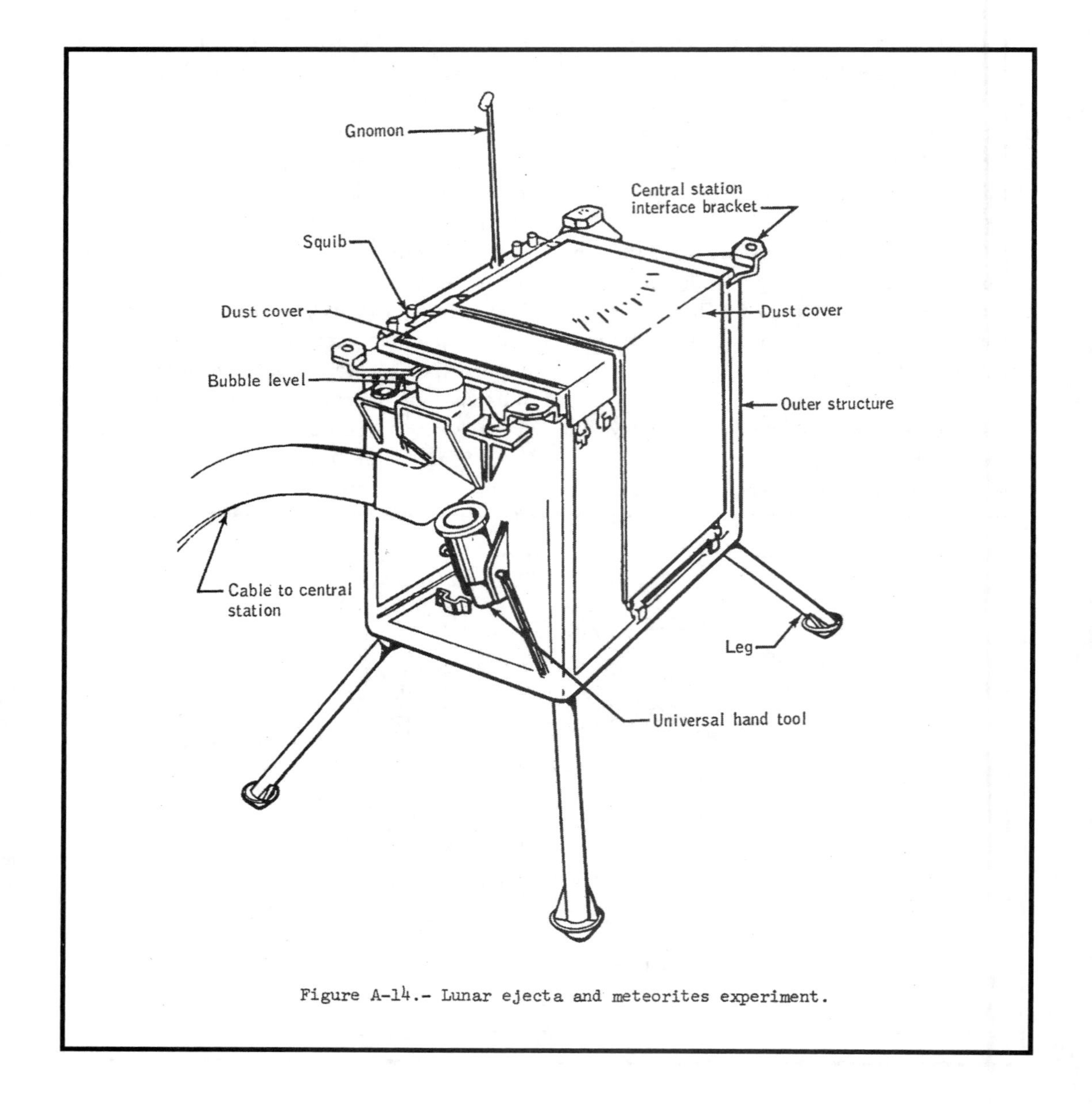

Figure A-14.- Lunar ejecta and meteorites experiment.

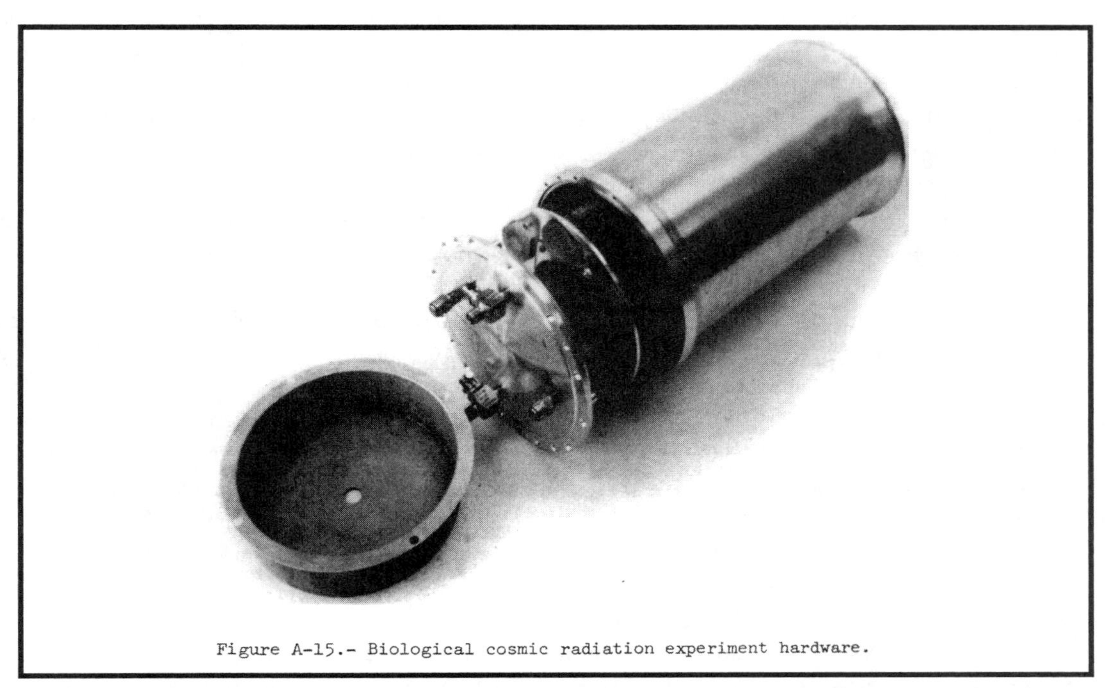

Figure A-15.- Biological cosmic radiation experiment hardware.

specimens added were colpoda cucullus (protozoa cycts - unicellular organisms) and tribolium castaneum (eggs of a flour beetle). The radiation detector was a salt substance made of silver chloride.

Biological cosmic radiation experiment (M-212)

The biological cosmic radiation experiment (fig. A -15), flown for the first time, was a radiation experiment using mice. The hardware consisted of a hermatically sealed, cylindrical aluminum canister that contained seven perforated cylindrical tubes. Attached to one end of the canister were redundant pressure relief valves and two manually controlled purge valves. Six of the seven tubes were arranged around the inside wall of the canister. The seventh tube was centrally located in the six-tube circular arrangement. Five of the six tubes contained a mouse and its food supply. The sixth tube was flown empty because the oxygen generating capability of the environmental control system indicated a marginal supply of oxygen for 6 mice. The tube in the center contained potassium superoxide granules for life support and operated by converting the carbon dioxide from the mice into oxygen. A radiation dosimeter was implanted beneath the scalp of each mouse, and a separate dosimeter was included in the stowage locker next to the hardware. Two self-recording temperature sensors were located in two of the mouse tube end caps.

Perognathus longimembris (little pocket mice) were selected as the animal species because of their small size, no requirement for water, and the availability of a large quantity of background data. Five mice, from a large number of mice having surgically implanted subscalp radiation dosimeters, were selected for flight specimens three days prior to launch and loaded into the canister. The canister was placed aboard the command module 30 hours prior to lift-off. The experiment was completely passive and required no crew participation.

The experiment was secured in the command module with its longitudinal axis perpendicular to the thrust axis during launch and entry. Location of the experiment was in an area shielded no more than 50 gm/cm2.

Postflight analysis of the subscalp

dosimeters will provide information that will permit sectioning of the mouse brain and eyes in such a manner as to maximize the probability of locating abnormalities (lesions) caused by heavy cosmic particles. The subscalp dosimeters will indicate the type, intensity, and trajectory of heavy cosmic particles traversing and stopping in the animals' brains and eyes. The self-recording temperature sensors that were located in the experiment hardware and the radiation dosimeter located in the locker will provide environmental data necessary for interpretation of the data provided by examination of the mouse tissues.

A.4.3 Inflight Science

The scientific instrument module experiments, excluding the panoramic and mapping cameras and laser altimeter flown on Apollo 15 and 16, were replaced by three new experiments. The three new experiments, which were the ultraviolet spectrometer, infrared scanning radiometer, and the lunar sounder, collected data on the physical properties and characteristics of the moon. The ultraviolet spectrometer was used to measure the moon's atmospheric density and composition. The infrared scanning radiometer mapped the inner surface thermal characteristics and the lunar sounder's developed a geological model of the lunar surface. The S-band transponder that used the command and service module communications system was not changed. Descriptions of the equipment used for each of these experiments are contained in the following paragraphs.

Far ultraviolet spectrometer.- The far ultraviolet spectrometer was an electro-optical device that detected ultraviolet energy in the 1180 Angstroms region after being re-radiated from particulate in the lunar atmosphere. The spectrometer measured the intensity of radiation as a function of the wavelength. The instrument (fig. A-16) was a 0.5-meter focal-length Ebert spectrometer whose optical components consisted of an external baffle, entrance slit, Ebert mirror, scanning diffraction grating, exit slit mirrors, exit slit, and photoelectric sensor (photomultiplier tube). The external baffle was a multiple-angled horn that greatly attenuated the light originating from sources outside the specified field of view.

The Ebert mirror collimated the radiation admitted by the entrance slit and redirected the radiation onto the diffraction grating. The mirror also refocused the diffracted radiation onto the exit slit. The specified entrance and exit slit dimensions and the Ebert mirror dimensions

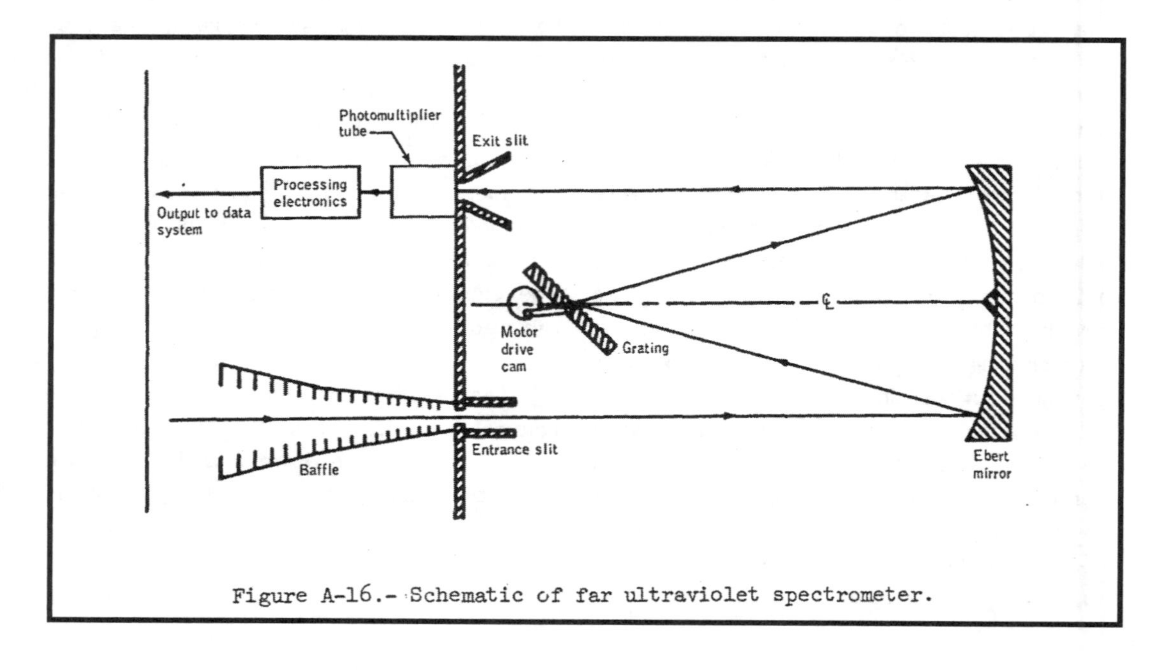

Figure A-16.- Schematic of far ultraviolet spectrometer.

were designed to maximize the energy transmission. The inclusion of the exit slit mirrors reduced the optical vignetting virtually to zero. The diffraction grating had an area of 100 square centimeters with 3600 grooves per millimeter.

The diffraction grating was cycled through the 1180 Angstroms to 1680 Angstroms wavelength range by a grating mechanism consisting of a motor, gear reducer drive, cam, cam follower, grating housing, and fiducial mark generator. A synchronous motor and gear reducer unit drove the cam in one direction at a uniform rate to repeat the wavelength scan cycle once per 12 seconds. As the cam rotated, the cam follower tilted the grating back and forth within the spectrum limits. A fiducial mark was generated by passing light from a gallium arsenide diode through a small hole in the cam to a photodiode detecter. The fiducial mark indicated the end of scan and its output synchronized the data word format.

The photomultiplier tube produced an electrical signal that was related to the intensity of the incident light. The electrical signal was processed by the electronics module for telemetry. The electronics module contained a pulse amplifier and discriminator, a high-voltage supply for the photomultiplier tube, a low-voltage dc/dc converter, pulse counting circuitry, and telemetry preconditioning circuitry.

A protective cover over the entrance of the external baffle shielded the photoelectric sensor from direct sunlight impingement and contamination from spacecraft effluent sources and reaction control system exhaust products. The cover was opened and closed from the command module.

Infrared scanning radiometer.- The infrared scanning radiometer basically consisted of a telescope with an electronic thermometer at its focal point. The device measured energy radiated from a given spot on the lunar surface. Hardware components of the system were a continuously rotating, time-referenced beryllium scan mirror; a Cassegrain telescope; a silicon hyperhemispheric immersion lens; a thermistor balometer (sensitive thermometer) ; three buffer amplifiers; and the attendant signal conditioning equipment.

Light entered the telescope from a moter-driven mirror. The mirror rotated so that the spot on the moon's surface, as seen by the telescope, was scanned across the ground track of the spacecraft. The light passed through the various mirrors, baffles, and lens of the telescope to the thermistor (detector). The detector, which was a solid state thermometer, related the radiant energy to an electrical signal that was amplified, telemetered, and recorded for post-mission conversion to lunar surface temperature data. The relationship of the orbiting spacecraft to the area scanned by the experiment is shown in figure A-17.

The field of view of the telescope was such that the surface resolution of the nadir was approximately 2 kilometers, temperatures were recorded over a range of -213° C to +127° C with an accuracy of one degree.

A sensor protective cover, operated from the command module when required, provided protection to the optical scanning unit from direct sunlight impingement and spacecraft contamination sources such as effluent liquids and reaction control system products.

Lunar sounder.- The lunar sounder equipment (fig. A-18) consisted of a coherent synthetic aperture radar system that operated at either 5 MHz, 15 MHz, or in the very high frequency band of 150 MHz and transmitted a series of swept frequency pulses.

The lunar sounder transmitted electromagnetic pulses from the spacecraft during lunar orbit. A small part of the pulse energy was reflected from the lunar surface and subsurface features, and subsequently detected by a receiver on the spacecraft. The optical recorder recorded the radar video output from the receiver on film. The film cassette was retrieved during the transearth extravehicular activity. The recordings were time referenced for later data reduction and analysis. The shape, amplitude, and

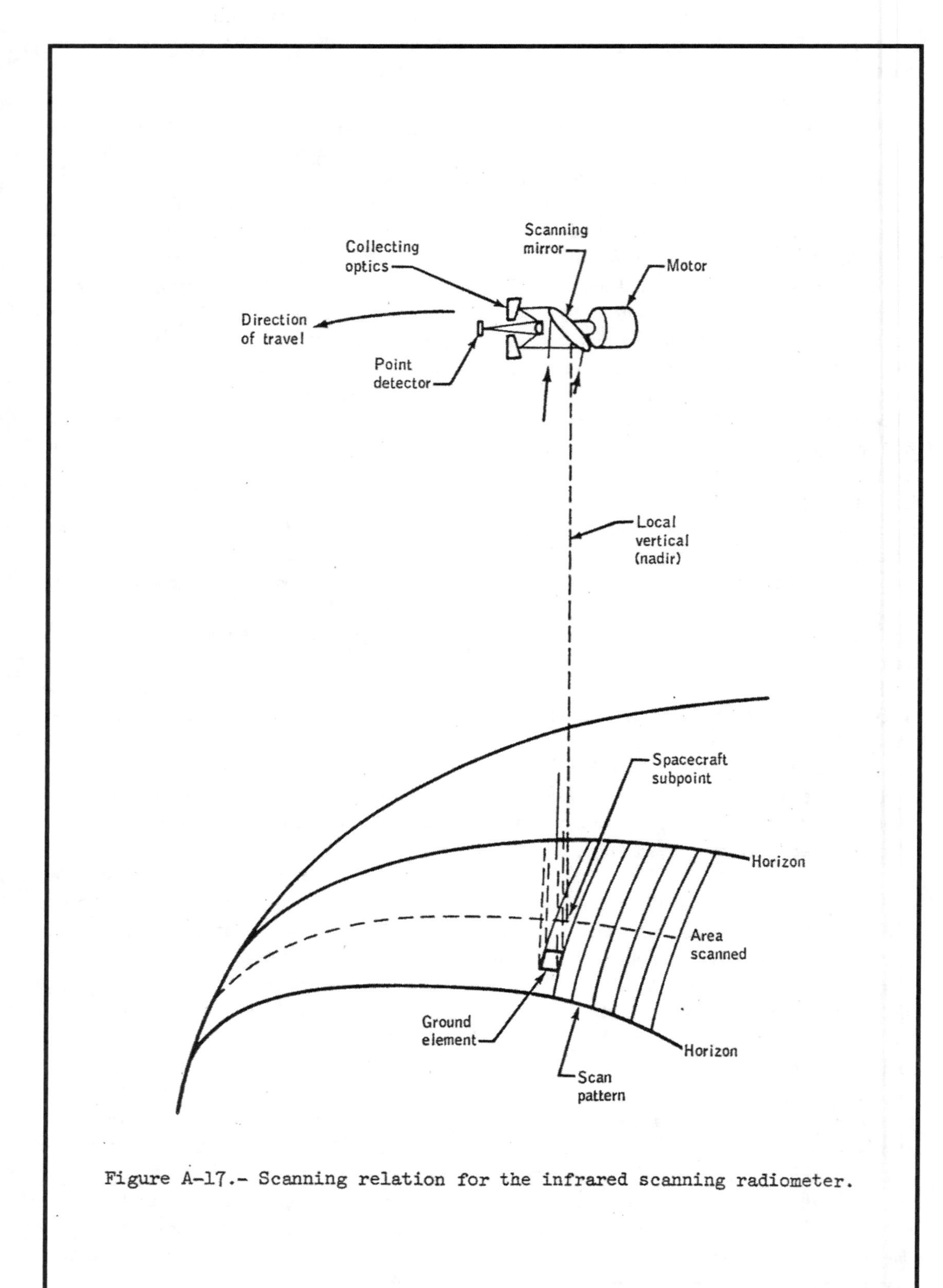

Figure A-17.- Scanning relation for the infrared scanning radiometer.

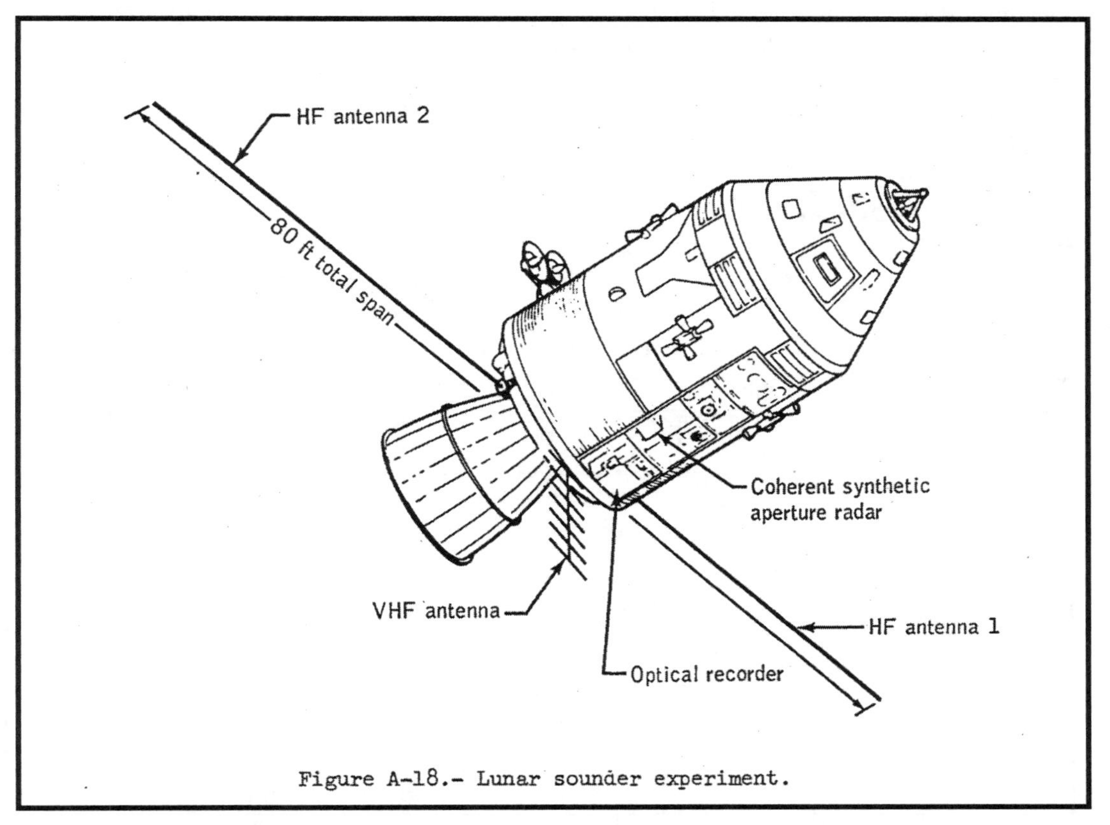

Figure A-18.- Lunar sounder experiment.

time of arrival of the detected pulses relative to the shape, amplitude, and time of transmission provided data which may be interpreted in terms of the three dimensional distribution of scattering centers. This information together with calculated electromagnetic parameters will be used to produce a structural model of the moon to a depth of one kilometer.

Two retractable antennas were used by the high-frequency radar and a Yagi antenna was used by the very high frequency radar. The high-frequency antenna assembly was deployed at the moldline of the service module so that the antennas, when deployed, were at 90° to the service module center axis and 180° apart. The overall span of both antennas, when deployed, was 80 feet. The time required for deployment or retraction was approximately 2 minutes. The very high frequency antenna was mounted on the aft shield and consisted of a spring-loaded 274-centimeter boom with seven cross-bars. In the deployed position, the antenna extended beyond the service module moldline and at right angles to the service module center axis. Deployment occurred automatically when the spacecraft /launch vehicle adapter panels were jettisoned.

Mapping camera and laser altimeter.- Improvements to the mapping camera and laser altimeter equipment were made to correct anomolous performances experienced on previous flights. The abnormal behavior of the mapping camera deployment mechanism was corrected by modifying the gear box seal and removing the locking rollers and excess lubricants. The degraded performance of the laser altimeter (low output power) was corrected by using a better grade of quartz in the flash tube and changing the pulse forming voltage. Contamination of optical surfaces within the laser module was reduced by placing a cover over the drive mechanism and using lubricant impregnated bearings.

Heat flow and convection demonstration.- A heat flow and convection demonstration device was flown. The apparatus was similar in operation and design to the one flown on Apollo 14. Crew participation and photographic coverage was required.

A.5 PHOTOGRAPHIC TASKS AND EQUIPMENT

The majority of the mission activities required photographic coverage. Cameras were located in the service module, command module, lunar module, and on the lunar surface to accomplish numerous tasks. Table A-I lists the cameras, lens, and films used to accomplish the specific tasks of most photographic activities.

A.6 MASS PROPERTIES

Mass properties for the Apollo 17 mission are summarized in table A-II. These data represent the conditions as determined from postflight analyses of expendable loadings and usage during the flight. Variations in the command and service module and lunar module mass properties are determined for each significant mission phase from lift-off through landing. Expendables usage

TABLE A-I.- PHOTOGRAPHIC EQUIPMENT

Subject	Camera type (a)	Lens	Film type (b)
Scientific Instrument Module photographic tasks	PC	24 inch	LBW
	MC	3 inch	BW (3400)
	SC	3 inch	BW (3401)
Solar and Lunar surface photographic tasks from the command module	HEC	80-mm	VHBW
	DAC	18-mm	VHBW
	35	55-mm	VHBW
	HEC	250-mm	VHBW
	HEC	250-mm	CEX
Lunar sounder	CRT	12.3 cm	TVBW
ALSEP/central station	HEDC	E0-mm	HCEX
Radioisotape thermoelectric generator	HEDC	60-mm	HCEX
Heat flow	HEDC	60-mm	HCEX
Lunar geology	HEDC	60-mm	BW (3400)
	HEDC	60-mm	RW (3401)
Soil Mechanics	HEDC	60-mm	BW (3410)
	HEDC	60-mm	HCEX
Traverse gravimeter	HEDC	60-mm	HCEX
Lunar ejecta and meteorites	HEDC	60-mm	HCEX
Lunar surface electrical properties	HEDC	60-mm	HCEX
Lunar seismic profiling	HEDC	60-mm	HCEX
Lunar atmospheric composition	HEDC	60-mm	HCEX
Lunar surface gravimeter	HEDC	60-mm	HCEX

[a] camera nomenclature:

PC Panoramic camera
MC Mapping camera
SC Stellar carver
HC Electric camera
HEDC Electric data camera
DAC Data acquisition camera
35 35-mm camera
ORCRT Optical recorder cathode ray tube

[b] Film nomenclature:
LBW Low speed black and white
BW Medium speed black and white (3400 and 3401)
VHBW Very high speed black and white
CEX Color exterior
TVBW Television black and white
HCEX High-speed color exterior

are based on reported real-time and postflight data as presented in other sections of this report. The weights and center-of-gravity of the individual modules (command, service, ascent stage, and descent stage) were measured prior to flight and inertia values were calculated. All changes incorporated after the actual weighing were monitored, and the mass properties were updated.

TABLE A-II.- MASS PROPERTIES

Event	Weight, lb	Center of gravity, in.			Moment of inertia, slug-ft²			Product of inertia, slug-ft²		
		X	Y	Z	I_{XX}	I_{YY}	I_{ZZ}	I_{XY}	I_{XZ}	I_{YZ}
Command and service module/lunar module										
Lift-off	116 265	843.2	3.1	2.5	73 297	1 238 778	1 239 235	4232	10 205	2700
Earth orbit insertion	107 161	804.5	3.3	2.7	72 424	768 200	768 679	6487	11 585	2687
Transposition and docking:										
Command & service module	66 593	934.1	5.0	4.8	36 228	80 286	82 004	-2083	179	2349
Lunar module	36 274	1233.1	-0.7	0.8	25 885	26 237	27 100	-517	228	-417
Total docked	103 267	1041.0	3.0	3.4	62 355	576 019	578 682	-11 363	-5740	2046
Lunar orbit insertion	102 639	1041.6	2.9	3.5	61 929	574 405	577 170	-11 171	-5997	2170
Descent orbit insertion	76 354	1084.0	1.9	2.1	43 315	442 736	448 763	-8708	-1464	-925
Separation	74 762	1085.2	1.8	2.0	43 876	434 027	439 371	-7924	-1758	-1050
Command and service module circularization	37 960	944.1	3.8	3.4	21 403	60 315	64 736	-2419	1253	-604
Command and service module plane change	37 464	944.4	3.8	3.4	21 100	60 134	64 466	-2427	1276	-563
Docking:										
Command and service modules	36 036	945.8	3.6	3.4	20 336	59 726	63 433	-2349	1257	-621
Ascent stage	5878	1165.7	4.8	-2.9	3289	2304	2690	-101	-17	-359
Total after docking:										
Ascent stage manned	41 914	976.6	3.8	2.5	23 670	114 848	118 899	-2148	-267	-1091
Ascent stage unmanned	41 896	974.4	3.5	2.5	23 584	110 226	114 280	-2598	-60	-957
After ascent stage jettison	36 619	946.6	3.7	3.1	20 453	59 785	63 465	-2249	1069	-640
Transearth injection	36 394	946.7	3.8	3.1	20 265	59 638	63 352	-2260	1086	-583
Transearth extravehicular activity	26 933	971.7	1.1	3.7	15 494	45 505	44 686	-841	791	-1154
Command and service module prior to separation	26 659	972.4	1.1	3.9	15 184	44 977	44 214	-773	796	-1042
After separation:										
Service module	13 507	907.7	2.3	2.0	9217	14 954	14 724	-364	454	-1033
Command module	13 152	1038.8	-0.2	5.9	5936	5289	4786	56	-385	9
Entry	13 140	1038.8	-0.2	5.8	5929	5282	4767	56	-383	9
Main parachute deployment	12 567	1037.3	-0.2	6.0	5724	4879	4406	57	-331	12
Landing	12 120	1035.4	-0.2	6.0	5678	4583	4092	55	-337	12
Lunar module										
Earth lift-off	36 277	184.2	0.4	-1.0	25 890	27 352	26 201	63	558	163
Separation	36 771	185.2	0.4	-0.4	27 342	28 574	27 285	56	857	145
Second descent orbit insertion	36 746	185.1	0.4	-0.5	27 328	28 529	27 234	57	858	142
Powered descent initiation	36 686	185.0	0.4	-0.5	27 295	28 401	27 086	59	858	142
Lunar landing	18 305	208.5	0.6	-0.8	15 711	16 661	17 970	16	915	170
Lunar lift-off	10 997	243.7	0.0	3.0	6776	3415	5985	66	174	-32
Orbit insertion	6042	256.6	0.0	5.5	3371	2917	2116	66	82	-32
Terminal phase initiation	5970	256.3	0.0	5.5	3336	2904	2076	66	84	-29
Docking	5878	256.1	0.0	5.6	3259	2885	2016	66	85	-28
Jettison	5277	255.3	-0.4	3.3	3128	2764	1953	77	73	-29

APPENDIX B - SPACECRAFT HISTORIES

The history of command and service module (CSM-114) operations at the manufacturer's facility, Downey, California, is shown in figure B-1, and the operations at Kennedy Space Center, Florida in figure B-2. The history of the lunar module (LM-12) at the manufacturer's facility, Bethpage, New York, is shown in figure B-3, and the operations at Kennedy Space Center, Florida, in figure B-4.

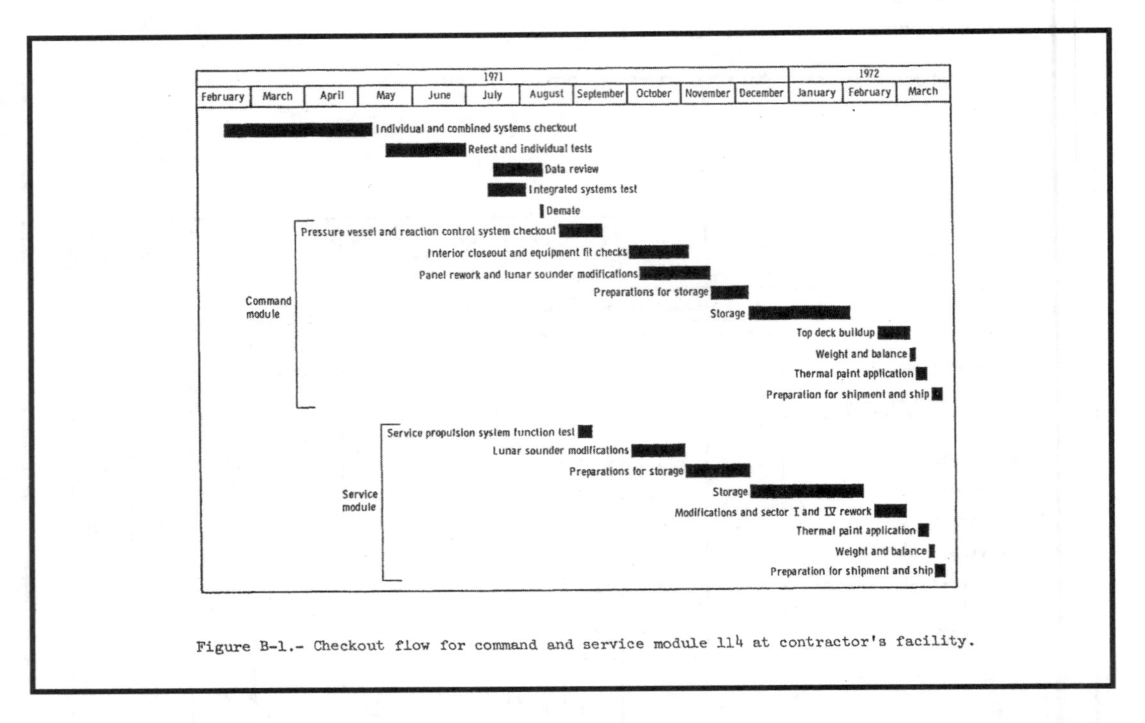

Figure B-1.- Checkout flow for command and service module 114 at contractor's facility.

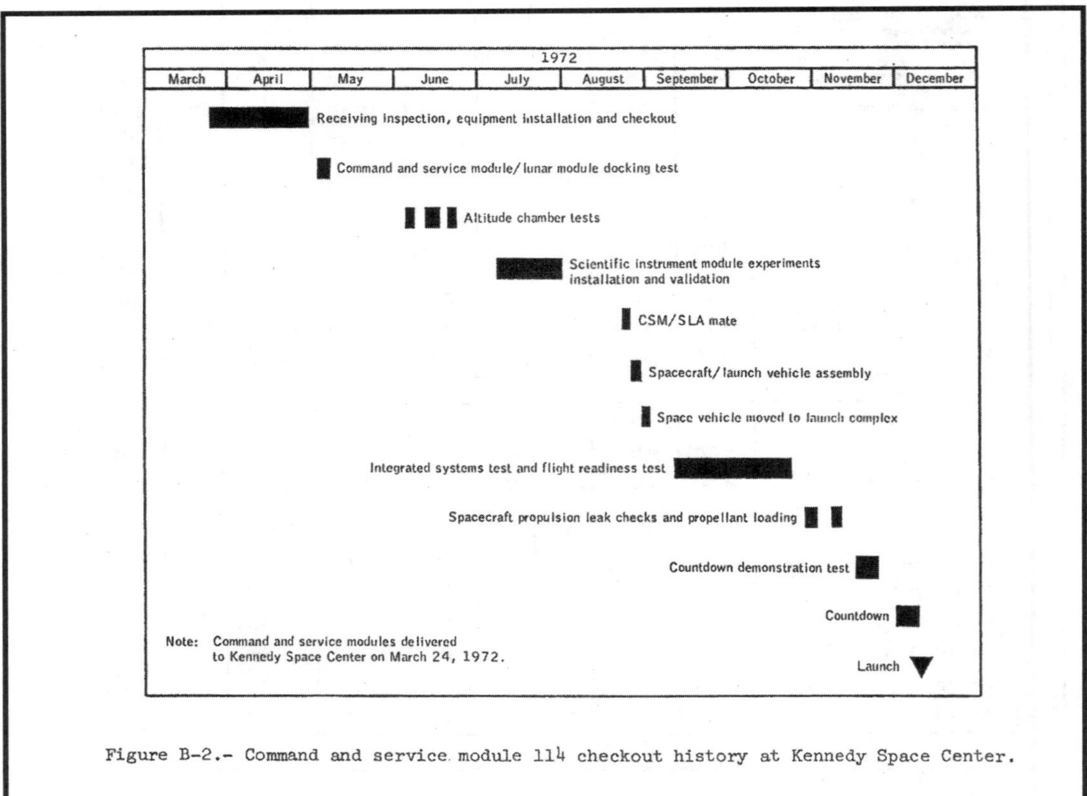

Figure B-2.- Command and service module 114 checkout history at Kennedy Space Center.

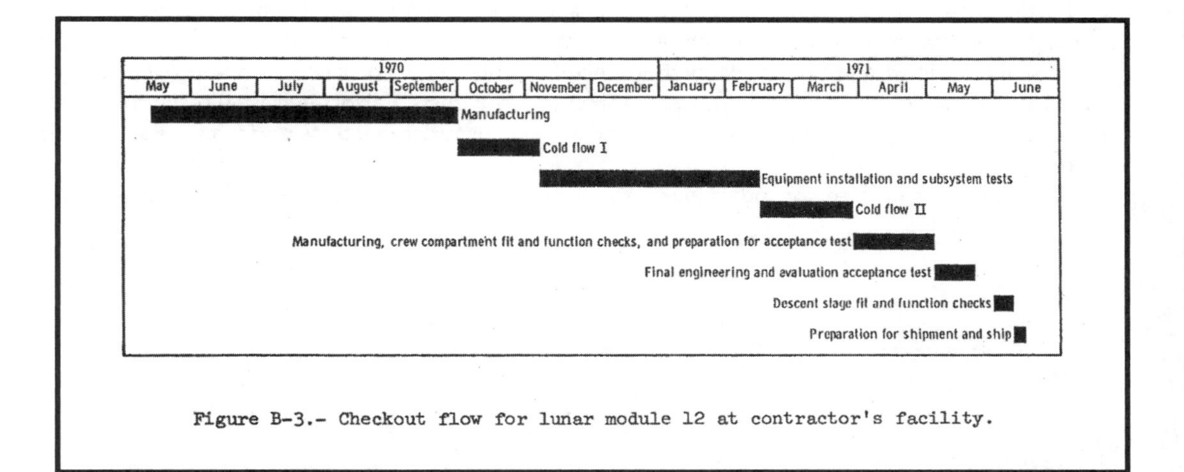

Figure B-3.- Checkout flow for lunar module 12 at contractor's facility.

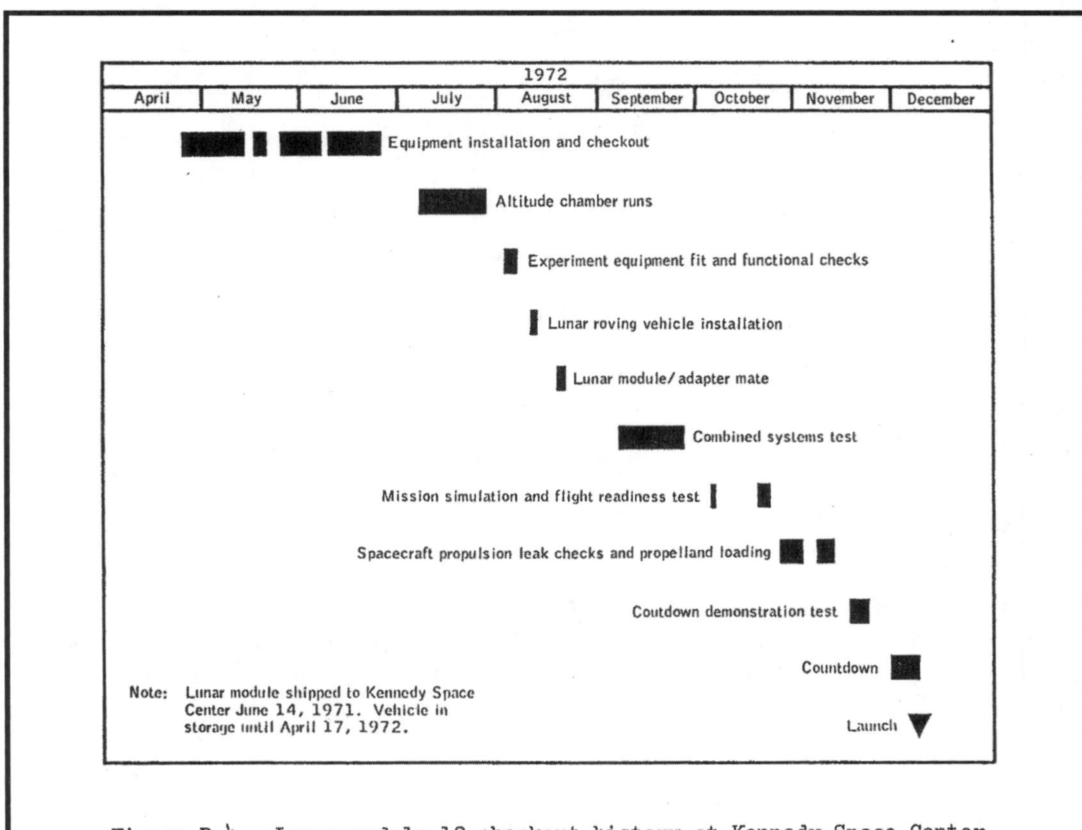

Figure B-4.- Lunar module 12 checkout history at Kennedy Space Center.

APPENDIX C - POSTFLIIGHT TESTING

Postflight testing and inspection of the command module and crew equipment for evaluation of the inflight performance and investigation of the flight irregularities were conducted at the contractor's and vendor's facilities and at the Manned Spacecraft Center in accordance with approved , Apollo Spacecraft Hardware Utilization Requests (ASHUR's). The tests performed as a result of inflight problems are described in table C-I and discussed in the appropriate systems performance secitons of this report.

Tests being conducted for other purposes in accordance with other ASHUR's and the basic contract are not included.

TABLE C-I.- POSTFLIGHT TESTING SUMMARY

ASHUR no.	Purpose	Tests performed	Results
Guidance and Navigation			
114015 114508	Determine cause of unexplained accelerometer bias shifts in the entry monitor system.	Perform bench tests and analysis.	Bench testing did not reveal any malfunction. The inflight problem could not be duplicated.
Electrical Power			
114016	Determine cause of panel 306 mission timer being 15 seconds slow.	Perform functional tests.	Timer operated normally during all testing, maintaining correct time to the second.
114017	Determine cause of spurious master alarms without accompanying matrix light and momentary main A undervolt light.	Perform functional tests in commend module. Remove panel and inspect wiring, solder joints, connectors, and switches. Disassemble and inspect suspect switches and oxygen pressure meter for contamination.	Tests in command module did duplicate the anomaly. Inspection of wiring and connectors and examination of suspect switches did not disclose any abnormal conditions. Testing is not complete.
Communications			
114020	Determine cause for loss of several inatrumentatton parameters daring a two-minute period.	Functionally test the pulse code modulation equipment in the command module. Remove assembly from command module and test for presence of contamination	Spacecraft tests did not reveal the cause, Testing is not complete,
Crew Equipment			
114018	Determine cause for gray tape not sticking properly.	Perform pull testing on tape from same lot as that on Apollo 17.	Bonding properties were within specification.
114021	Determine why set screws on 18-mm lens interferred with mounting 16-mm camera on right angle mirror.	Examine returned hardware.	Examination is not complete.

APPENDIX D - DATA AVAILABILITY

There was no data processed for Apollo 17 for postmission evaluation. The following table contains the times that experimental data were made available to the principal investigators for scientific analysis.

Time, hr:min	
From	To
81:23	104:10
110:51	115:40
120:25	132:30
136:36	161:15
187:41	192:25
193:30	194:25
195:51	208:35
212:10	227:42
229:11	253:37
256:35	299:32

APPENDIX E - MISSION REPORT SUPPLEMENTS

Table E-I contains a listing of all reports that supplement the Apollo 7 through Apollo 17 mission reports. The table indicates the present status of each report not yet completed and the publication date of those which have been published.

TABLE E-I.- MISSION REPORT SUPPLEMENTS

Supplement number	Title	Publication date/status
	Apollo 7	
1	Trajectory Reconstruction and Analysis	May 1969
2	Communication System Performance	June 1969
3	Guidance, Navigation, and Control System Performance Analysis	November 1969
4	Reaction Control System Performance	August 1969
5	Cancelled	
6	Entry Postflight Analysis	December 1969
	Apollo 8	
1	Trajectory Reconstruction and Analysis	December 1969
2	Guidance, Navigation, and Control System Performance and Analysis	November 1969
3	Performance of Command and Service Module Reaction Control System	March 1970
4	Service Propulsion System Final Flight Evaluation	September 1970
5	Cancelled	
6.	Analysis of Apollo 8 Photography and Visual Observations	December 1969
7	Entry Postflight Analysis	December 1969
	Apollo 9	
1	Trajectory Reconstruction and Analysis	November 1969
2	Command and Service Module Guidance, Navigation, and Control System Performance	November 1969
3	Lunar Module'Abort Guidance System Performance Analysis	November 1969
4	Performance of Command and Service Module Reaction Control System	April 1970
5	Service Propulsion System Final Flight Evaluation	December 1969
6	Performance of Lunar Module Reaction Control System	August 1970
7	Ascent Propulsion System Final Flight Evaluation	December 1970
8.	Descent Propulsion System Final Flight Evaluation	December 1970
9	Cancelled	
10	Stroking Test Analysis	December 1969
11	Communications System Performance	December 1969
12	Entry Postflight Analysis	December 1969
	Apollo 10	
1	Trajectory Reconstruction and Analysis	March 1970
2	Guidance, Navigation, and Control System Performance Analysis	December 1969
3	Performance of Command and Service Module Reaction Control System	August 1970
4	Service Propulsion System Final Flight Evaluation.	September 1970

Supplement number	Title	Publication date/status
5	Performance of Lunar Module Reaction Control System	August 1970
6	Ascent Propulsion System Final Flight Evaluation	January 1970
7	Descent Propulsion System Final Flight Evaluation	January 1970
8	Cancelled	
9	Analysis of Apollo 10 Photography and Visual Observations	August 1971
10	Entry Postflight Analysis	December 1969
11	Communications System Performance	
	Apollo 11	
1	Trajectory Reconstruction and Analysis	May 1970
2	Guidance, Navigation, and Control System Performance Analysis	September 1970
3	Performance of Command and Service Module Reaction Control System	December 1971
4	Service Propulsion System Final Flight Evaluation	October 1970
5	Performance of Lunar Module Reaction Control System	December 1971
6	Ascent Propulsion System Final Flight Evaluation	September 1970
7	Descent Propulsion System Final Flight Evaluation	September 1970
8	Cancelled	
9	Apollo 11 Preliminary Science Report	December 1969
10	Communications System Performance	January 1970
11	Entry Postflight Analysis	April 1970
	Apollo 12	
1	Trajectory Reconstruction and Analysis	September 1970
2	Guidance, Navigation, and Control System Performance Analysis	September 1970
3	Service Propulsion System Final Flight Evaluation	December 1971
4	Ascent Propulsion System Final Flight Evaluation	Publication
5	Descent Propulsion System Final Flight Evaluation	Publication
6	Apollo 12 Preliminary Science Report	July 1970
7	Landing site Selection Processes	Final Review
	Apollo 13	
1	Guidance, Navigation, and Control System Performance Analysis	September 1970
2	Descent Propulsion System Final Flight Evaluation	October 1970
3	Entry Postflight Analysis	Cancelled
	Apollo 14	
1	Guidance, Navigation, and Control System Performance Analysis	January 1972
2	Cryogenic Storage System Performance Analysis	March 1972

Supplement number	Title	Publication date/status
3	Service Propulsion System Final Flight Evaluation	May 1972
4	Ascent Propulsion System Final Flight Evaluation	May 1972
5	Descent Propulsion System Final Flight Evaluation	Publication
6	Apollo 14 Preliminary Science Report	June 1971
7	Analysis of Inflight Demonstrations	January 1972
8	Atmospheric Electricity Experiments on Apollo 13 and 14 Launches	January 1972
	Apollo 15	
1	Guidance, Navigation and Control System Performance Analysis	October 1972
2	Service Propulsion System Final Flight Evaluation	Preparation
3	Ascent Propulsion System Final Flight Evaluation	October 1972
4	Descent Propulsion System Final Flight Evaluation	October 1972
5	Apollo 15 Preliminary Science Report	April 1972
6	Postflight Analysis of the Extravehicular Communications System - Lunar Module Communications Link	January 1972
7	Analysis of Command Module Color Television Camera	Cancelled
	Apollo 16	
1	Guidance, Navigation and Control System Performance Analysis	November 1972
2	Service Propulsion System Final Flight Evaluation	Preparation
3	Ascent Propulsion System Final Flight Evaluation	Preparation
4	Descent Propulsion System Final Flight Evaluation	Preparation
5	Apollo 16 Preliminary Science Report	Preparation
6	Microbial Response and Space Environment Experiment (S-191)	Preparation
7	Analysis of Fluid Electrophoresis Demonstration	Preparation
	Apollo 17	
1	Apollo 17 Preliminary Science Report	Preparation
2	Cosmic Radiation Effects on Biological Systems Exposed on Apollo 16 and 17	Preparation
3	Further Observations on the Pocket Mouse Experiment (BIOCORE-M212)	Preparation
4	Results of Apollo Light Flash Investigations	Preparation
5	Results of Apollo 17 Heat Flow and Convection Demonstration	Preparation
6	Apollo 17 Calibration Results for Gamma Ray Spectrometer Silver Iodide Crystal	Preparation
7	Meteorological Observation and Forecasting During the Apollo 17 Mission	Preparation

APPENDIX F - GLOSSARY

Ackermann steering	A type of linkage that provides a high wheel response for a small steering motion.
Agglutinate	A deposit of originally molten ejecta.
Albedo	Relative brightness, defined as the ratio of radiation reflected from a surface to the total amount incident upon it.
Aldosterone	A steroid hormone extracted from the adrenal cortex that is very active in regulating the salt and water balance in the body.
Anorthositic gabbro	A granular textured igneous rock regarded as having solidified at considerable depth. It is composed almost entirely of a soda-lime feldspar.
Antidiuretic	Suppressing the secretion of urine.
Arrhythmia	any variation from the normal rhythm of the heart beat.
Atomic	Having its atoms in an uncombined form (atomic hydrogen).
Basalt	Any fine-grained dark-colored igneous rock that is the extrusive equivalent of gabbro.
Benard	A circulation phenomenon associated with convective heating in liquid flow.
Bimodal	Possessing two statistical modes.
Bouguer corrections	Empirical adjustments to relative gravity values for variations in the density of material from the density of material from the reference point (lunar module site).
Breccias	A rock consisting of sharp fragments embedded in a five-grained matrix.
Bungee	A latch loading mechanism for locking the docking ring latches.
Bungee strap	Elasticized cord used as a fastener.
Clast	A discrete fragment of rock or mineral included in a larger rock.
Competent lava	A mass strong enough to transmit thrust effectively and sustain the weight of overlying strata.
Continuum of morphologies	Continuous charges in geological and topographical structure.
Corticoids	A term applied to hormones of the steroid adrenal convex or any other natural or synthetic compound having a similar activity.
Cortisone	A colorless crystalline steroid hormone of the adrenal cortex.
Craterlets	Little craters.
Dacite domes	Mounds present in volcanic fields formed by the volcanic extrusion of dacite, a type of rock that is highly viscous.
Dikelet	A small dike 2 to 3 centimeters in width.
Down-sun	In the same direction as the solar vector and used in specifying lunar surface photography requirements.
Earthshine	Illumination of the moon's surface by sunlight reflected from the earth's surface and atmosphere.
Ecliptic	The great circle of the celestial sphere that is the apparent path of the sun among the stars or of the earth as seen from the sun.
Ejects	Material thrown out from volcanic action or meteoroid impact.
Endocrine	Secreting internally; applied to organs whose function is to secrete into the blood.
Ephemeris	A table of the computed positions of celestial bodies at regular intervals.
Euthanatized	Put to death in a painless humane fashion.
Feldspar	A group of minerals, closely related in crystalline form, that are the essential constituents of nearly all crystalline rocks. They have a glassy luster and break rather easily in two directions at approximately right angles to each other.

Fillet	Debris (soil) piled against a rock.
Fissure	Crack.
Flatulence	Of or having gas in the stomach or intestines.
Free air correction	Adjustments to relative gravity values for distances above the reference level (lunar module site).
Herpetic	An inflammatory virus disease of the skin.
Homeostatic	Stability in the normal body states of the organism.
Hornfelsic	Resembling a fine-grained silicate rock.
Hummocks	Low rounded hills or knolls.
Hypogravic	Pertaining to low gravity, particularly the low gravity experienced in space.
Ilmenite	A mineral, rich in titanium and iron, and usually black with a submetallic luster.
In situ	In its original place or natural locale.
Insensible water	Not appreciable by or perceptible to the senses. Irradiated The application of rays, such as ultraviolet rays, to a substance to increase its vitamin efficiency.
Isophote	A line or surface on a chart forming the locality of points of equal illumination or light intensity from a given source.
J-Mission	A classification of Apollo lunar explorations missions for which provisions were made for extended lunar surface stay time, surface vehicular mobility and communications, and more extensive science data acquisition. Apollo 15, 16, and 17 were of this classification.
Krytox oil	Fluorinated hydrocarbon lubricant.
Libration	Any of several phenomena by which an observer on earth, over a period of time, can observe more than one hemisphere of the moon.
Limb	The outer edge of the apparent disc of a celestial body such as the moor_ or earth, or a portion of the edge.
Lineament	The distinguishing topographic or characteristic feature of a mass.
Lithic	Stone like.
Lithology	The character of a rock formation or of rock found in a geological area expressed in terms of its physical characteristics.
Lunation	The period of one revolution of the moon about the earth with respect to the sun. A period of 29 days 12 hours 44 minutes 2.8 seconds.
Magnetite	Black isometric mineral of the spinel group. An non-important iron ore that is strongly attracted by a magnet and sometimes possesses polarity.
Mantle	The inner part of a terrestrial-like planet lying between the crust and the core.
Mascon	Large mass concentrations beneath the surface of the moon.
Maskelynite	A feldspar found in meteorites.
Massif	A large fault block of mountainous topography displaced as a unit without internal change.
Microdensitometry	An instrument that measures optic density. Morphology The external structure of rocks in relation to the development of erosional forms or topographic features.
Mottled	Marked with blotches or spots of different colors or shades.
Nadir	The point on the celestial sphere that is vertically downward from the observer.
Newton rings	Colored rings due to light interference effects from reflected or transmitted light rays that are seen about the contact of a convex lens with a plane surface, or of two lenses differing in curvature.

Norite	A variety of granular igneous rock composed essentially of a calcium-sodium series that usually contains a ferromagnesium mineral.
Orthopyroxene	A non-metallic mineral (see pyroxene).
Orthostatic	Pertaining to or caused by standing erect.
Penumbra	A partially illuminated region on either side of a completely dark shadow cast by an opaque body such as a planet or satellite.
Peristalsis	Wavelike motion of the walls of hollow organs consisting of alternate contractions and dilations of transverse and longitudinal muscles.
Plagioclase	A feldspar mineral composed of varying amounts of sodium and calcium with aluminum silicate.
Platelets	Small flattened masses.
Potassium gluconate	A salt of crystalline acid.
Pre-Imbrian	Older than the stratigraphic time of Imbrian age.
P-wave	A deflection in an electrocardiographic tracing that represents auricular appendix activity of the heart.
Pyroxene	A mineral occurring in short, thick prismatic crystals or in square cross section; often laminated; and varying in color from white to dark green or black (rarely blue).
Radon atmosphere	A radioactive gaseous environment of radon formed by the radioactive decay of radium.
Ubiquitous	The state, fact or capacity of being everywhere at the same time.
Vagotonia	A hypertension condition in which the vagus nerve ' dominates the general functioning of the body organs.
Vagus nerve	A wandering nerve. Its length and wide distribution are in the brain.
Vesicles	A small cavity in a mineral or rock, ordinarily produced by expansion of vapor in a molten mass.
Vesicular	A mineral or rock containing small cavities produced ordinarily by the expansion of vapor in the molten mass.
Vestibular	Inner ear cavity.
Vignetting	Pertaining to the progressive reduction in the intensity of illumination falling on a photographic film toward the edges of the frame due to the obstruction of oblique light beams.
Vug	A small cavity in a rock.
Zap pitting	Impacts from particles traveling at extremely high velocities that leave small glass-lined depressions in rocks.
Zodiacal light	A faint glow extending around the entire zodiac, but showing most prominently in the area of the Sun.

REFERENCES

1. Johnson Space Center: Apollo 17 Preliminary Science Report. (To be published as a NASA Special Publication.)
2. Marshall Space Flight Center: Saturn V Launch Vehicle Flight Evaluation Report - AS-512 Apollo 17 Mission. MPR -SAT-FE-73-1. February 28, 1973.
3. Boeing Company: AS-512 Final Postflight Trajectory. Unnumbered document (to be published).
4. NASA Headquarters: Mission Implementation Plan for Apollo 17 Mission. Unnumbered document. Revision 3. October 1972.
5. Manned Spacecraft Center: Mission Requirements SA-512/CM-114/LM-12, J-3 Type Mission. MSC-05180, Change D. March 16, 1972 (reprinted September 18, 1972).
6. Manned Spacecraft Center: Apollo 15 Mission Report. MSC-05161. December 1971.
7. Manned Spacecraft Center: Apollo 16 Mission Report. MSC-07230. August 1972

SUBJECT: Lunar Orbiter Mission V:
DATE: July 26, 1967 Potential AAP Landing Sites
Case 232
FROM: Farouk El-Baz

ABSTRACT

Sixteen sites, concentrated mainly in the northern half of the lunar front face, constitute the potential AAP landing sites proposed for Lunar Orbiter Mission V. Two sites fall within craters (Copernicus and Alphonsus). The rest, however, are at least partly in mare material or mare-like blankets, especially at or near contacts with highland terrains.

Specific points of interest in these sites are
shown on reproductions of Mission IV photographs. This is done to allow a clear representation of the general character of the sites, albedo variations, approach paths, landing points and priorities among the points of interest themselves.

It is concluded that the 16 sites represent a good variety of lunar surface units and structural features directly or indirectly related to major scientific questions concerning the moon. This report should be of use during the mission and of value in future planning of AAP manned landing missions.

MEMORANDUM FOR FILE

I. INTRODUCTION

The Lunar Orbiter Project Office has approved, with minor modifications, the Mission V Plan submitted by the Mission V Planning Group and Subgroup on June 14, 1967. (Surveyor Orbiter Utilization Committee Minutes, June 14, 1967.) Following is a description of those sites considered at present to be potential AAP landing sites.

II. LOCATION

As it stands, the Mission Plan includes 16 sites to be considered as potential AAP landing sites. Figure 1 is an index map of their location. For each site a serial number is given followed by the photographic orbit number. The map indicates that these sites are distributed as follows:

Portion of lunar front face	No of Sites
Upper left quarter of moon	8
Upper right quarter of moon	6
Lower right quarter of moon	0
Lower left quarter of moon	2
Total	16

What appears to be a bias for the upper half of the lunar front face is actually due to the following:

1. A relatively flat, smooth surface is considered, at present, a prime requirement for AAP. This means that the approach paths and landing points ought to be mainly in mania and/or mare-like blankets. About 70% of the mania on the lunar front face fall within the upper half of the moon.
2. Scientifically rich portions of the maria are those at or close to contacts with highland terrain. Judging from the above-mentioned fact, a longer strip of these contacts is found in the upper half of the moon's near side.
3. Orbital conflicts with Apollo sites, film budget and other constraints have also played a role in limiting the number of potential AAP landing sites in a specific area.

III. PHOTOGRAPHIC COVERAGE

Eleven sites are to be photographed by a set of fast-4 frames.* The rest constitute three groups:
A. Fast-8 sets are planned for two cases where continuous coverage of a large area seemed demanding. These are Copernicus (11/63) and Marius Hills (16/83).
B. Slow-4** sets were preferred over fast-4 sets in areas where continuous coverage is not as critical as the extent of the areas to be covered (Hadley Rille, 6/50 and Copernicus Secondaries, 10/61).
C. One frame was allowed in the case of Copernicus CD site (8/59). It is a particular case. The one frame coverage would, under usual circumstances, exclude the site from the list of potential AAP landing sites, due to the lack of stereo photograhy. However, it was felt that the site possesses unique characteristics and therefore should be kept in as a potential landing site. This is also supported by the fact that a good portion of the information about it would be conveyed by the one frame.

*i.e. four frames the high resolution photography contiguous.

**i.e. four frames the low resolution photography contiguous.

IV. CHARACTERISTICS

It is believed that a thorough description of the selected sites could only be given using photographs. This way, the reader becomes much more familiar with the general character of the site as well as the parameters to be considered in selecting specific points of interest. In addition, presenting the photographs may serve as a cross check of the information during more detailed examinations of the sites.
For this reason Lunar Orbiter IV Photographs were used in describing the sites. Figures 3 through 10.

Figure 2 serves as a key to the method used in preparing Figures 3-10. An approach path of about 10 km is indicated as an east-west line in each site. This line serves also as a scale. A suggested landing point is represented in each site by an "x" at the left end of the approach path. Specific points of scientific interest are marked by numbers (1,2,3, etc.) in order of decreasing priority.

It must be stated however that no universal agreement exists concerning the presetned data. What is provided constitutes the author's own understanding of the details within each area of the 16 sites. Factors such as approach paths and landing points are likely to change after closer examination of the sites following the Mission. The specific points of interest in each site and their priority lists are given for use prior to the Mission. They may or may not change when Mission V photographs are studied in the future.

Figure 11 is a summary description of the characteristics of the individual sites with special emphasis on the geological viewpoint. This is supplemented by Figure 12, which represents a cumulative graph of the frequency of lunar surface units and structural features in the 16 sites. Both figures should be of value in:
1. Real-time planning of a non-nominal mission in case of need to select replacements sites.
2. Future selection of sites for actual AAP manned landing missions.
From the two figures it is clear that certain features of the lunar surface are not represented in the 16 sites. For example there is no coverage of pitted highland plains, a crater pair, areas showing clear indications of erosion and/or deposition of material, rays in highland terrains, escarpments with possible exposure of major stratigraphic units, etc. It is also clear that some features are represented in many more sites than others. As an example, there are nine sites which include fresh (young) craters as opposed to only one incidence of ghost craters.

V. CONCLUSION

In spite of the various constraints (mainly orbital. conflicts and frame budget), Mission V plan includes a good variety of sites to be considered as potential AAP landing sites. The sixteen sites selected for this consideration portray a wide spectrum of lunar surface features and structures directly or indirectly connected with major genetic problems related to the moon and its surface features. The photographic coverage of these sites, as planned, should allow us to draw conclusions as to the significance of the areas to be considered in future lunar exploration missions.

1012-FEB-ljb Farouk El-Baz

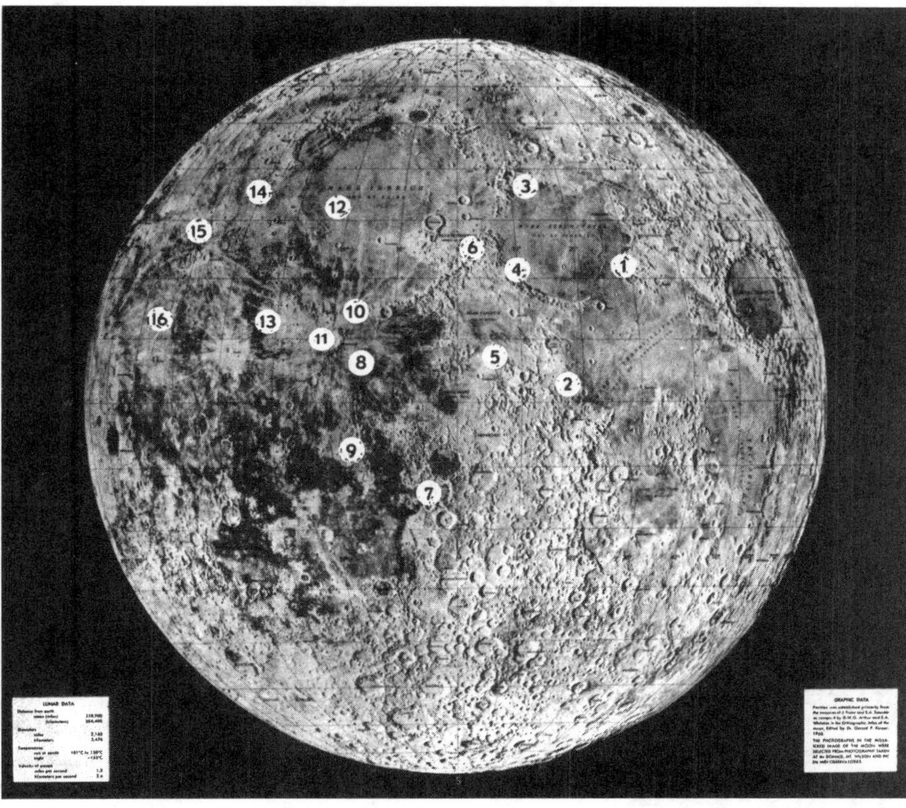

FIGURE 1 INDEX MAP OF LUNAR ORBITER V AAP LANDING SITES

NO.	DESIGNATION
1/35	LITTROW
2/41	DIONYSIUS
3/45	S.ALEXANDER
4/46	SULP.GALLUS
5/47	HYGINUS R.
6/50	HADLEY R.
7/53	ALPHONSUS
8/59	COPER. CD
9/60	FRA MAURO
10/61	COPER.SECON.
11/63	COPERNICUS
12/65	IMBR. FLOWS
13/69	TOBIAS MAYER
14/76	JURA-GRUIT.
15/82	ARISTARCHUS
16/83	MARIUS HILLS

PHOTOGRAPH

2
1
X ———
3
F
4

DESCRIPTION

SITE DESIGNATION NO./ORBIT
COORDINATES NO. OF FRAMES
LO IV SUBFRAME

1,2,3 ETC.: MAJOR POINTS OF INTEREST ARRANGED IN ORDER OF DECREASING IMPORTANCE (PRIORITY).

——— POSSIBLE APPROACH PATH = 10 KM
X POSSIBLE LANDING POINT
1-4 FEATURES OF INTEREST
F FRAMELET SEPARATION (SCALE):
= 12 KM IN EQUATORIAL REGIONS
= 16 KM IN POLAR REGIONS

FIGURE 2 - EXPLANATION OF FIGURES 3-10

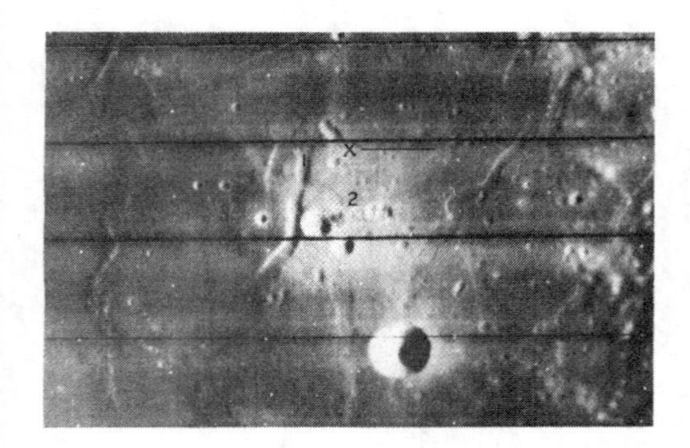

LITTROW RILLES 1/35
29^{0} 20' E 22^{0} 12' N F4
LO-IV H-78-R

1. DARK MARE MATERIAL CUT BY RILLE, BEDDING MAY BE EXPOSED
2. BRIGHT HALO CRATERS OF IRRIGULAR SHAPE (YOUNG VOLCANICS?)

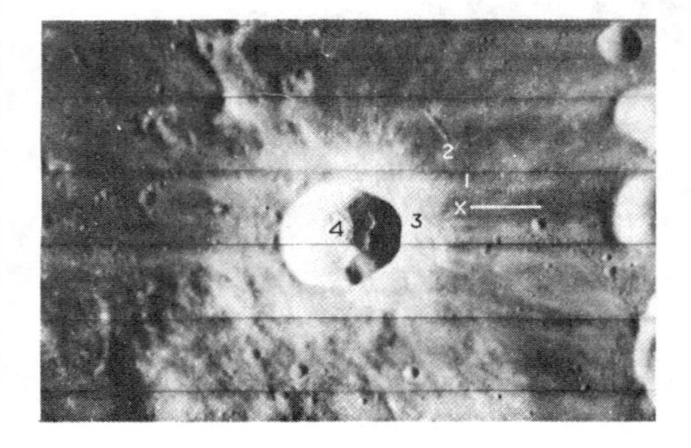

DIONYSIUS 2/41
18^{0} 00' E 2^{0} 42' N F4
LO-IV H-90-1

1. ALTERNATING DARK AND LIGHT RAYS
2. LINEAR RILLE CUTTING RAYS
3. BRIGHT BLANKET AND SHARP RIM
4. FLOOR OF FRESH CRATER

FIGURE 3

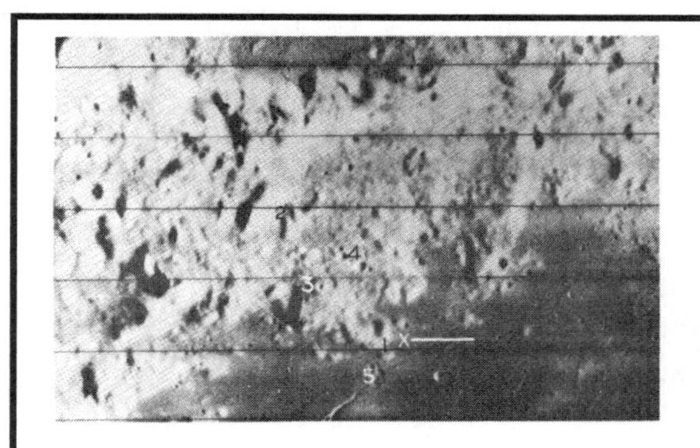

SOUTH OF ALEXANDER 3/45
13° 30' E 38° 30' N F4
LO-IV H-98-2

1. MARE-HIGHLAND CONTACT
2. TWO HIGHLAND TERRAINS CONTACT
3. DOMES AND SCARPS IN HIGHLAND
4. SMALL CRATERS IN HIGHLAND
5. RILLE AND CRATER CLUSTER

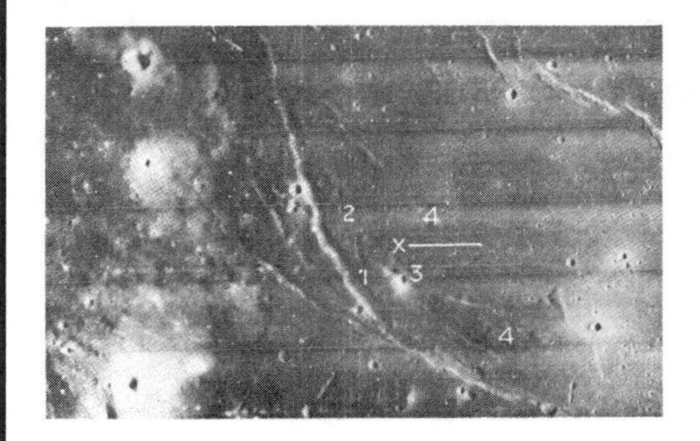

SULPICIUS GALLUS RILES 4/46
9° 20' E 21° 00' N F4
LO-IV H-97-1

1. RILLE COMPLEX IN MARE
2. DARK MATERIAL COVERING MARE & TERRA
3. BRIGHT HALO CRATER WITH PROBABLE EXPOSURE OF BEDS
4. DOMES IN MARE MATERIAL

FIGURE 4

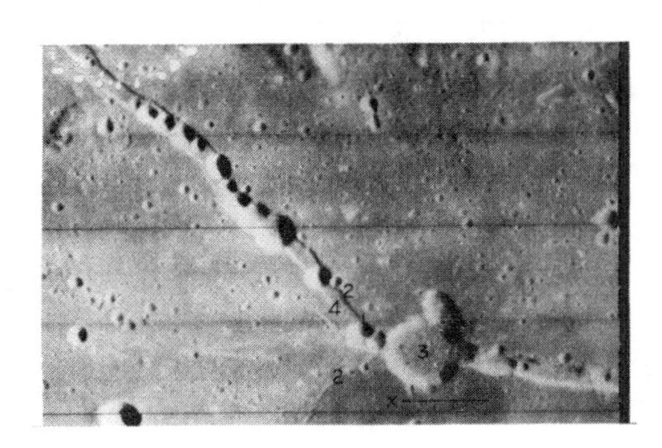

HYGINUS RILLE 5/47
6° 00' E 8° 15' N F4
LO-IV H-102-2

1. CRATER CHAIN - RILLE CONTACT
2. RILLE - MARE LINE
3. CRATER FLOOR
4. RILLE FLOOR

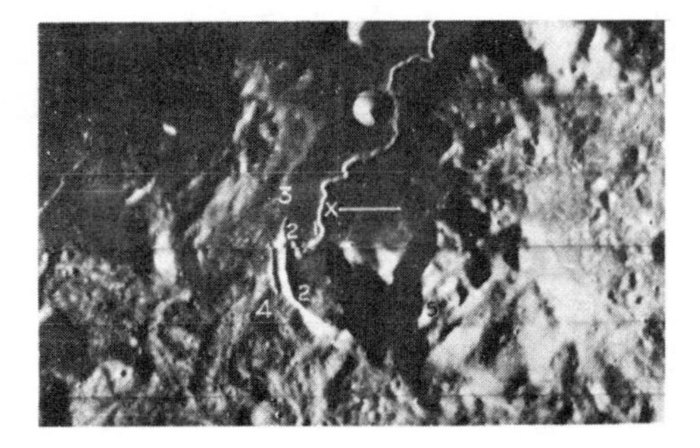

HADLEY RILLE 6/50
3° 00' E 26° 12' N S4
LO-IV H-102-3

1. SINUOUS RILLE IN MARE
2. TERMINATION OF RILLE
3. MARE - HIGHLAND CONTACT
4. APENNINE BENCH MATERIAL
5. EVEN-RUGGED HIGHLAND CONTACT

FIGURE 5

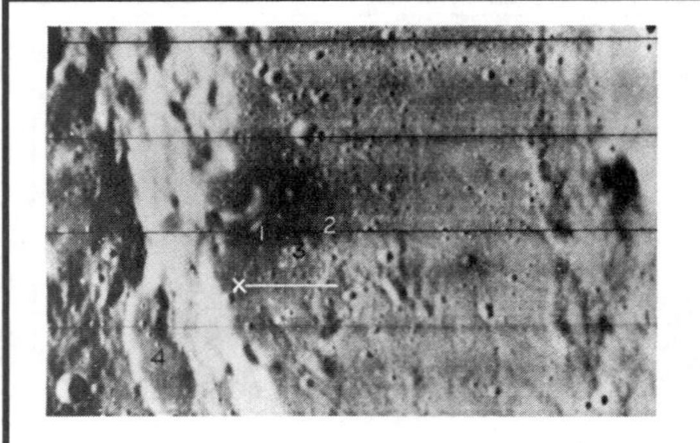

ALPHONSUS 7/53
4° 10' W 13° 40' S F4
LO-IV H-108-2

1. DARK HALO CRATER
2. RILLE IN CRATER FLOOR
3. DARK-LIGHT MATERIAL CONTACT
4. ELONGATE FEATURE ON RIM

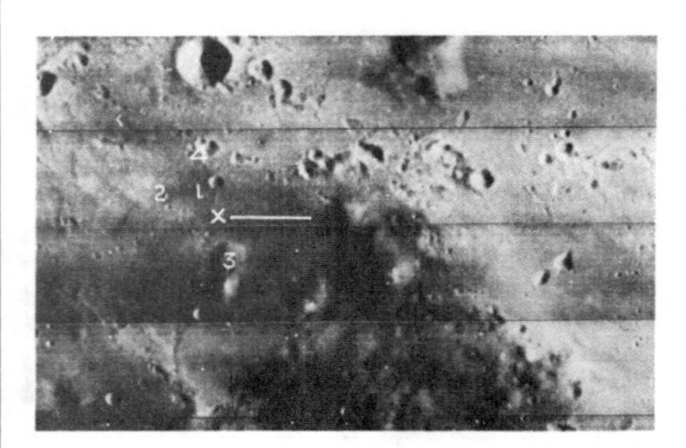

COPERNICUS CD 8/59
14° 45' W 6° 25' N 1
LO-IV H-121-1 AND 2

1. DARK MANTLING MATERIAL
2. DARK-LIGHT MATERIAL CONTACT
3. DOME (WITH SUMMIT CRATER)
4. SECONDARY CRATERS

FIGURE 6

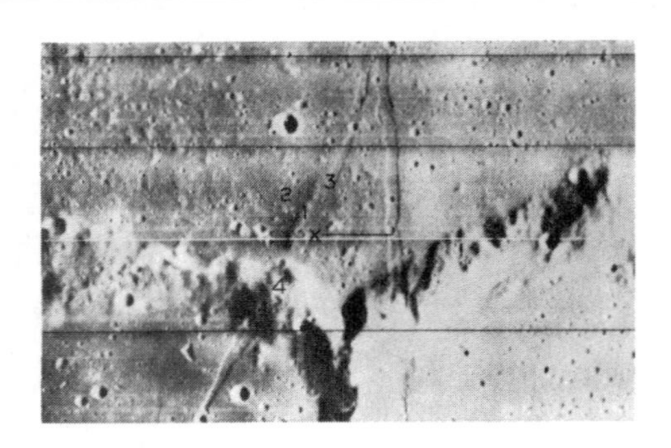

FRA MAURO 9/60
16° 45' W 7° 00' S F4
LO-IV H-120-3

1. DOMES IN FLOOR
2. DARK-LIGHT MATERIAL CONTACT
3. FILLING OF RILLE
4. RILLE IN RIM

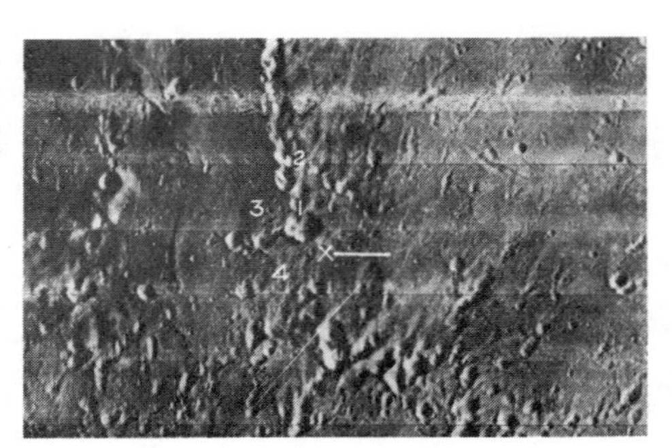

COPERNICUS SECONDARIES 10/61
16° 15' W 14° 40' N S4
LO-IV H-121-2

1. CRATER CHAIN
2. RAYS FROM CHAIN
3. SMOOTHER TERRAIN
4. ROUGHER TERRAIN

FIGURE 7

COPERNICUS 11/63
20° 18' W 10° 25' N F8
LO-IV H-121-2

1. SMOOTH AND HUMMOCKY MATERIAL
2. DOMES IN CRATER FLOOR
3. SINUOUS RILLES IN FLOOR
4. BEDDING IN RIM
5. CENTRAL PEAKS AND CONES

IMBRIUM FLOWS 12/65
22° 00' W 32° 40' N F4
LO-IV H-127-L

1. CONTACT BETWEEN TWO FLOWS
2. SMALL CRATERS IN MARE
3. BRIGHT HALO CRATER

FIGURE 8

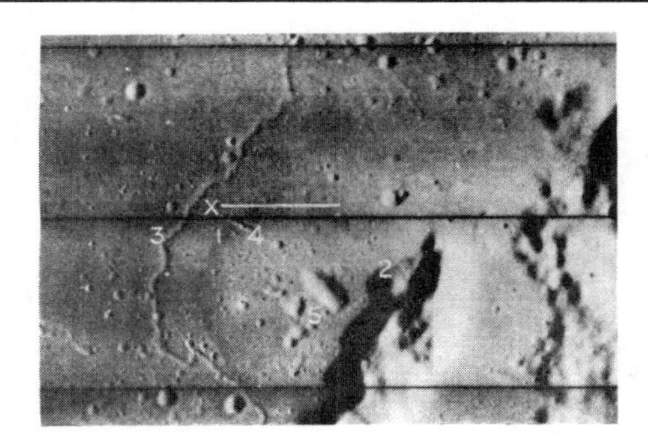

TOBIAS MAYER DOME 13/69
30° 55' W 13° 10' N F4
LO-IV H-133-2

1. DOME - MARE CONTACT
2. DOME - TERRA CONTACT
3. SIMUOUS RILLE
4. CRATER CHAIN
5. SUMMIT CRATERS

JURA-GRUITHUISEN 14/76
41° 30' W 35° 55' N F4
LO-IV H-145-1

1. PANCAKE CRATER INTERIOR
2. DOME IN HIGHLAND
3. MARE - HIGHLAND CONTACT
4. SUMMIT CRATER
5. DOME IN MARE
6. SIMUOUS RILLE IN MARE
7. WRINKLE RIDGE

FIGURE 9

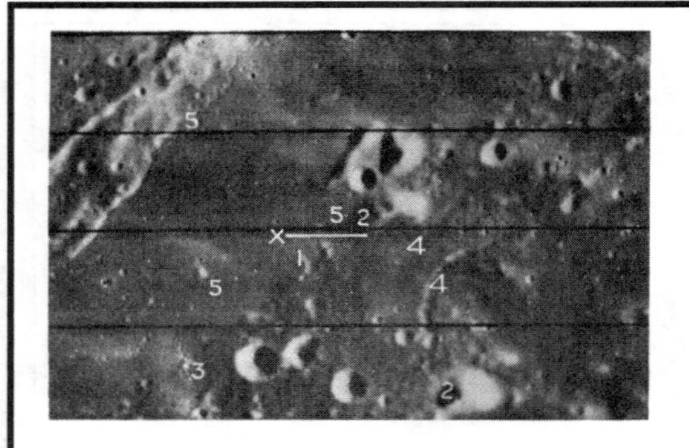

ARISTARCHUS PLATEAU 15/82
52^0 45' W 28^0 00' N F4
LO-IV H-158-L

1. MARE - TERRA MANTLE CONTACT
2. CONES AND DOMES
3. CRATER CHAIN
4. RILLE IN TERRA
5. RILLE IN MARE

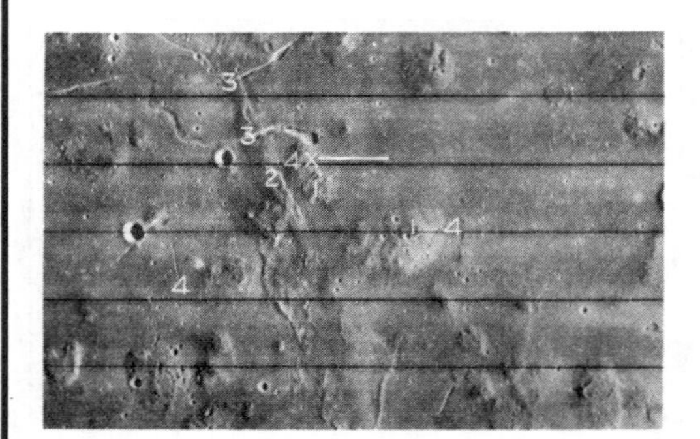

MARIUS HILLS 16/83
56^0 00' W 13^0 45' N F8
LO-IV H-157-C

1. DOMES IN MARE
2. FLOW FRONTS
3. SINUOUS RILLE
4. CRATER CHAINS

FIGURE 10

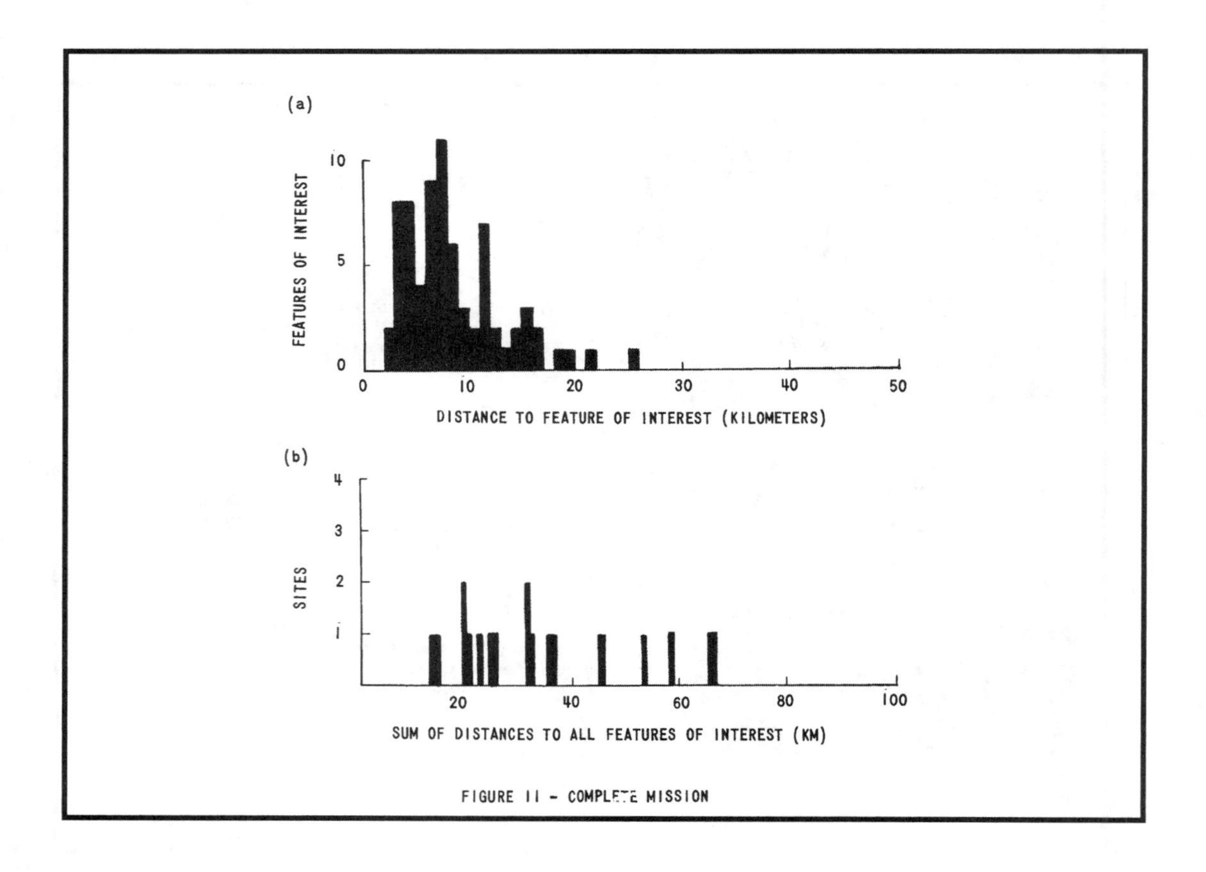

FIGURE 11 - COMPLETE MISSION

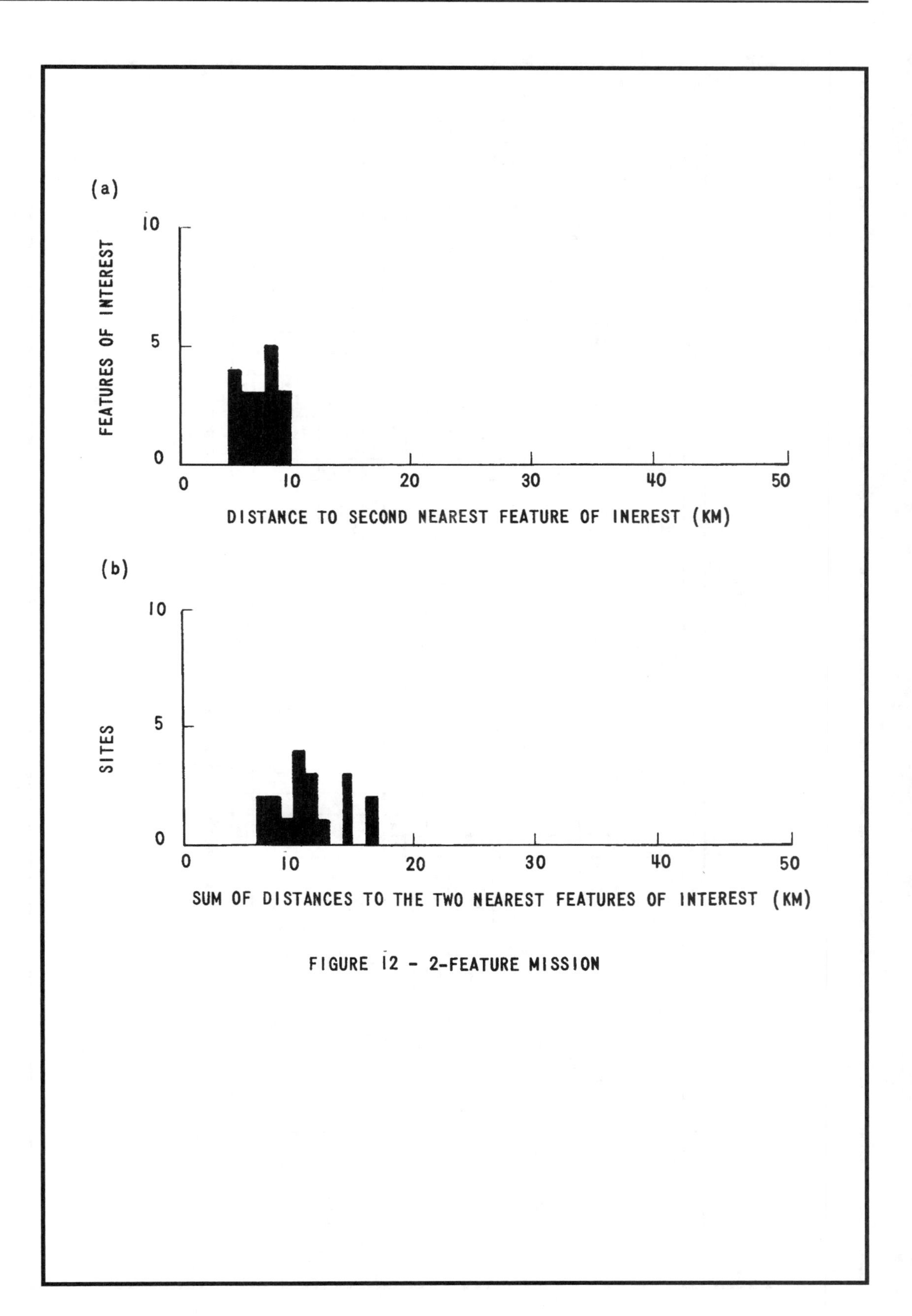

FIGURE 12 - 2-FEATURE MISSION

MAJOR UNITS			MARIA				HIGHLAND			CRATERS																		
CHARACTERISTICS			ALBEDO		SURFACE		SURFACE			NUMBER			GEOMETRY				SITE		HALO		RIM		STRUCTURE			AGE		
NUMBER	ORBIT	DESIGNATION OF SITE	LIGHT	DARK	SMOOTH	ROUGH	EVEN	PITTED	RUGGED	SINGLE	CLUSTER	CHAIN	CIRCULAR	ELONGATE	POLYGONAL	IRREGULAR	IN MARE	IN HIGHLAND	BRIGHT	DARK	STEPPED	EJECTA	CENTER PEAK	CENTER RIDGE	HUMMOCKY	FRESH	FLOOR FILLED	GHOST
1	35	LITTROW		X		X				X	X					X	X		X							X		
2	41	DIONYSIUS								X					X		X					X			X	X		
3	45	S. ALEXANDER	X	X	X		X		X	X	X		X				X	X								X		
4	46	SULP. GALLUS		X		X	X			X			X				X		X							X		
5	47	HYGINUS R.	X		X							X	X	X			X									X		
6	50	HADLEY R.		X	X		X		X																			
7	53	ALPHONSUS								X			X					X		X	X		X	X			X	
8	59	COPER. CD	X			X				X	X		X			X	X			X						X		
9	60	FRA MAURO	X			X	X				X				X		X					X			X		X	X
10	61	COPER. SECON.					X					X				X		X							X			
11	63	COPERNICUS								X			X				X	X			X	X	X		X	X		
12	65	IMB. FLOWS		X	X					X			X				X		X							X		
13	69	TOBIAS MAYER	X		X				X	X		X	X	X			X	X										
14	76	JURA-GRUIT.	X	X	X		X		X	X			X			X	X	X										
15	82	ARISTARCHUS		X	X		X			X		X	X			X	X	X								X		
16	83	MARIUS HILLS	X			X						X					X											

MAJOR UNITS			STRUCTURAL FEATURES																					
CHARACTERISTICS			STRATA				FAULTS			RILLES				RAYS				RIDGES		DOMES		ANOMALIES		
NUMBER	ORBIT	DESIGNATION OF SITE	BEDDING	CONTACTS	EROSION	DEPOSITION	SCARP	SLUMPAGE	DISPLACEMENT	LINEAR	SINUOUS	IN MARE	IN HIGHLAND	BRIGHT	DARK	IN MARE	IN HIGHLAND	FLOW	WRINKLE	IN MARE	IN HIGHLAND	COLOR	THERMAL	RADAR
1	35	LITTROW	X						X	X		X												
2	41	DIONYSIUS								X		X		X	X	X							X	
3	45	S. ALEXANDER		X						X		X									X			
4	46	SULP. GALLUS	X						X	X		X								X				
5	47	HYGINUS R.							X	X		X											X	
6	50	HADLEY R.		X			X				X	X												
7	53	ALPHONSUS					X			X														
8	59	COPER. CD												X		X		X		X				
9	60	FRA MAURO								X											X			
10	61	COPER. SECON.																						
11	63	COPERNICUS	X					X			X		X	X		X					X			
12	65	IMB. FLOWS																				X		
13	69	TOBIAS MAYER									X	X								X				
14	76	JURA-GRUIT.		X							X	X							X	X	X			
15	82	ARISTARCHUS		X						X	X	X	X								X			
16	83	MARIUS HILLS									X	X							X	X				

FIGURE 11 - CHARACTERISTICS OF THE SIXTEEN POTENTIAL AAP LANDING SITES PROPOSED FOR LUNAR ORBITER MISSION V

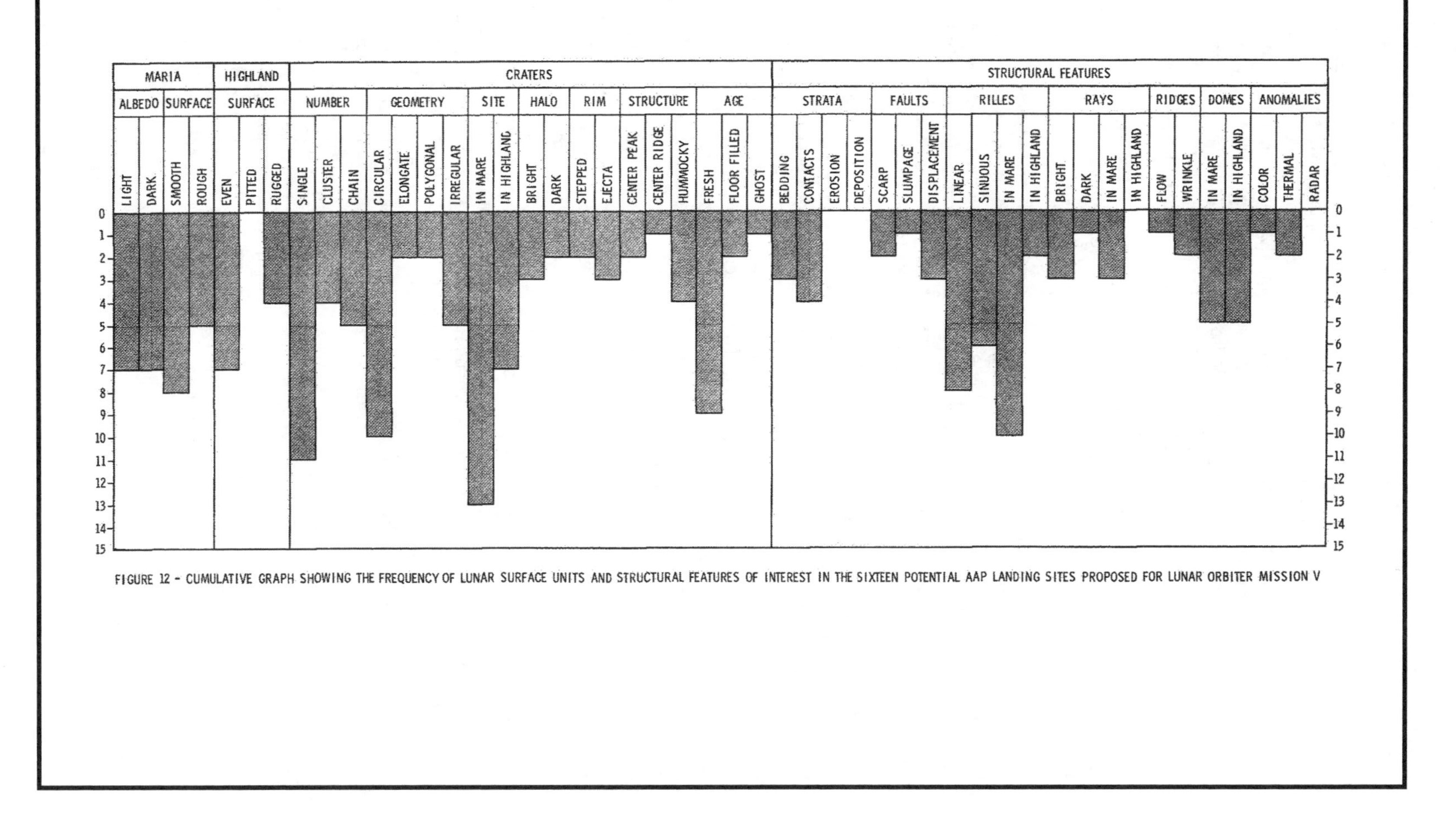

FIGURE 12 - CUMULATIVE GRAPH SHOWING THE FREQUENCY OF LUNAR SURFACE UNITS AND STRUCTURAL FEATURES OF INTEREST IN THE SIXTEEN POTENTIAL AAP LANDING SITES PROPOSED FOR LUNAR ORBITER MISSION V

SUBJECT: Mobility Requirements for AAP
DATE: August 21, 1967
Lunar Surface Missions Case 232
FROM: C. J. Byrne

ABSTRACT

The sixteen AAP lunar landing targets of Lunar Orbiter V are analysed to determine mobility requirements for surface missions. In order to visit all features of interest, a mobility system with a total effective range of 132 km and an operating radius of 25 km would be required. If only two features were to be visited on each landing mission, a total effective range of 30 km and an operating radius of 8 km would be sufficient.

I. INTRODUCTION

The sixteen AAP targets of Lunar Orbiter V form the pool of potential landing sites for AAP missions. In a recent paper, F. El Baz examined Lunar Orbiter IV photography of these sites (about 60 meters resolution), tentatively choosing a touchdown point and features of scientific interest for ground missions*. The touchdown points were chosen for apparent smoothness of touchdown and approach path and for a central location close to several features of interest. The ground missions would require a mobility system to extend an astronaut's operating range. The purpose of this paper is to examine the sites to determine the minimum required range capabilities of such a system. It is recognized that examination of high resolution data from Lunar Orbiter V and more detailed mission planning will result in a refinement of this study.

II. DISTANCE STATISTICS

For each landing site, the distances from the chosen touchdown point to each feature of scientific interest were measured on unrectified Lunar Orbiter IV photographs (reproduced from the memorandum of F. El Baz in Figures 1 to 10). The distances were scaled from the 10 km line in the figures. The following special rules were observed:

1. Where two touchdown points are described for a single site, the northern one is designated (a) and the southern one (b).
2. Where several features of the same type are present, only the distance to the feature nearest to the touchdown point is measured.

*El Baz, F., Lunar Orbiter Mission V: Potential AAP Landing Sites, Bellcomm Memorandum for File, July 26, 1967.

The measurements are summarized in Table I. In addition to the measurements of each feature of interest, the sum of all distances for a particular site is determined and the sum of distances for the two nearest features on each site is determined. Two sets of histograms have been plotted, for two types of missions (see Figures 11 and 12). In the complete missions, all features of interest are visited for a particular landing site. A star pattern is used; one excursion, out and back, is carried out for each feature. In the 2-feature missions only the two nearest features are visited.

III. CONCLUSIONS

It should be emphasized that the conclusions that can be drawn at this stage are tentative because the missions are based on low resolution data. Furthermore only the complete mission was envisioned in designating points of interest and the touchdown point. A definitive study must await detailed mission planning based on high resolution photographs. However, to provide interim estimates of mobility requirements, the following inferences can be drawn from the histograms and associated site photography:

1. A mobility system with a 25 km maximum one-way effective range (50 km round trip) could visit all points of interest. A system with a 16 km maximum one-way effective range could visit 95% of all features of interest. The eliminated sites would be mostly in highlands.
2. The total round-trip effective range required to visit all features is 132 km.
3. For the 2-feature mission, an 8 km maximum one-way effective range is sufficient for all sites.
4. The total round trip effective range required for 2-feature missions at all sites is 30 km.

From these tentative conclusions, one would be led to carry two levels of vehicle performance through study; one with a total effective range of about 150 km and an operating radius of 25 km and the other with a total effective range of about 30 km and an operating radius of 8 km. Of course, safety margins, an allowance for non-straight paths due to obstacle avoidance and exploration deviations and guidance errors must be added. These two sets of requirements need not imply separate vehicle designs; they may only imply different power or fuel recharging and operational strategies. A third possible mission type, requiring mobility from one landing site to another, is not covered in this study.

C. Byrne

ACKNOWLEDGEMENTS

The author is indebted to Mrs. K. Jackson for the measurements, calculations, and preparation of histograms.

TABLE A Distances of Points of Scientific Interest From The Touchdown Point

Name	Site No.	Point No.			Distance (KM)
Littrow Rilles	l/35	1			6.4
		2			7.2*
			Sum	all	13.6
			Sum	2	13.6
Dionysius	2/41	1			3.7
		2			7.4
		3			5.5*
		4			14.8
			Sum	all	31.4
			Sum	2	9.2
South of Alexander	3/45	1			2.9
		2			25.4
		3			15.8
		4			14.1
		5			7.5*
			Sum	all	65.7
			Sum	2	10.4
Sulpicius Gallus	4/46	1			5.7
		2			7.1
		3			3.9
		4			5.3*
			Sum	all	22.0
			Sum.	2	9.1
Hyginius Rille	5/47	1			8.8
		2			6.1
		3			6.4*
		4			11.1
			Sum	all	32.4
			Sum	2	12.5
Hadley Rille	6/50	1			4.1
		2			6.8*
		3			7.2
		4			18.6
		5			21.3
			Sum	all	60.0
			Sum	2	10.9
Alphonsus	7/53	1			6.1
		2			10.9
		3			7.2*
		4			11.2
			Sum	all	35.4
			Sum	2	13.3
Copernicus CD	8/59	1			4.0
		2			7.0
		3			5.3*
		4			7.3
			Sum	all	23.6
			Sum	2	9.3
Fra Mauro	9/60	1			3.3
		2			6.6*
		3			7.4
		4			8.1
			Sum	all	25.4
			Sum	2	9.9
Copernicus Secondaries	10/61	1			8.1
		2			15.7
		3			13.3
		4			7.6
			Sum	all	44.7
			Sum	2	15.7
Copernicus A	11/63 A	1			3.4
		2			4.8*
		3			11.0
			Sum	all	19.2
			Sum	2	8.2
Copernicus B	11/63 B	1			4.1
		2			3.4*
		3			3.1
		4			8.9
		5			11.7
			Sum	all	31.2
			Sum	2	6.5
Imbrium Flows	12/65	1			3.4
		2			5.7*
		3			10.0
			Sum	all	19.1
			Sum	2	9.1
Tobias Mayer Dome	13/69	1			2.2
		2			14.1
		3			4.7
		4			4.1
		5			11.3
			Sum	all	36.4
			Sum	2	6.9
Jura Gruithuisen A	14/76 A	1			7.9*
		2			10.8
		3			15.4
		4			11.2
		5			7.5
		6			12.08
			Sum	all	64.88
			Sum	2	15.4
Jura Gruithuisen B	14/76 B	1			3.6*
		2			6.8
		3			2.8
			Sum	all	13.2
			Sum	2	6.4
Aristarchus Plateau	15/82	1			3.3
		2			10.0
		3			17.6
		4			15.3
		5			7.0*
			Sum	all	53.2
			Sum	2	10.3
Marius Hills	16/32	1			3.6*
		2			5.6
		3			8.8
		4			2.4
			Sum	all	20.4
			Sum	2	6.0

*Denotes second shortest distance.

Date: April 5, 1971
Distribution: from N.W. Hinners
Subject: A Reconsideration of Copernicus As an Apollo Candidate Site — Case 340

ABSTRACT

The leading candidate sites for an Apollo 17 mission have been Copernicus, Marius Hills, and a new highland site. The new flight schedule and consideration of other targets that one might photograph on Apollo 15 and 16 are such that Copernicus might, by default, become the prime candidate. In this paper I have reviewed the Copernicus central peaks mission objectives, originally established before lunar sample return, in the light of recent interpretations of Orbiter, Apollo 11, 12,14, 15 (expected) and Luna 16 data. Appropriate consideration is also given to the proposed Apollo 17 experiments. I conclude that:

(1) there is now sufficient indirect data regarding the age of Copernicus (>0.8, <1.5 billion years) that obtaining a sample for precise age-dating is no longer of high priority;
(2) there is a high probability that we have already sampled sub-mare material similar to that expected at the peaks;
(3) to a first approximation, the role of impact associated volcanism at Copernicus is now reasonably well understood;
(4) the site may be less favorable than alternate sites for geophysical studies planned for an Apollo 17 mission; and
(5) the more significant unresolved problems associated with meteoroid impacts (flux as a function of time, base-surge mechanics) and lunar erosion rates could be better studied on the ejecta blanket of a much younger crater.

Within the scope of the existing Apollo Program and realizing that Apollo 17 is the last planned Apollo flight, the above conclusions lead to my recommendationthat Copernicus be considered a low priority candidate for Apollo 17 relative to a new highland site. Every effort should now be made to acquire a new highland site, far removed from the circum-Imbrium region, by re-examining existing photography and by judicious planning of Apollo 15 and 16 photography.

MEMORANDUM FOR FILE

The Apollo Site Selection Board (ASSB), at its meeting on September 24, 1970, selected Hadley-Apennine as the Apollo 15 site for the July, 1971 launch date (ASSB minutes). Descartes was identified as the prime candidate for Apollo 16, then scheduled for January, 1972, contingent upon the acquisition of satisfactory landing site photography from the Apollo 14 mission. Implicit in the decision was the consideration that should adequate photography not be acquired, Copernicus (central peaks) would provide an acceptable alternate. Lastly, Copernicus, Marius Hills, and a new highland site (to be selected from Apollo 14 and/or 15 photography) were recognized as leading candidates for Apollo 17, then scheduled for June, 1972. Since the September 1970 ASSB meeting, the launch schedule has been changed such that Apollo 16 (Descartes) is now scheduled for March, 1972 and Apollo 17 for December, 1972. The immediate result is that Marius Hills is no longer a viable Apollo 17 candidate as it is not accessible under Apollo 15 mission design constraints in the December, 1972 - February, 1973 time period (K. E. Martersteck, Bellcomm, personal communication).

Regarding a new highland site for Apollo 17, there do not appear to be good prospects for obtaining one in the region covered by Apollo 14 photography if Apollo 16 goes to Descartes. First, the useful photography obtained on Apollo 14 is for the most part in geologic terrain similar to that at the prime Descartes site. Second, an Apollo 17 mission to, say, Theophilus, which is covered by the photography and which may be in a different terrain, would result in an almost exact duplication of the Apollo 16 orbital remote sensing and photography. Non-mare areas available for photography on Apollo 15 are limited to a small region W to S of Mare Crisium. and in the Haemus and Apennine Mountain regions. No further comment can now be made on the photogeology of the Crisium region since it is in the process of being evaluated. It appears, however, that the Haemus Apennine Mountain area would be a geological duplication of the Apollo 14 and 15 sites. For certain a mission there would result in major duplication of the Apollo 15 orbital science coverage.

It may be possible, now that there is a nine month interval between Apollo 16 and 17, to obtain Apollo 17 site photography on Apollo 16. In such a situation, however, it would still be necessary to have a high priority back-up site. Neglecting the back-up site for the moment, and assuming Descartes on Apollo 16, it appears that it would be necessary to obtain coverage ~10° south of the Descartes site in order to photograph large areas of a different geologic terrain although there appear to be small patches of such between Descartes and Alphonsus. The available plane change capability and the characteristics of the region accessible to site photography on a

Descartes mission are under investigation.

The immediate consequence of the above brief review of the status of site selection is that the path-of-least-resistance is to regard Copernicus as the prime Apollo 17 candidate site, assuming that it is not used as the Apollo 16 site. In view of that tendency and realizing that the final selection of the Apollo 16 site has not been made, it is useful to review the attributes of Copernicus at this time.

The proposed landing site at Copernicus is the central peaks region. The scientific objectives for a central peaks mission were established before Apollo 11 and, according to El-Baz (1968), are:

1. to sample the central peaks, believed to be a rebound structure formed immediately after the impact which created Copernicus. It is thought that the material of the peaks came from original depths of ~5-10 km and that such samples would consist largely of the sub-mare crustal material;
2. to obtain the time of the impact in order to establish a point on the meteoroid flux versus time curve and to study (recent) erosion rates by analyzing samples from a "young-looking" feature, namely a bright-rayed crater; and
3. to sample the volcanic-appearing crater floor material in order to study the problem of impact-associated volcanism.

Information from recent studies of Orbiter photography, Apollo 11, 12 and 14 sample analyses, combined with the expected return from the Apollo 15 mission to Hadley Apennine,and Luna 16 sample analyses, is now sufficient to allow us to re-examine the above objectives.

1. Central peaks sub-mare crustal material

A schematic cross-section of Copernicus, based upon the photogeologic map of Schmitt et al., (1967) is shown in figure 1. The impact is believed to have cratered through a section of lunar crust consisting of, top to bottom, thin layers totaling as much as a kilometer of mare fill, a layer (~l km) of Imbrium ejecta (Fra Mauro formation) and several kilometers of pre-Fra Mauro lunar crust. Some unknown mix of the aggregate section was ejected during the impact, some of which formed the ray system radial to Copernicus. One such ray crosses the Apollo 12 site and is thought to be the source of the exotic light-colored soil [commonly known as KREEP for high potassium (K), rare earth element (REE) and phosphorous (P)] found at the site (Hubbard et al., 1971). Further, KREEP-type material is very similar to samples returned from Apollo 14 which landed in known Fra Mauro material (R. Brett, MSC, personal communication).

Assuming for the moment that the KREEP at Apollo 12 is indeed Copernicus-ejected Fra Mauro, the question at hand regards its original depth beneath Imbrium, particularly as compared to that at the Apollo 14 site. While one cannot confidently specify those depths in the pre-Imbrian crust, the cratering mechanics models (Roberts, 1968) indicate that the pre-impact shallower material is concentrated, relative to deeper material, in the ejecta found furthest from the impact. On this basis, Fra Mauro sampled at Apollo 14 contains more material from slightly shallower depths than that sampled at Apollo 12. How much shallower is impossible to say and about all one can do is to obtain a qualitative feel by noting, based on the work of Eggleton (1962), that the estimated Fra Mauro thickness at Apollo 14 is 100-200 meters versus the ~1000 meters or so at Copernicus. Alternatively, but less likely, suppose now that the ray material is not Copernicusejected Fra Mauro but the subjacent crust. It is highly likely that the subjacent crust is indistinguishable from Fra Mauro formation. This statement, of prime importance to the evaluation of the scientific value of a Copernicus peaks mission, is based upon the following:

a. Both Apollo 11 (Mare Tranquillitatis) and 12 (Oceanus Procellarum) soils have isotopic model ages of ~4.4 - 4.6 billion years (Papanastassiou and Wasserburg, 1971) as does the soil returned from Mare Fecunditatis by Luna 16 (Vinogradov, 1971). Several lines of evidence indicate that those soils must contain an old (~4.4 - 4.6 billion years) exotic "rock component with a radioactivity such that it isotopically dominates the contribution of the local basaltic rock to the soil. Hubbard et al. (1971) have presented evidence that the KREEP (Fra Mauro) material is that component, at least in the Apollo 12 soils and most likely in the Apollo 11 soil. It is reasonable to conjecture that a KREEP-type component is also present in the Luna 16 samples. The Rb-Sr data of Hubbard et al. (1971) is consistent with the KREEP material being 1,4.4 billion years old. Consider now that the Apollo 11 and 12 sites are separated by ~1700 km and that the Apollo 12 and Luna 16 sites are ~2500 km apart. A study of the photogeologic map indicates that no visible Copernicus ray comes near the Apollo 11 site and that the nearest Fra Mauro formation is over 200 km away (Morris and Wilhelms, 1967). The implication is that there is

a "local" source of KREEP-type material near the Apollo 11 and Luna 16 sites, i.e., there is a widespread lunar occurrence of that old component. This is consistent with the postulate of Papanastassiou and Wasserburg (1971) that a lunar-wide large scale differentiation occurred shortly after or coincident with the time of lunar origin ~4.6 billion years ago. One might expect such a process to result in a sensibly homogeneous composition. The arguments above lead me to conclude
that the old pre-Fra Mauro crust in the Copernicus region should not differ greatly, either compositionally or chronologically, from that present in the nearby Imbrium region.

b. There is an absence of detectable major layering in the upper 20-30 km of lunar crust near the Apollo 12 site (Latham et. al., 1970). Assuming an analogous situation exists beneath Mare Imbrium, and remembering (a) above, one would expect a Copernicus central peak sample from 5-10 km deep to be lithologically similar to Apollo 14 Fra Mauro material derived from the upper several kilometers of the Imbrium basin. More likely yet is it apt to be similar to Fra Mauro material mantling the Apennines and/or to material constituting the Apennine block itself. This follows since the base of the Apennines, averaging ~2-3 km beneath the crests, presumably is a sample of the pre-Imbrium crust at that depth.

c. The Copernicus central peaks material appears (figure 2) to be composed of at least two units of distinctly different mechanical strength. One of these is a resistant unit tending to form ridges and is the most probable source of the boulders seen at the base of the hills. The other unit appears to be mechanically weak, eroding and moving downslope at a greater rate than the resistant units. I interpret the resistant units as basaltic or gabbroic dikes intruded into fragmental pre-Imbrian sub-surface material contemporaneously with the Procellarum basaltic flooding on the surface. (Crater counts by Gault, 1970, indicate that the Copernicus impact occurred on a Procellarum surface).

2. "Age-dating" the Copernicus impact and the study of erosion rates

When Copernicus was initially proposed as a site, it was thought by many that it was an extremely young feature, possibly only tens of millions of years old (see e.g., Gault, 1970) and that analysis of Copernicus samples would enable one to establish recent cratering and erosion rates. By correlating Apollo 11 and 12 data on radio-isotope-derived absolute ages with crater statistics applicable to those sites, Hartmann (1970) has shown that the cratering rate averaged over the past two to three billion years has been relatively low and on the average relatively uniform (but see below). Prior to 3 billion
years ago, however, it was significantly higher. Thus the impact that created Copernicus appears to be a random event in the last several billion, relatively quiet, years. Similarly, studies on Apollo 11 and 12 rocks exposed by small cratering events to the surface environment have indicated that average erosion rates over the "recent" past 30 million years, and probably over the past 2-3 billion, have been extremely low - several angstroms/year (see e.g., Hörz, et al., 1971).

Despite the above, it would be scientifically useful to know the age of the Copernicus impact event. However, it is probable that we already know it to within ±350 million years:

a. revised cratering rates, based upon Apollo 11 and 12 data, indicate that the Copernicus event occurred ~1.3 billion years ago if one judges by the areal density of large craters (>2 km) (Hartmann, 1970).

b. Using the technique of Soderblom (1970a, b), which depends upon measuring the erosion of craters by the statistically significant large numbers of micrometeoroids, indicates that the flux registered on the Copernicus ejecta blanket is ~0.25 of that on the Apollo 12 surface. Assuming a constant micrometeoroid flux since the time the Apollo 12 mare surface formed ~3.3 billion years ago (Papanastassiou and Wasserburg, 1971) then puts the age of Copernicus at ~0.8 billion years. Allowing for a slightly decreasing erosion rate in more recent times would favor a slightly older age.

c. Uranium-thorium-lead dating of Apollo 12 soil which contains probable Copernicus ejecta, indicates that the ages are discordant (i.e., the various U-Th-Pb ages do not agree). One way to account for such discordancy is to assume a multi-stage uraniumthorium-lead evolution. Considering such, Cliff et al. (1971) note that the soil must have lost lead at least as recently as 1.5 billion years ago. More recent work by Silver (CIT, personal communication, 1971) puts such a loss at 0.8 - 0.9 billion years ago, consistent with the findings of Tatsumoto et al. (1971) who show that the soil components fall on the chord joining the 4.63 and 0.9 billion year points on the concordia curve. Both Silver and Tatsumoto at al believe that the most likely "event" causing the lead loss is the Copernicus impact.

Considering the three essentially independent age dating techniques described above one comes to the conclusion that Copernicus was formed >0.8 and <1.5 billion years ago, during a period of relatively quiet lunar history. (One can argue about how independent (a) and (b) are but they depend on the meteoroid flux at mass points separated by over nine orders of magnitude).

Before leaving the subject of cratering and erosion rates, recall that some pre-Apollo estimates of lunar ages were drastically low (Gault, 1970). Those estimates made use of recent meteoroid flux measurements thus indicating as one possibility that the present flux is anomalously high. Weak support for such an interpretation comes from both Hartmann's (1970) speculation that recent asteroidal collisions may now be increasing the flux of particles and from the clustering of cosmic ray exposure ages of stone meteorites around 5-20 million years but ranging up to -.100 million years (Wood, 1968). If the increase is real, one way to pin it down is to sample a recently formed lunar surface, i.e., of age comparable to the time the flux increase became significant. One such surface is the ejecta blanket at Tycho which, based on Hartmann's data (1970), has about 1/7th the crater density of the Apollo 12 area, corresponding to an age of -.400 (+200) million years. The technique of Soderblom (1970a, b) leads to an estimate of an age of Tycho as <~ 200 million years.

3. Impact Associated Volcanism

The large expanses of circular and irregular basin mare fill are not directly associated with impact_ craters. Pre-Apollo studies (e.g., Baldwin, 1963) indicated that a time interval existed between mare basin formation and the filling while Apollo sample analysis indicates that the basalts are mostly likely a product of partial melting in the deep lunar interior (e.g., see Gast et al., 1970 and Haskin et al., 1970). Alternatively, studies of the soil isotopic and elemental abundances indicate (Papanastassiou and Wasserburg, 1970) that the basalts were not formed by melting of local soil as might be expected in direct: impact melting (with the caveat that the soil now exposed is not necessarily the same as that existing at that location prior to the formation of the basalts). One cannot deny that an indirect association between mare flooding arid basin formation may well exist with the large basin-forming impacts fracturing the crust, such fractures then providing conduits for lavas. The association of mare-type filling ir. the low lying regions of the concentric ring valleys around the relatively unfilled Mare Orientale basin supports this interpretation. It is also possible that brecciation of the surface layers (or acquisition of a covering of fragmented ejecta) might decrease the thermal diffusivity sufficiently to allow an anomalously high subsurface temperature increase.

A specific question at Copernicus centers about whether or not the floor filling material and the smooth interior and exterior terrace pools are volcanic in origin (versus ejecta fallback) and when such volcanism might have occurred relative to the impact. Recent crater size-frequency analyses by Greeley and Gault (1970) show, first, that the fissured volcanic appearing floor and exterior terraces
seem to have the same crater density and size distribution as the ejecta blanket indicating that to within some small time interval the main floor filling and rim volcanism were contemporaneous with the impact (see figure 3). Second, a floor "fissure-flow" has a lower (and non-equilibrium) crater density than the ejecta blanket, indicating that it is a relatively young feature relative to the time of impact while the interior terraces have greater non-equilibrium crater densities. The details of the crater size distribution, compared with terrestrial volcanic regions, indicates rather conclusively that in addition to impact craters, the terraces contain a population of collapse craters of non-impact origin associated with the drainage of fluid lava. In summary one sees that although we do not have a precise date for the quantitatively minor interior terraces and fissure flows, we do have a sensibly good first order idea of the time and sequence of crater-modifying volcanic events at Copernicus.
There are other impact-related problems, raised by findings on Apollo 11 and 12: many of the breccias appear to be products of simple physical compaction beneath local, small impact-produced craters (Chao, et al., 1971). Others, however, appear to have been involved in a "thermal" and physically violent event purported to be the "base-surge" of a large impact (McKay et al., 1970). Base-surge induced mixing of lunar surface materials thus may be significant in horizontal mixing of lunar surface materials. That such mixing occurs is an established fact as evidenced both by the old model age of the soils (see above) and by the chemical anomalies (Hubbard et al., 1971). However, base-surge deposits, as interpreted from Orbiter photography and as seen at terrestrial explosion craters, are radial to craters and best seen and studied on the ejecta

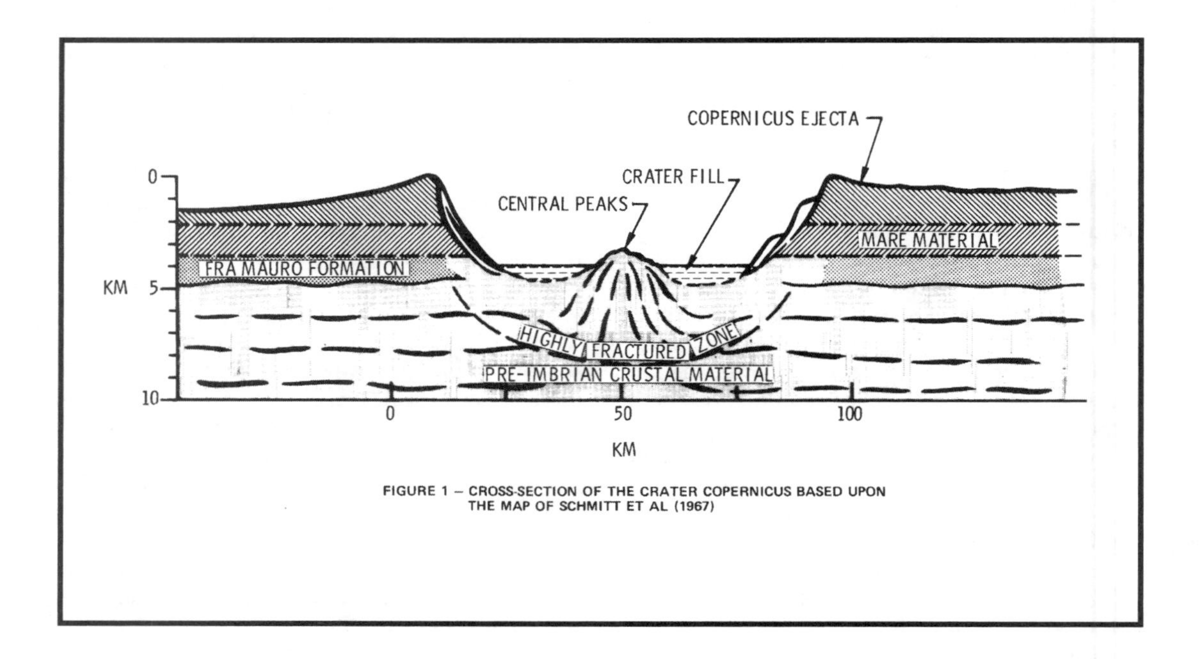

FIGURE 1 – CROSS-SECTION OF THE CRATER COPERNICUS BASED UPON THE MAP OF SCHMITT ET AL (1967)

FIGURE 2 – COPERNICUS CENTRAL PEAKS SHOWING RESISTANT UNITS (A) INTERPRETED AS BASALTIC DIKES INTRUDED INTO PRE-IMBRIAN FRAGMENTAL MATERIAL (B). (ORBITER 2, PHOTOGRAPH 162).

FIGURE 3 – CRATER COPERNICUS SHOWING TYPICAL AREAS INVESTIGATED BY GREELEY AND GAULT (1971): EJECTA (A), EXTERIOR TERRACE (B), INTERIOR TERRACE (C), WALL (D), AND FLAT FLOOR (E). THE FLOOR "FISSURE FLOW" IS NOT OBSERVABLE ON THIS PHOTO. (ORBITER 5, 157M, FRAMELET WIDTH ~ 3.8 KM).

blanket (Fisher and Waters, 1969). Alternatively, the worst place to study the phenomenon is in the geometric center of a crater - the central peak region.

The last-point to belabor re Copernicus is its suitability as a site for Apollo 17 traverse geophysics and the emplacement of science stations (ALSEP's). The chaotic nature of the crater floor is evidently not conducive to doing good traverse geophysics and the relative nearness (~400 km) to the Apollo 12 and 14 sites is not favorable for either establishing wide-spread networks or for seeing a genuinely new part of the moon. Similar arguments in the past led the geophysicists to place Copernicus in a low priority (see Apollo Site Selection Board Minutes, Meeting of September 24, 1970).

Summary and Recommendation

The leading candidate sites for an Apollo 17 mission have been Copernicus, Marius Hills, and a new highland site. The new flight schedule and consideration of other targets that one might photograph on Apollo 15 and 16 are such that Copernicus might, by default, become the prime candidate. In this paper I have reviewed the Copernicus central peaks mission objectives originally established before lunar sample return, in the light of recent interpretations of Orbiter, Apollo 11, 12, 14, 15 (expected) and Luna 16 data. Appropriate consideration is also given to the proposed Apollo 17 experiments. I conclude that:

(1) there is now sufficient indirect data regarding the age of Copernicus (>0.8, <1.5 billion years) that obtaining a sample for precise age-dating is no longer of high priority;

(2) there is a high probability that we have already sampled sub-mare material similar to that expected at the peaks;

(3) to a first approximation, the role of impact-associated volcanism at Copernicus is now reasonably well understood;

(4) the site may be less favorable than alternate sites for geophysical studies planned for an Apollo 17 mission; and

(5) the more significant unresolved problems associated with meteoroid impacts (flux as a function of time, base-surge mechanics) and lunar erosion rates could be better studied on the ejecta blanket of a much younger crater.

Within the scope of the existing Apollo Program and realizing that Apollo 17 is the last planned Apollo flight, the above conclusions lead to my recommendation that, Copernicus be considered a low priority candidate for Apollo 17 relative to a new highland site. Every effort should now be made to acquire a new highland site, far removed from the circum-Imbrium region, by re-examining existing photography and by judicious planning of Apollo 15 and 16 photography.

N.W. Hinners

REFERENCES

Apollo Site Selection Board, Minutes of the Apollo Site Selection Board Meeting Held on September 24, 1970, Apollo Program Office, Office of Manned Space Flight, National Aeronautics and Space Administration, Washington, D. C., 1970.

Baldwin, R., The Measure of the Moon, 488-pp., The University of Chicago Press, Chicago, 1963.

Chao, E. C. T., J. A. Boreman, and G. A. Desborough, Unshocked and Shocked Apollo 11 and 12 Microbreccias Characteristics and Some Geologic Implications. Second Lunar Science Conference, Houston, Texas, 1971.

Cliff, R. A., C. Lee-Hu and G. W. Wetherill, Rb-Sr and U-Th-Pb Measurements on Apollo 12 Material, Second Lunar Science Conference, Houston, Texas, 1971.

Eggleton, R. E., Thickness of the Apenninian Series in the Lansberg Region of the Moon, in Astrogeol. Studies Ann. Prog. Rept., U. S. Geological Survey Open-File Report, pp. 19-31, August 1961 - August 1962.

El-Baz, F., Geologic Characteristics of the Nine Lunar Landing Mission Sites Recommended by the Group for Lunar Exploration Planning, Technical Report TR-68-340-1, Bellcomm, Inc., Washington, D. C., 1968.

Fisher, R. V. and A. C. Waters, Bed Forms in Base-Surge Deposits. Lunar Implications. Science 165, 1349, 1969.
Gast, P. W., N. J. Hubbard and H. Wiesmann, Chemical Composition and Petrogenesis of Basalts from Tranquillity Base, Geochim. Cosmochim. Acta, Suppl. 1, 2, 1143, 1970.

Gault, D. E., Saturation and Equilibrium Conditions for Impact Cratering on the Lunar Surface; Criteria and Implications, Radio Science 5, 273, 1970.

Greeley, R. and D. E. Gault, Endogenetic Craters Interpreted from Crater Counts on the Inner Wall of Copernicus, Science 171, 477, 1971.

Hartmann, W. K., Lunar Cratering Chronology, Icarus, 13, 299, 1970.

Haskin, L. A., R. O. Allen, P. A. Helmke, T. P. Paster, M. R. Anderson, R. C. Korotev, and K. A. Zweifel, Rare-earths and Other Trace Elements in Apollo 11 Lunar Samples, Geochim. Cosmochim. Acta, Suppl. I, 2, 1213, 1970.

Horz, F., J. A. Hartung, and D. E. Gault, Lunar Microcraters, Second Lunar Science Conference, Houston, Texas, 1971.

Hubbard, N. J., C. Meyer, Jr., P. W. Gast, and H. Wiesmann, The Composition and Derivation of Apollo 12 Soils, Earth Planet. Sci. Letters 10, 341, 1971.

McKay, D. S., W. R. Greenwood and D. A. Morrison, Origin of Small Lunar Particles and Breccia from the Apollo 11 Site, Geochim. Cosmochim. Acta, Suppl. 1, 1, 673, 1970.

Morris, E. C. and D. E. Wilhelms, Geologic Map of the Julius Caesar Quadrangle of the Moon, MAP I-510, U. S. Geological Survey, Washington, D. C., 1967.

Latham, G., M. Ewing, J. Dorman, F. Press, N. Toksoz, G. Sutton, R. Meissner, F. Duennebier, Y. Nakamura, R. Kovach, and M. Yates, Seismic Data from man-Made Impacts on the Moon, Science 170, 620, 1970.

Papanastassiou, D. A. and G. J. Wasserburg, Rb-Sr Ages from the Ocean of Storms, Earth Planet. Sci. Letters 8, 269, 1970.

Papanastassiou, D. A. and G. J. Wasserburg, Lunar Chronology and Evolution from Rb-Sr Studies of Apollo 11 and 12 Samples, Pre-print, submitted to Earth Planet. Sci. Letters, 1971.

Roberts, W. A., Shock Crater Ejecta Characteristics, in Shock Metamorphism of Natural Materials, edited by B. M. French and N. M. Short, 644 pp., Mono Book Corp., Baltimore, 1968.

Schmitt, H. H., N. J. Trask and E. M. Shoemaker, Geologic Map of the Copernicus Quadrangle of the Moon, MAP I-515, U. S. Geological Survey, Washington, D. C., 1967.

Soderblom, L. A., The Distribution and Ages of Regional Lithologies in the Lunar Maria, Ph.D. Thesis, California Institute of Technology, 1970a.

Soderblom, L. A., A Model for Small-Impact Erosion Applied to the Lunar Surface, J. Geophys. Res. 75, 2655, 1970b.

Tatsumoto, M., R. J. Knight and B. R. Doe, U-Th-Pb Systematics of Apollo 12 Lunar Samples, Second Lunar Science Conference, Houston, Texas, 1971.

Vinogradov, A. P., Preliminary Data on Lunar Ground Brought to Earth by Automatic Probe "Luna 16", Second Lunar Science Conference, Houston, Texas, 1971.

Wood, J. A., Meteorites and the Origin of Planets, 117 pp., McGraw-Hill, New York, 1968.

Distribution List
Complete Memorandum to
Complete Memorandum to
NASA Headquarters
Ames Research Center
R.J. Allen by - MAL
D. E. Gault
D.A. Beattie - MAL
R. Greeley
D. O. Beck - MA
R. P. Bryson - MAL
Bellcomm, Inc.
J. K. Holcomb - MAO
R. A. Bass
E. W. Land - MAO
A. P. Boysen, Jr.
C. M. Lee - MA
J. O. Cappellari, Jr.
A. S. Lyman - MR
F. El-Baz
W. T. O'Bryant - MAL
D. R. Hagner
R. A. Petrone - MA
W. G. Heffron
L. R. Scherer - MAL
J. J. Hibbert
T. B. Hoekstra
Manned Spacecraft Center
M. Liwshitz
P. R. Brett - TN
J. L. Marshall
A. J. Calio - TA
K. E. Martersteck
E. A. Cernan - CB
J. Z. Menard
P. W. Gast - TN
G. T. Orrok
A. W. Patteson - TJ
P. E. Reynolds
G. W. Ricks - FM
P. F. Sennewald
H. H. Schmitt - CB
R. V. Sperry
J. R. Sevier - PD
A. W. Starkey
M. G. Simmons - TA
J. W. Timko
D. W. Strangway - TN
A. R. Vernon
J. W. Young - CB
R. L. Wagner

D. B. Wood
California Institute of Technology
All Members-2015
L. T. Silver
Central Files

Dept. 1024 Files
U. S. Geological Survey
Library
H. Masursky - Flagstaff
G. A. Swann - Flagstaff
Abstract Only to
D. E. Wilhelms - Menlo Park

Bellcomm, Inc.
U. of Texas
J. P. Downs
W. Muehlburger
D. P. Ling

M. P. Wilson
Princeton University
R. Phinney

U. of Arizona
W. Hartmann

APOLLO 17 SCIENTIFIC INSTRUMENT MODULE

Equipment for conducting orbital experiments is mainly located in the scientific instrument module (SIM) bay of the service module (SM), which accommodates more than 1,000 pounds of such equipment. Additional experiment equipment and all necessary control and displays are located in the command/service module (CSM). A single foot restraint in the SIM bay will be used for EVA during transearth coast to transfer film cassettes from the SM to the CM.

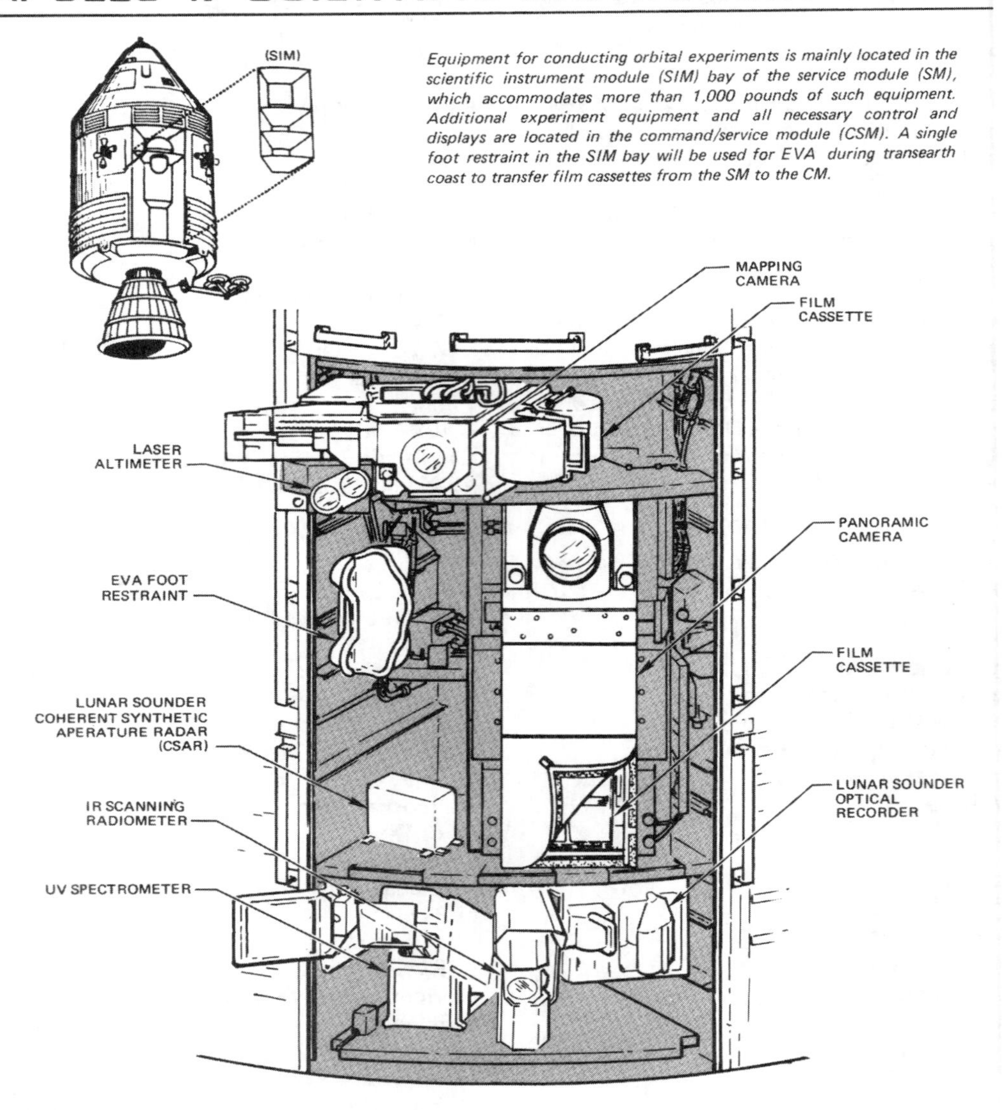

APOLLO LUNAR LANDING SITES

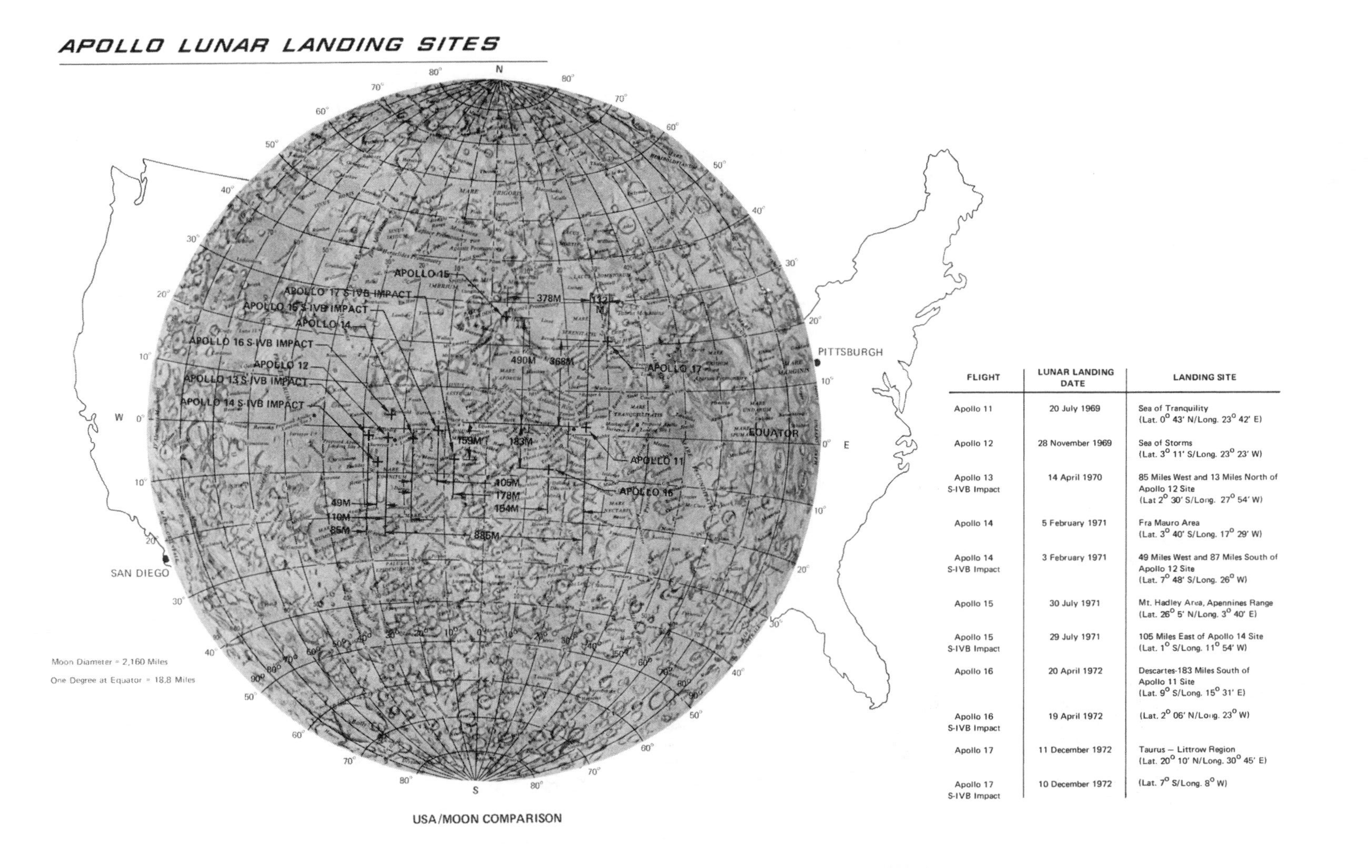

USA/MOON COMPARISON

FLIGHT	LUNAR LANDING DATE	LANDING SITE
Apollo 11	20 July 1969	Sea of Tranquility (Lat. 0° 43' N/Long. 23° 42' E)
Apollo 12	28 November 1969	Sea of Storms (Lat. 3° 11' S/Long. 23° 23' W)
Apollo 13 S-IVB Impact	14 April 1970	85 Miles West and 13 Miles North of Apollo 12 Site (Lat 2° 30' S/Long. 27° 54' W)
Apollo 14	5 February 1971	Fra Mauro Area (Lat. 3° 40' S/Long. 17° 29' W)
Apollo 14 S-IVB Impact	3 February 1971	49 Miles West and 87 Miles South of Apollo 12 Site (Lat. 7° 48' S/Long. 26° W)
Apollo 15	30 July 1971	Mt. Hadley Area, Apennines Range (Lat. 26° 5' N/Long. 3° 40' E)
Apollo 15 S-IVB Impact	29 July 1971	105 Miles East of Apollo 14 Site (Lat. 1° S/Long. 11° 54' W)
Apollo 16	20 April 1972	Descartes-183 Miles South of Apollo 11 Site (Lat. 9° S/Long. 15° 31' E)
Apollo 16 S-IVB Impact	19 April 1972	(Lat. 2° 06' N/Long. 23° W)
Apollo 17	11 December 1972	Taurus – Littrow Region (Lat. 20° 10' N/Long. 30° 45' E)
Apollo 17 S-IVB Impact	10 December 1972	(Lat. 7° S/Long. 8° W)

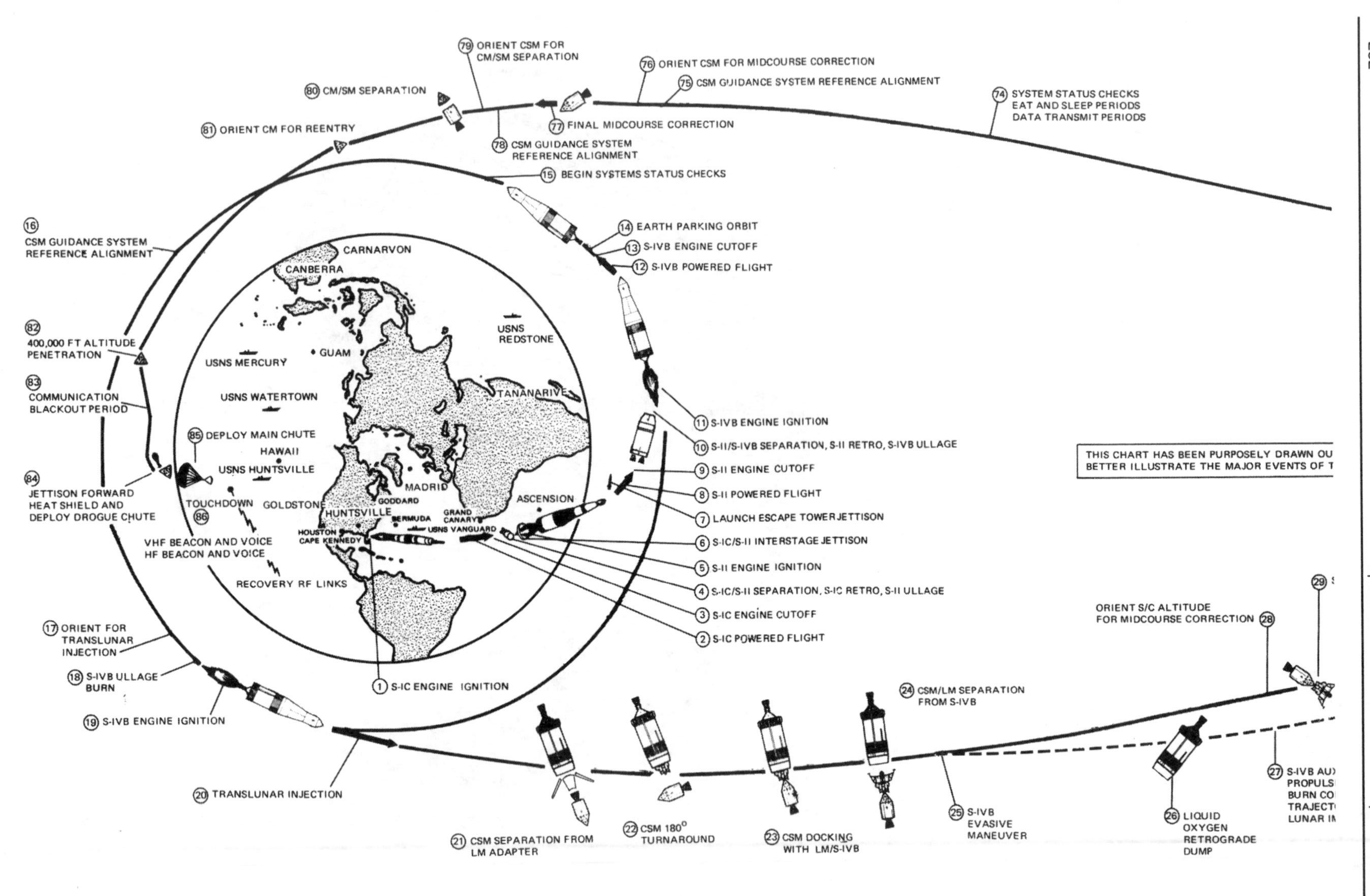
79 ORIENT CSM FOR CM/SM SEPARATION
76 ORIENT CSM FOR MIDCOURSE CORRECTION
75 CSM GUIDANCE SYSTEM REFERENCE ALIGNMENT
80 CM/SM SEPARATION
74 SYSTEM STATUS CHECKS EAT AND SLEEP PERIODS DATA TRANSMIT PERIODS
81 ORIENT CM FOR REENTRY
77 FINAL MIDCOURSE CORRECTION
78 CSM GUIDANCE SYSTEM REFERENCE ALIGNMENT
15 BEGIN SYSTEMS STATUS CHECKS
16 CSM GUIDANCE SYSTEM REFERENCE ALIGNMENT
14 EARTH PARKING ORBIT
13 S-IVB ENGINE CUTOFF
12 S-IVB POWERED FLIGHT
CARNARVON
CANBERRA
USNS REDSTONE
82 400,000 FT ALTITUDE PENETRATION
GUAM
USNS MERCURY
83 COMMUNICATION BLACKOUT PERIOD
USNS WATERTOWN
TANANARIVE
11 S-IVB ENGINE IGNITION
85 DEPLOY MAIN CHUTE
10 S-II/S-IVB SEPARATION, S-II RETRO, S-IVB ULLAGE
HAWAII
THIS CHART HAS BEEN PURPOSELY DRAWN OU
BETTER ILLUSTRATE THE MAJOR EVENTS OF T
USNS HUNTSVILLE
84 JETTISON FORWARD HEAT SHIELD AND DEPLOY DROGUE CHUTE
9 S-II ENGINE CUTOFF
MADRID
8 S-II POWERED FLIGHT
ASCENSION
TOUCHDOWN 86
GOLDSTONE
HUNTSVILLE
GODDARD
BERMUDA
GRAND CANARY
USNS VANGUARD
7 LAUNCH ESCAPE TOWER JETTISON
HOUSTON
CAPE KENNEDY
VHF BEACON AND VOICE
HF BEACON AND VOICE
6 S-IC/S-II INTERSTAGE JETTISON
5 S-II ENGINE IGNITION
RECOVERY RF LINKS
4 S-IC/S-II SEPARATION, S-IC RETRO, S-II ULLAGE
3 S-IC ENGINE CUTOFF
ORIENT S/C ALTITUDE FOR MIDCOURSE CORRECTION 28
2 S-IC POWERED FLIGHT
17 ORIENT FOR TRANSLUNAR INJECTION
18 S-IVB ULLAGE BURN
1 S-IC ENGINE IGNITION
24 CSM/LM SEPARATION FROM S-IVB
19 S-IVB ENGINE IGNITION
20 TRANSLUNAR INJECTION
27 S-IVB AU PROPULS BURN CO TRAJECT LUNAR IM
25 S-IVB EVASIVE MANEUVER
26 LIQUID OXYGEN RETROGRADE DUMP
21 CSM SEPARATION FROM LM ADAPTER
22 CSM 180° TURNAROUND
23 CSM DOCKING WITH LM/S-IVB

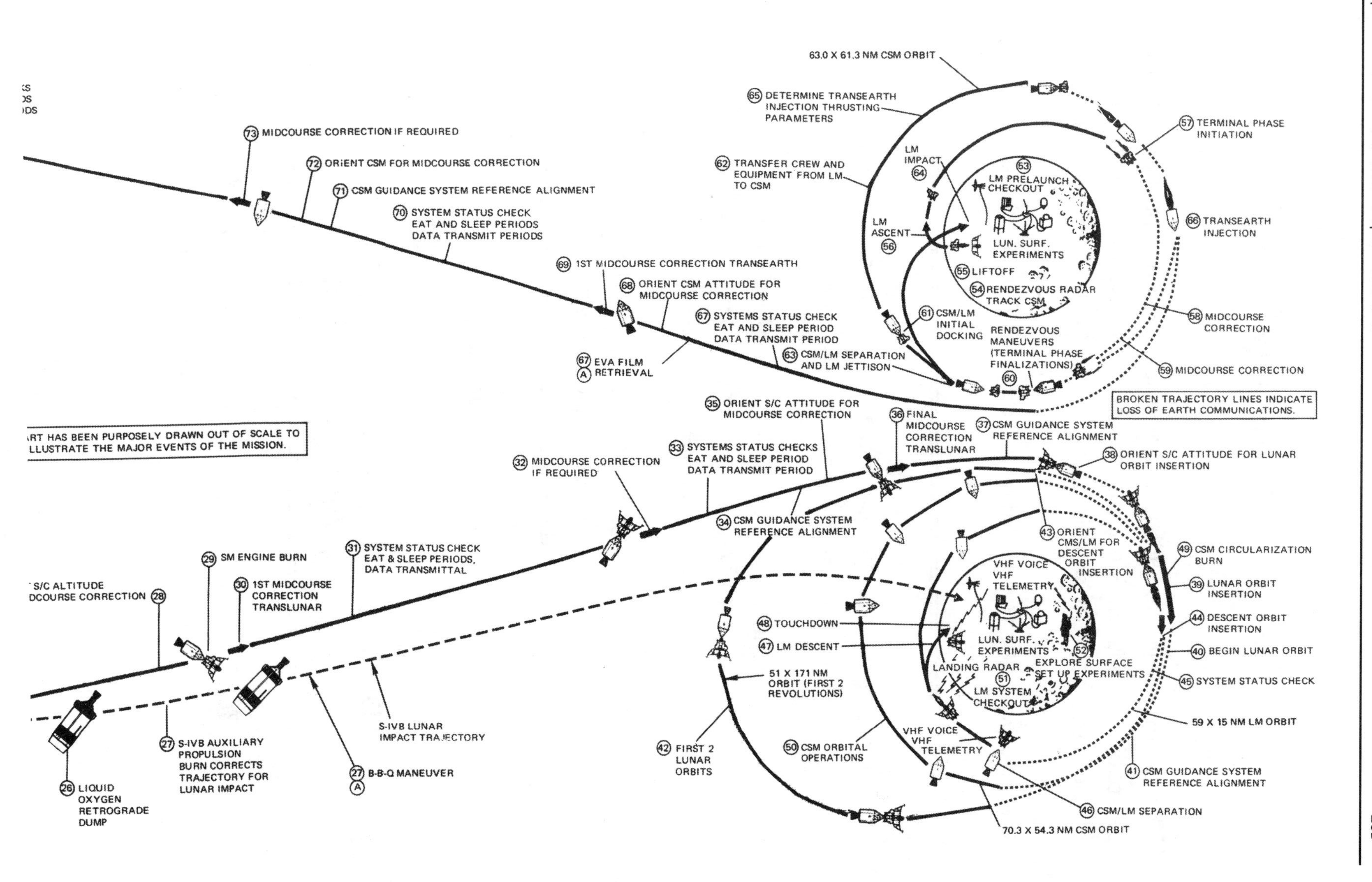

63.0 X 61.3 NM CSM ORBIT
65 DETERMINE TRANSEARTH INJECTION THRUSTING PARAMETERS
57 TERMINAL PHASE INITIATION
73 MIDCOURSE CORRECTION IF REQUIRED
72 ORIENT CSM FOR MIDCOURSE CORRECTION
71 CSM GUIDANCE SYSTEM REFERENCE ALIGNMENT
70 SYSTEM STATUS CHECK EAT AND SLEEP PERIODS DATA TRANSMIT PERIODS
62 TRANSFER CREW AND EQUIPMENT FROM LM TO CSM
LM IMPACT 64
53 LM PRELAUNCH CHECKOUT
LM ASCENT 56
LUN. SURF. EXPERIMENTS
66 TRANSEARTH INJECTION
69 1ST MIDCOURSE CORRECTION TRANSEARTH
55 LIFTOFF
54 RENDEZVOUS RADAR TRACK CSM
68 ORIENT CSM ATTITUDE FOR MIDCOURSE CORRECTION
67 SYSTEMS STATUS CHECK EAT AND SLEEP PERIOD DATA TRANSMIT PERIOD
61 CSM/LM INITIAL DOCKING
RENDEZVOUS MANEUVERS (TERMINAL PHASE FINALIZATIONS) 60
58 MIDCOURSE CORRECTION
63 CSM/LM SEPARATION AND LM JETTISON
67 A EVA FILM RETRIEVAL
59 MIDCOURSE CORRECTION
BROKEN TRAJECTORY LINES INDICATE LOSS OF EARTH COMMUNICATIONS.
35 ORIENT S/C ATTITUDE FOR MIDCOURSE CORRECTION
36 FINAL MIDCOURSE CORRECTION TRANSLUNAR
37 CSM GUIDANCE SYSTEM REFERENCE ALIGNMENT
RT HAS BEEN PURPOSELY DRAWN OUT OF SCALE TO
LLUSTRATE THE MAJOR EVENTS OF THE MISSION.
33 SYSTEMS STATUS CHECKS EAT AND SLEEP PERIOD DATA TRANSMIT PERIOD
38 ORIENT S/C ATTITUDE FOR LUNAR ORBIT INSERTION
32 MIDCOURSE CORRECTION IF REQUIRED
34 CSM GUIDANCE SYSTEM REFERENCE ALIGNMENT
43 ORIENT CMS/LM FOR DESCENT ORBIT INSERTION
31 SYSTEM STATUS CHECK EAT & SLEEP PERIODS, DATA TRANSMITTAL
29 SM ENGINE BURN
49 CSM CIRCULARIZATION BURN
VHF VOICE VHF TELEMETRY
S/C ALTITUDE
DCOURSE CORRECTION 28
30 1ST MIDCOURSE CORRECTION TRANSLUNAR
39 LUNAR ORBIT INSERTION
48 TOUCHDOWN
44 DESCENT ORBIT INSERTION
LUN. SURF. EXPERIMENTS
47 LM DESCENT
52 EXPLORE SURFACE SET UP EXPERIMENTS
40 BEGIN LUNAR ORBIT
LANDING RADAR
51 LM SYSTEM CHECKOUT
51 X 171 NM ORBIT (FIRST 2 REVOLUTIONS)
45 SYSTEM STATUS CHECK
S-IVB LUNAR IMPACT TRAJECTORY
27 S-IVB AUXILIARY PROPULSION BURN CORRECTS TRAJECTORY FOR LUNAR IMPACT
59 X 15 NM LM ORBIT
VHF VOICE VHF TELEMETRY
42 FIRST 2 LUNAR ORBITS
50 CSM ORBITAL OPERATIONS
27 A B-B-Q MANEUVER
41 CSM GUIDANCE SYSTEM REFERENCE ALIGNMENT
26 LIQUID OXYGEN RETROGRADE DUMP
46 CSM/LM SEPARATION
70.3 X 54.3 NM CSM ORBIT

Special Bonus DVD-ROM

The enclosed disc is a DVD-ROM disc, it will not play on a standard set-top DVD player. A computer and DVD-ROM drive are required to view the content of this disc. A web browser is recommended.

On the disc are the following videos in MPG format:

Unique private 8mm footage of the launch of Apollo 17.

Unique private 8mm footage of the training of the Apollo 17 crew in Hawaii.

An exclusive 100 minute interview with Apollo 17 Commander Eugene Cernan.

Also included are:

In PDF format the complete Apollo 17 Science Report. Over 700 MB of information.

All 3647 still images taken during the Apollo 17 Mission organised by reel with their image numbers.